# ORIGIN STORY

ALSO BY HOWARD MARKEL

AUTHOR

*The Secret of Life: Rosalind Franklin, James Watson, Francis Crick, and the Discovery of DNA's Double Helix*

*Literatim: Essays at the Intersections of Medicine and Culture*

*The Kelloggs: The Battling Brothers of Battle Creek*

*An Anatomy of Addiction: Sigmund Freud, William Halsted, and the Miracle Drug Cocaine*

*When Germs Travel: Six Major Epidemics That Invaded America Since 1900 and the Fears They Have Unleashed* (2004; revised second edition, with new introduction and afterword, 2020)

*Quarantine! East European Jewish Immigrants and the New York City Epidemics of 1892* (1997; revised second edition, with new introduction, 2021)

COAUTHOR

*The Practical Pediatrician: The A–Z Guide to Your Child's Health, Behavior, and Safety*

*The Portable Pediatrician* (1992; second edition, 2000)

*The H. L. Mencken Baby Book*

COEDITOR

*The Influenza Epidemic of 1918–1919: A Digital Encyclopedia* (2012; second edition, 2016; third edition, 2023)

*Formative Years: Children's Health in the United States, 1880–2000*

*Caring for Children: A Celebration of the Department of Pediatrics and Communicable Diseases at the University of Michigan*

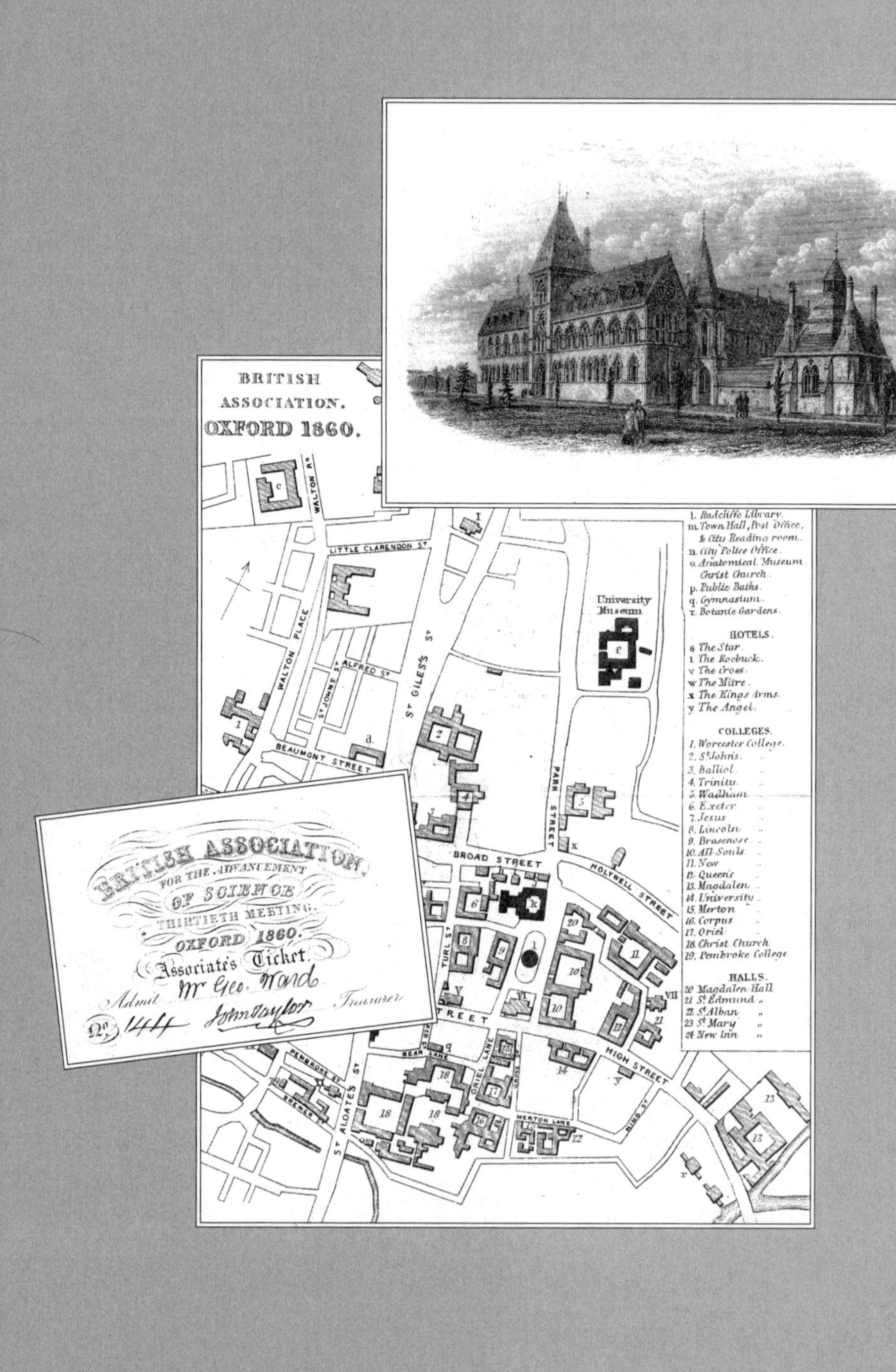

BRITISH ASSOCIATION.
OXFORD 1860.
WALTON RD
LITTLE CLARENDON ST
WALTON PLACE
ST JOHNS ST
ALFRED ST
ST GILES'S ST
BEAUMONT STREET
PARK STREET
BROAD STREET
HOLYWELL STREET
TURL ST
HIGH STREET
BEAR LANE
ORIEL LANE
MERTON LANE
PEMBROKE ST
BREWER ST
ST ALDATES ST
KING ST
University Museum
l. Radcliffe Library.
m. Town Hall, Post Office, & City Reading room.
n. City Police Office.
o. Anatomical Museum Christ Church.
p. Public Baths.
q. Gymnasium.
r. Botanic Gardens.
HOTELS.
s The Star.
t The Roebuck.
v The Cross.
w The Mitre.
x The Kings Arms.
y The Angel.
COLLEGES.
1. Worcester College.
2. St Johns.
3. Balliol.
4. Trinity.
5. Wadham.
6. Exeter.
7. Jesus.
8. Lincoln.
9. Brasenose.
10. All Souls.
11. New.
12. Queens.
13. Magdalen.
14. University.
15. Merton.
16. Corpus.
17. Oriel.
18. Christ Church.
19. Pembroke College.
HALLS.
20 Magdalen Hall.
21 St Edmund "
22 St Alban "
23 St Mary "
24 New Inn "
BRITISH ASSOCIATION
FOR THE ADVANCEMENT
OF SCIENCE
THIRTIETH MEETING.
OXFORD 1860.
Associate's Ticket.
Admit Mr Geo. Ward
No. 144 John Taylor Treasurer

# ORIGIN STORY

## THE TRIALS OF CHARLES DARWIN

Howard Markel

Printed in the United States of America
First Edition

For information about permission to reproduce selections from this book, write to Permissions, W. W. Norton & Company, Inc., 500 Fifth Avenue, New York, NY 10110

For information about special discounts for bulk purchases, please contact W. W. Norton Special Sales at specialsales@wwnorton.com or 800-233-4830

Manufacturing by Lakeside Book Company
Book design by Chris Welch
Production manager: Lauren Abbate

Library of Congress Cataloging-in-Publication Data Available

ISBN 978-1-324-03674-6

W. W. Norton & Company, Inc.
500 Fifth Avenue, New York, N.Y. 10110
www.wwnorton.com

W. W. Norton & Company Ltd.
15 Carlisle Street, London W1D 3BS

1 2 3 4 5 6 7 8 9 0

*In memory of*

*Horace W. Davenport, Ph.D., D.Sc. (Oxon),*

*who played such an important role in my personal evolution*

# CONTENTS

## PART III: FRIENDS AND FOES



## PART IV: OXFORD

# PREFACE

For millennia, Judeo-Christian coreligionists have accepted Genesis's account of how the earth and its inhabitants were created. That authority eroded with the 1859 publication of Charles Darwin's *On the Origin of Species*. For many, Darwin's theory of natural selection serves as a scientific foundation for how life and biodiversity began. Yet more than a century and a half after its initial appearance—and despite overwhelming evidence—the validity of Darwin's *Origins* remains a topic of debate in public arenas around the world.

The first, formal, and very public contest over this book occurred at the 30th annual meeting of the British Association for the Advancement of Science (BAAS). It was held on the campus of Oxford University, on June 30, 1860. Astonishingly, Darwin was absent from the proceedings. Stricken by "fits of flatulence" and numerous other problems, Darwin was ill for most of his adult life.[1] At the time, he was also mourning the death of his eighteen-month-old son, contending with the serious illness of one of his daughters, and exhausted from racing to finish his magnum opus before competitors beat him to the published page. So, instead of defending his life's work at the BAAS, Darwin was nursing his ailing gastrointestinal tract at a water cure hospital in Surrey.[2]

Fortunately, Darwin's "Bulldog"—the formidable and eloquent zoologist

Thomas H. Huxley—served as the author's surrogate. In front of a room of one thousand spectators, Huxley bested the Bishop of Oxford, Samuel Wilberforce, in a deliberation over Darwinian theory and man's relation to apes. With the passage of time, the Huxley-Wilberforce exchange has metamorphosed from an academic quarrel into a parable depicting *the* critical moment when scientific truth defeated religious faith. More recently, the Oxford debate has been attacked by historians as an ossified, tall tale because it was not a "real debate," nor did it immediately settle the divide between science and religious doctrine.[3] Upon closer scrutiny, such reminders of self-appointed umpires are pedestrian, if not outright obvious. Of course, the 1860 BAAS meeting did not reconcile the question of life's "mystery of mysteries" after only a few hours of argument! Across the broad expanse of scientific inquiry, there are relatively few light switch moments where everything changes suddenly.[4] Science evolves—no pun intended—over long periods of time.

The Oxford row was not a giant asteroid slamming into the earth with a sudden force; it neither killed all the dinosaurs nor wreaked planetary havoc. But it does provide an early and important view of the shifting moral authority from the church to the intellectual power of science.[5] As Thomas Huxley's son Leonard shrewdly concluded, decades after the fact: "The importance of the Oxford meeting lay in the open resistance that was made to authority, at a moment when even a drawn battle was hardly less effectual than acknowledged victory."[6]

In other words, if the fracas at Oxford was an opening salvo in a war that has never ended, it is because our quest for better explanations of our world—and beyond—has never ended, either.

# AUTHOR'S NOTE

*With respect to the words "monkey" and "ape"—they represent different species; one obvious difference between these creatures is that the former has tails while the latter does not. Gorillas and chimpanzees, for example, are apes and, consequently, have no tails. It is interesting that during an era fixated on taxonomy, variety, and species classification, Victorians often interchanged the labels of apes and monkeys in their everyday conversations. Herein, I have quoted what phrases and terms the historical actors uttered at the time, without correction and with my apologies to all the primates.*

# ORIGIN STORY

INTRODUCTION

# A Temple of Science

Now and then, in the course of the century, a great man of science, like Darwin . . . has been able to isolate himself, to keep himself out of the clamorous claims of others, to stand "under the shelter of the wall," as Plato puts it, and so to realize the perfection of what was in him, to his own incomparable gain, and to the incomparable and lasting gain of the whole world.

—*Oscar Wilde, 1891*[1]

There are two sides to every question; there are often half a dozen.

—*Vyvyan Holland (Oscar Wilde's younger son), 1954*[2]

On the bright summer morning of June 30, 1860, a crowd congregated outside Oxford University's Museum of Natural History. The occasion was the annual meeting of the British Association for the Advancement of Science (BAAS), one of the world's leading scientific societies. The attendees queued up early to secure their seats for a special plenary session discussing Charles Darwin's new book, *On the Origin of Species by Means of Natural Selection, or the Preservation of Favoured Races in the Struggle for Life*. The author—a simultaneously reclusive and intellectually audacious squire from Kent—claimed to have solved "that mystery of mysteries," how species originated, adapted, and modified into new ones.[3] Without fear of exaggeration, the 490 pages comprising Darwin's thesis forever changed our understanding of the life sciences and the natural world.

Exterior, Oxford Natural History Museum.

Interior of the museum.

The expectant throng could hardly have found a better place for their Darwinian deliberations. Inspired by the Victorian polymath John Ruskin's dictum of "truth to nature," the new museum—a neo-Gothic pile of granite and limestone, trimmed with sandstone—was designed to impress and instruct.[4] Four years earlier, in 1856, an Oxford professor named Charles Daubeny told the BAAS membership that when completed, the museum's central court would serve as "the Sanctuary of the Temple of Science, intended to include all those wonderful contrivances by which the Author of the Universe manifests himself to His creatures." The rooms comprising the periphery of this grand space were for "lectures and researches connected with all branches of Physical Science, [and] may represent the chambers of the ministering Priests, engaged in worshipping at her altar, and in expounding her mysteries."[5]

The completed exhibition hall was illuminated by natural sunlight, thanks to a heavy glass and iron roof—which, in its first iteration, almost crashed to the ground. Learning from their structural miscalculations, the architects made sure the second version was fully supported by a battalion of cast-iron pillars and and arches. The columns were ornamented with wrought iron leaves, representing a metallic arboretum of trees found

Interior corbels, Museum of Natural History.

across the British Isles—from bushy holly to the mighty oak. Aside from their weight-bearing nature, the pillars cloistered the main gallery into three wide aisles, accommodating the reassembled remains of woolly mammoths, whales, and dinosaurs, along with rows of oak-trimmed glass display cases filled with fossils, rocks, taxidermized animals, and that rare bird, the dodo.

The two-story Main Court was bounded by a rectangle of 126 columns, each one composed of a different British rock, labeled with its identity and source. Sculpted into their capitals and corbels was a Pre-Raphaelite wonderland of plants, demonstrating all the botanical orders. For accuracy, the stonemasons used specimens clipped from the university's botanical garden. Interspersed between the rocky columns and arches was a parade of marble statues lined up like Latter-Day Saints—including Aristotle, Gali-

Man in front of statue of Lord Bacon at Oxford Museum of Natural History, circa 1860.

Angel at entry.

leo, Bacon, Newton, and Linnaeus—with a few empty niches that would eventually be filled with sculptures of some of the most eminent scientists expected to be present that morning.[6]

At the pinnacle of the Museum's arched entrance was a bas-relief of an angel. In one hand, she holds a Bible turned to Genesis; in the other, a culture plate depicting a cell in the act of division and reproduction.[7] Even Darwin—a Cambridge man to his core—contributed unwittingly to the building's facade. To the right of the portal is a capital adorning a first-floor window. Instead of a heraldic seal or a gargoyle, the masons carved a pair of four-legged sea monsters with voracious mouths and long tails, affectionately known to this day as "bear-whales." This stony statement was a humorous reference to page 184 of the *Origin of Species*, in which Darwin playfully proposed what might happen if "a race of bears" ever took to

Bear-whales.

living in the sea: "I can see no difficulty in a race of bears being rendered, by natural selection, more and more aquatic in their structure and habits, with larger and larger mouths, till a creature was produced as monstrous as a whale."[8] No work of revisionist history, the "bear-whales" assumed their petrified position as the museum's facade was being completed in early 1860—a few months after the initial publication of *Origin*.[9]

At noon, an eminent physiologist from New York University named John William Draper was scheduled to deliver a lecture on "The Intellectual Development of Europe, Considered with Reference to the Views of Mr. Darwin and Others, That the Progression of Organisms Is Determined by Law." His argument was that "man in civilization does not occur accidentally, or in a fortuitous manner, but is determined by immutable law."[10] Draper's turn at the podium was originally slotted for a later day in the program. But his talk was moved up after a verbal dustup on June 28 between Thomas Huxley, a comparative anatomist who was Darwin's most energetic advocate, and Richard Owen, a paleontologist who despised everything about Darwin and his theories. Although Darwin had not yet publicly committed to details concerning the descent of man, Huxley was already arguing that man's closest relatives in the "tree of life" were apes. Owen, who "had probably dissected more apes than any man," maintained that such notions diminished the Judeo-Christian belief in the ultimate authority of human origins, Genesis 1:26–27—which proclaimed that God created men and women "in His own image."[11] Slamming the lectern with clenched fists, Owen remonstrated that his research had "definitively" shown that the genus *Homo*, or man, was not merely "a distinct order, but of a distinct subclass of the Mammalia." Humans, he decreed in print only a few years earlier, were "as different from a chimp as the ape was from a platypus."[12] To which Darwin once quipped, "I cannot swallow Man making a division as distinct from a Chimpanzee, as [a platypus] from a Horse: I wonder what a Chimpanzee [would] say to this."[13]

Among those waiting outside the museum were the crème de la crème of British science. These men were attired in black morning coats, gray-striped trousers, and tall silk hats. Off to one side gathered a group of long-robed clergymen—their necks wrapped in white chokers and their fingers already flapping against a theory they deemed as heresy. Assembled nearby was a herd of old nags—self-styled natural philosophers, scholars, and armchair scientists who bridled, stomped, and shied at the mere mention of Darwin's name. The creepy dons were reported to have escorted a bevy of lasses, forming "the union under the same roof of grave octogenarian professors with ladies not out of their teens, of deep and learned discussions with the frothiest of small talk, of the natural magic which experimental philosophy can display with the still more wondrous witchery of bright eyes, graceful forms, and becoming dresses—provocative of that feeling of the ridiculous, which has so close a connection with contrast and 'the unexpected.'" Indeed, the commentator, whose name is lost to history, found it difficult to maintain his composure, let alone a straight face, when spying "the fair fingers which have rarely turned the leaves of anything more serious than one of Mr. James's novels or Mozart's Symphony in D employed upon pages where the simplest and most intelligible terms are such as organic compounds . . . and the isomers of cumole."[14]

A pack of college boys milled about the museum's lush lawn. They sported straw boaters, jackets of many collegiate colors, a spectrum of opinions, and a love for good (and bad)-natured raillery. One was John Richard Green, who hailed from a high Anglican church, Tory family and attended Jesus College. Later that year, he completed his bachelor's degree and joined the diaconate. Assigned to several curates in London until hanging up his cassock in 1869, Green would become one of Great Britain's most distinguished historians before his premature death at forty-five. In 1885, the journal *Nature* eulogized him as "the first historian who appreciated the function of science in a State."[15] A few days after Draper's

John Richard Green.

lecture, Green wrote a letter to his friend "Dax"—the future geologist and archaeologist W. Boyd Dawkins.

Green described walking to the Museum with John David Jenkins, a fellow at Jesus College. The Welsh-born Jenkins served as a missionary in South Africa for six years until ill health sent him back to Oxford in 1858, where he studied the hidebound history of Christianity and ministered to railway workers. During breakfast at hall that morning, Jenkins "proposed going 'to hear the Bishop of Oxford smash Darwin.'" Jenkins was referring to the same bishop who ordained him in 1851—the Right Reverend Samuel Wilberforce, an accomplished theologian, member of the House of Lords, fellow of the Royal Society, and one of the most facile and popular orators of his day. "Smash Darwin! Smash the Pyramids," Green replied, "in great wrath, and muttering something about 'impertinence.'" To which Jenkins patiently explained that "'the bishop was a first-class in mathematics, you know, and so has a right to treat on scientific matters,' which of course silenced [Green's] cavils."[16]

Wilberforce made his star turn at the meeting primarily because it was

The Right Reverend Samuel Wilberforce.

held on his ecclesiastical turf; but as an amateur naturalist and ornithologist, he was also an active member of the BAAS and a patron of the new museum. A few months earlier, the bishop blessed the new Oxford museum as a pantheon of natural theology—a thick British brew of biblical lore and empirical observations of plants, trees, animals, and rocks that explained a Creator-planned world, with man and woman being God's greatest, and most perfect, works. This concept was most famously articulated by a Christian philosopher, clergyman, and Cambridge alumnus named William Paley. In 1802, he published an influential book entitled *Natural Theology, or Evidences of the Existence and Attributes of the Deity from the Appearances of Nature.*[17] Bishop Wilberforce was hardly the only enthusiastic reader of Paley's popular tome. On November 22, 1859—two days before *Origin* hit the booksellers' shelves—Charles Darwin admitted his fondness for Paley's book when he was still a student at Cambridge University: "I do not think I hardly ever admired a book more than Paley's *Natural Theology*. I could almost formerly have said it by heart."[18]

ſ

THE MUSEUM's heavy oak doors swung open at about a quarter to noon. The spectators rushed inside. As they made their way up the steps to the long, narrow library, there was a clatter of high button shoes and heavy leather boots, the rustling of velvet trousers, and the hollow clinks of crinoline hoop skirts. Many of the chairs provided an obstructed view of the hastily outfitted speakers' stage. Above them was a freshly painted, hammer beam roof with curved, arched, and specially cut timbers that appeared to hold up the ceiling at acute angles. The quarter-sawn oak bookshelves lining the chamber were newly varnished and still sticky. As a result, tens of thousands of books remained boxed and placed at one end of the room instead of being put out for perusal. Such arrangements were of little consequence, as geologist Charles Lyell reported a few days later, for "the excitement was tremendous."[19]

Hardly a debate in the tradition of politicians who spout off sound bites and search for "gotcha" moments, the deliberations held after Draper's lecture were typical of many academic symposiums: the professor's equivalent of a blood sport, where facts, figures, and experiments were the weapons of choice. Although Darwin referred to it as "the Battle of Oxford," historians are quick—and correct—to remind that this event was not "the Galileo moment of the 19th century."[20] Nonetheless, most of the men and women present knew that the gathering represented a confrontation between the faith and future of the Christian church and the origin and nature of modern biology.[21]

Darwin's deputies—Thomas H. Huxley and the botanist Joseph D. Hooker—had already taken their seats and were anxious to begin. So, too, was the speaker, John W. Draper, and John Stevens Henslow, the president, or chairman, of the session. Finally, the portly, smug Bishop of Oxford, Samuel Wilberforce, made a purposefully delayed entrance—slowly walking down the aisle and shaking the hands of his supporters before settling into a place up front. At this point, the audience craned their collective necks, many murmuring the same question: *Where was Darwin*?

# Part I

# DOWN

I fear that my head will stand no thought, but I would sooner be the wretched contemptible invalid, which I am, than live the life of an idle squire.

—*Charles Darwin, June 2, 1857*

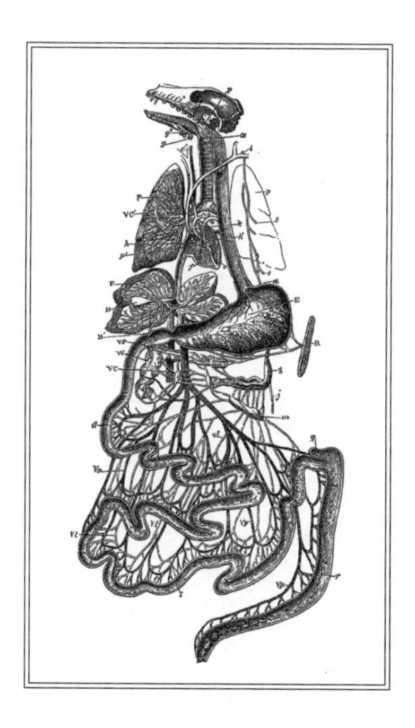

# 1

# The Letter

> I am extremely glad to hear that you are attending to distribution in accordance with theoretical ideas. I am a firm believer, that without speculation there is no good & original observation. Few travelers have [at]tended to such points as you are now at work on; & indeed the whole subject of distribution of animals is dreadfully behind that of Plants.
>
> —*Charles Darwin to Alfred Wallace, December 22, 1857*[1]

The eighteenth of June 1858 was not Charles Darwin's best day. His emotional and gastrointestinal barometers plummeted soon after the mailman delivered him a letter that was postmarked Ternate. A tiny island in what was then known as the Dutch East Indies, Ternate was rich in volcanic eruptions, earthquakes, and crustal warping; it also boasted an untouched ecosystem of plants and animals. On the front of the taut envelope was an easily deciphered cursive writing Darwin recognized to be from the hand of Alfred Russel Wallace, an intrepid explorer, collector, and natural historian.

Darwin had not sailed in search of unspoiled lands and creatures for nearly two decades. To compensate for his sedentary seclusion, he fed his insatiable appetite for ever more data by corresponding with far healthier explorers of the realm. Darwin wrote to a battalion of biologists, naturalists, geologists, animal breeders, farmers, gardeners, botanists, and anyone else who might slake his thirst for new information. He gathered all their facts, opinions, observations, descriptions, and figures. He copied down the most salient details into his notebooks and processed them through his

Alfred Russel Wallace.

encyclopedic mind. He conducted this work from one of the most important command stations in the history of biology: his study in Down House. Darwin described the method of his cluttered desk in an 1857 letter to his cousin, William D. Fox: "I am like Croesus overwhelmed by my riches in facts & I mean to make my book as perfect as ever I can."[2]

Throughout the nineteenth century, most scientists funded their research on their own time, from whatever revenue they raised by taking a teaching post or church position or from independent sources of income. Darwin was fortunate in having inherited large sums of money from his parents, beginning with his mother's share in the Wedgwood pottery works. Darwin not only descended from Wedgwoods; he married one—his first cousin Emma was the granddaughter of the fine china dishware manufacturer's founder, Josiah Wedgwood. Darwin's father, Robert, wisely invested both the Wedgwood profits and the revenue from his medical practice. These sums formed the trust funds he bequeathed to his children—and their heirs. Charles, too, was adept at nurturing his investments; he was partial to railroad stocks, one of the most profitable and expansive industries of the Victorian era. This substantial personal wealth allowed him to escape the dirt and noise of London in 1842 for an idyllic family home and one-man research center in bucolic Kent.

ʃ

DARWIN'S BIOLOGICAL brainstorming by correspondence coincided with the development of Great Britain's version of the World Wide Web—the Royal Mail, which then consisted of nearly ten thousand offices, branches, processing centers, and outposts across the expansive British Empire. During the mid-1830s, Rowland Hill, the social reformer and educator, successfully campaigned for a uniform Penny Post, prepaid by consumers using his creation, the postage stamp. One needed only to attach a penny stamp on a postcard to travel the world. Hill later became the Chief Post Office Secretary to the Postmaster General, from 1854 to 1864, and is often credited with modernizing Queen Victoria's Royal Mail into a vast physical enterprise that never ceased growing across the nineteenth century. By the 1850s, Britons came to rely on the Royal Mail whether engaging in long- or short-distance communications with family, friends, colleagues, business associates, and lovers. This process became even faster during the

Exterior, General Post Office, St. Martin-Le-Grand, London.

1860s, thanks to "the fact the railway-mail service has now assumed quite gigantic proportions" and the development of steam-powered, ocean-going vessels.[3]

In 1862, Hill reported that most cities and towns across the British Isles enjoyed upward of four deliveries per day, except for London, which formerly had "but six deliveries per diem" until that number was expanded to eleven every day but Sunday.[4] The last post arrived at 7:00 p.m., meaning that businessmen could send a letter in the morning and get a response by the close of the day. Individuals brought their letters and parcels directly to their local post offices or placed them in one of the 13,370 pillar mailboxes carrying the monogram of Queen Victoria and strategically positioned throughout the British Isles.[5]

These cast-iron boxes—originally painted bronze, sage, or olive green and, after 1874, the now familiar bright red—were the brainchild of Anthony Trollope, the traveling surveyor, Assistant Secretary of the Royal Mail, and popular novelist. Once a box was emptied, the postman rushed its contents to a general post office, where an avalanche of mail was sorted,

Interior, Sorting Room of the General Post Office, St. Martin-Le-Grand, London, circa 1856.

Mail pillar
("V.R." for Victoria Regina).

carried, and delivered. Between 1854 and 1880, the army of postal workers increased from twenty-one thousand to forty-nine thousand and made annual treks of over 149,000 miles, carrying bags that collectively weighed more than 4,300 tons. During the same period, the annual number of letters mailed nearly doubled in volume. In 1860, for example, there were 564 million letters processed; in 1880, 1.2 billion; and by 1900, 2.3 billion.[6]

Technology mattered, too. In the 1860s, immense mechanical sorters sped up the process of getting each letter to the right place and person. Letters, books, newspapers, manuscripts, government documents, checks, and money orders all connected the British Empire—where the sun never set. This net of post offices and human messengers was, in essence, a precursor to today's Internet, emails, texts, sites, e-books, and social media posts.

The so-called snail mail of Charles Darwin's era was every bit as important to his work as the barnacles and mollusks he dissected in his study. His correspondence spanned from England to Egypt, India, North America, South America, Europe, Africa, and Asia. Each day he wrote and received dozens of letters as well as manuscripts, newspapers, magazines, books, and scientific journals. Pouches of mail to and from Down House were so frequent that Darwin hooked up a mirror onto the window of his study so that he could see the postman coming up the walk. Darwin's letters constituted a slow-motion, international conversation—a dialogue separated by time and space—but a conversation, nonetheless, that was essential to his exploration of nature.[7] As he exclaimed in 1862, "a letter is something living."[8]

ʃ

ALFRED RUSSEL WALLACE and Charles Darwin had been corresponding for over a year when Wallace's mood-darkening letter arrived at Down on June 18, 1858. As a young man, Wallace attended lectures at the London Mechanics' Institute (now Birkbeck College), University of London, and briefly taught mapmaking and surveying at the Collegiate School in Leicester. Wallace was quite the dandy fellow, and he favored brightly colored waistcoats with matching lapels and pants, snowy-white linen shirts, and velvet frock coats, complemented by wire-rimmed glasses and a bushy brown beard.

Inspired by Darwin and Alexander von Humboldt—the early nineteenth-century Prussian naturalist and explorer of South America—Wallace traveled to Brazil in 1848 to explore the Amazon rainforest. He funded his exotic travels by collecting and selling rare specimens to upper-crust collectors back home in England.[9] Disaster struck in July 1852, when his ship caught fire and most of his notebooks and crates of specimens were lost at sea. Wallace somehow managed to get home safely, rescuing some of his precious notes and drawings. For the next eighteen months, he lived in London on the insurance money he collected for the loss of his expedition specimens. He also cobbled together two books on his adventures and the natural history of the Amazon River valley, the first maps tracing the course of the Rio Negro—the largest tributary of the Amazon—as well as several scientific papers.[10] His work was impressive enough for the Royal Geographical Society to support a 14,000-mile trip through Borneo and the Malay Archipelago from 1854 to 1862. It was in this pristine wilderness where his observations of variation within the same species led to a theory of transmutation and evolution. Without knowing Darwin's inner demons, Wallace wrote to the one person in the world who might understand the intense competition for survival that goes on in the natural world every day.[11]

While holding the sealed envelope, Darwin could plainly see that Wallace's latest packet—posted three months earlier, in March 1858—was too thin to contain a dead, preserved animal and too thick to contain a simple note. Once opened, he found several handwritten pages.[12] The folio comprised an essay entitled "On the Tendency of Varieties to Depart Indefinitely from the Original Type." Wallace wrote it while stuck on the Indonesian island of Halmahera in February 1858. As he later recalled, "I was suffering from a sharp attack of intermittent fever, and every day during the cold and succeeding hot fits had to lie down for several hours, during which I had nothing to do but to think over any subjects then particularly interesting to me."[13] In a fascinating coincidence of medical and natural history, Wallace's fever—most likely a by-product of malaria—proved inspirational. During his enforced bed rest, he read Thomas Malthus's *Essay on Population*, a "book [he] had read a dozen years before." He, like Darwin, was struck by the British economist's description of the checks and balances that kept a population from getting too large and running out of the resources necessary for life, such as food. After reviewing Malthus's discussion of "disease, famine, accidents, war, &c.," Wallace contemplated how these factors might act on wild and lower animals. "There suddenly flashed upon me," Wallace later recalled, "the idea of the survival of the fittest—that those individuals which every year were removed by these causes—termed collectively the 'struggle for existence'—must on the average and in the long run be inferior in some one or more ways to those which managed to survive."[14]

His resulting paper proposed "a general principle in nature which will cause many varieties to survive the parent species, and to give rise to successive variations departing further and further from the original type and which also produces, in domesticated animals, the tendency of varieties to return to the parent form."[15] Once his thesis was completed, the younger man asked Darwin to read the manuscript. He also requested him to forward the essay to the geologist Charles Lyell for his review and advice in publishing it.[16]

SIR CHARLES LYELL was a gentleman-scientist and, like Darwin, born with stacks of family money. Lyell's grandfather made his fortune supplying the Royal Navy at the port of Montrose in Scotland. As an Oxford man, Lyell read classics at Exeter College, sat through William Buckland's lectures on geology, and took a second-class honors bachelor's degree in 1819 and his master's in 1821. Typically, the label of a second-class degree doomed its recipient to business or, worse, the law. Thus, for a year or more, Lyell poured over legal texts at Lincoln's Inn, London, one of the four Inns of Court to which English barristers belong and are "called to the bar." He was predestined to be bored by that tangled and combative trade. Following his intellectual interests, Lyell studied geology by taking long

Charles Lyell.

walks in the countryside with only a bit of food for the journey, a small pick hammer to collect specimens, and a rucksack to carry them home for analysis. Lyell's wanderlust led to him writing several successful travel and geology books on Great Britain and the United States. In 1823, he left the law to become secretary of the prestigious Geological Society of London, and in 1826, he was elected into the prestigious Royal Society of London, the intellectual home of British science. During the 1830s, he taught geology at King's College, London.

There is a portrait of Lyell in London's National Portrait Gallery that captures this gentle and beloved genius to perfection. Painted by the pre-Raphaelite artist Lowes Cato Dickinson, the canvas depicts a warm, relaxed man with an almost egg-shaped head. He sports the requisite muttonchop side whiskers and is dressed in a grayish-brown frock coat draped over a dark suit jacket. Beneath those layers, he wears a beige waistcoat, dark-brown bowtie, and an ecru-colored shirt. Completing the sartorial ensemble are green velvet trousers. Seated in a red plush armchair, Lyell's thin legs are crossed. His hands are clasped and close to his chest. Below them, dangling on a black silk ribbon, is the monocle he used for reading. His face appears thoughtful, ruddy, and gentle in demeanor. Lyell's lips are pressed together as if he is prudently keeping himself in check before articulating his next thought. His soft, blue-gray eyes look off into the distance, ever eager for new geological vistas to conquer and explain. Lyell's wife loved the portrait so much that she told the curators of the National Portrait Gallery, "I afterwards asked Mr. Dickinson if he would make a copy of it, which he kindly did most carefully, and it really is a most exact replica."[17]

Thanks to his trust fund, book royalties, and speaker's fees, Lyell eventually resigned his post at King's and turned to full-time book writing and public appearances. His best-known work was the massive *Principles of Geology: Being an Attempt to Explain the Former Changes of the Earth's Surface by Reference to Causes Now in Operation.*[18] Appearing in three volumes between 1830 and 1833—at six shillings each—the book advanced the doctrine of uniformity, first espoused by James Hutton in his *Theory of the Earth* (1795); this dictum held that the natural laws and processes occurring in our present era have always operated in the same manner in the past. The term *uniformitarianism* was coined in 1832 by William Whewell, the scientific luminary and Master of Trinity College, Cambridge. Whewell used this long word while reviewing the second volume of Lyell's book.[19]

Since the planet was constantly in a state of gradual change, Lyell wrote, one could look to its geological strata to analyze their epoch and the events that formed them. Foreshadowing Darwin's revolutionary work on the origin of species, Lyell advanced the then-blasphemous theory that the earth was far older than what the Bible stated in Genesis and that there was no evidence of floods or other divine interventions as the source of mountains, valleys, or open fields; instead he proposed the view that the earth's topography was the result of slow and gradual changes caused by common forces, including the local weather and seismic activity. In this pursuit, Lyell was seen by many as the spiritual savior of geology; or as the historian of medicine and science Roy Porter once observed, "history showed that the study of the earth needed to be freed from Moses; but also from

John Stevens Henslow.

philosophers, and from man, if ideology was ever to yield to science."[20] The same might be said of Darwin.

IN 1831, after withdrawing from medical school at Edinburgh University, Darwin took his bachelor's degree at Christ's College, Cambridge, ranking 10th out of 178 candidates. His favorite teacher, John Stevens Henslow, recommended him to serve as naturalist without commission (that is, self-funded) on a Royal Navy expedition aboard the HMS *Beagle*.[21] The ship, under the command of Captain Robert FitzRoy, was preparing for an extensive voyage to South America. After consultations with family members, including his uncle Josiah Wedgwood II, Robert Darwin gave his son permission to go on this dangerous trip and pledged to "give him all the assistance in my power."[22] Less recalled is that Darwin was not the first choice for this position; both Henslow and one of Darwin's classmates, Leonard Jenyns, turned it down.

Darwin was almost rejected for his life-defining journey because Captain FitzRoy claimed to be able to judge a person's character "by the outline of his features." The irascible naval officer initially thought that Darwin's

Christ's College, Cambridge, circa 1900.

lumpy, long nose revealed a lack of determination and insufficient energy for the voyage. A second interview convinced FitzRoy to take the young man on despite his physiognomy.[23] During the five-year expedition, Darwin tolerated, admired, and feared the ill-tempered FitzRoy—whose volatile nature earned him the nickname "Hot Coffee."[24]

FitzRoy had a scholarly side, too, and presented Darwin with a copy of the first volume of Lyell's *Principles of Geology* as a shoving-off present. While exploring South America, whenever a new volume or edition was published, Darwin had them sent to him from London—another feat of the Royal Mail, considering the distance these books had to travel. Lyell's *Principles* proved to be a central guide for the young naturalist as he struggled to describe the evolution of life-forms, which he eventually argued were subject to gradual change over time.[25] As early as 1844, Darwin confessed, "I always feel as if my books came half out of Lyell's brains & that I never acknowledge this sufficiently . . . the great merit of the *Principles* was that it altered the whole tone of one's mind & therefore that when seeing a thing never seen by Lyell, one yet saw it partially through his eyes."[26]

Both Lyell's science and generosity of time and resources were essential to Darwin's scientific success after his voyage ended and he returned

HMS *Beagle*, exterior.

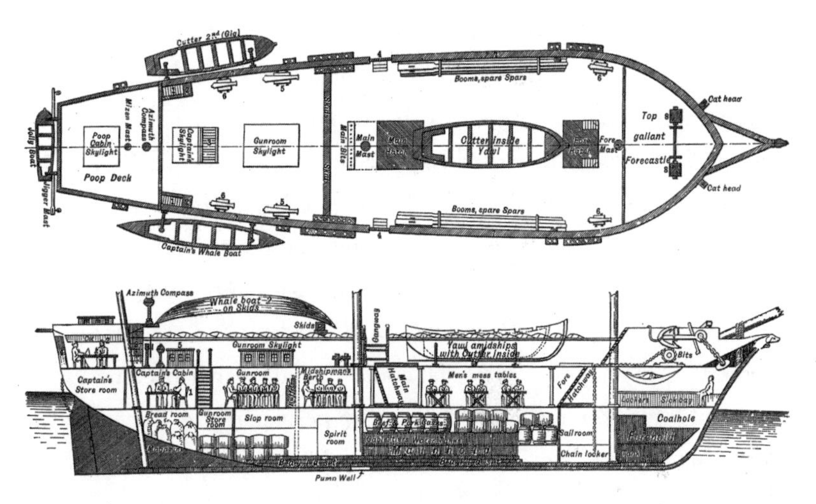


HMS *Beagle*, interior.

to London. Lyell put Darwin up for membership in the prestigious Geological Society and, soon after, for his old post as secretary of the Society. The latter position served as the perfect task for Darwin in that he was an able, generous editor of the reports that came his way; better still, the dispatches sent from rock collectors around the empire gave him an unparalleled view of earth science. The older man listened to Darwin's formative ideas, commiserated with him on the woes of his body, and offered friendly advice—but only when asked for it. Most important, Lyell read Darwin's manuscripts, gingerly edited him, and introduced him to his publisher, John Murray, who would soon become Darwin's publisher, too.

THE OTHER PERSON Darwin contacted about Wallace's paper was his best friend, Joseph Dalton Hooker. Like his father before him, Hooker was the director of the Royal Botanical Gardens at Kew.[27] Hooker was eight years junior to Darwin, but their résumés and scientific interests were remarkably similar. Both had trained in medicine, although neither made it his career because they detested the blood and guts of the craft. In 1839, three years after Darwin returned to England, Hooker began his own voyage aboard the HMS *Erebus*. Just before Hooker left Great Britain to tour the world, he and a shipmate who previously sailed with Darwin on the *Beagle* bumped into the naturalist in Trafalgar Square. The twenty-one-year-old Hooker was impressed by Darwin's "animated expression, heavy beetle brow, and delightfully frank and cordial greeting to his former shipmate."[28] While Hooker was at sea, their mutual friend, Charles Lyell, sent him page proofs of Darwin's *Journal of Researches*. Every night before retiring, Hooker read a page or two of it. He kept the manuscript under his pillow so that upon waking he could dip into what became Darwin's first book.[29] Hooker recognized it as the perfect model for what would become his memoir aboard the *Erebus*, under the command of Captain James Clark Ross.[30]

Four years later, on March 12, 1843, while Hooker was still exploring Antarctica, Darwin wrote to Joseph's father, William. He promised that if Joseph wrote up his research on the flora in that frosty clime, Darwin would "joyfully" lay the alpine plants he collected in Southern Patagonia

Joseph Dalton Hooker.

and Tierra del Fuego "at his disposal."[31] Thus initiated a decades-long conversation on natural selection, how species adapted across different latitudes and longitudes, environments, weather conditions, and new experimental methods. Hooker—who peered at plants from all over the globe through tiny "Franz Schubert" spectacles and grew wiry, muttonchop whiskers on his face, which eventually coalesced into a full white beard—worshipped the reclusive Darwin.

The attraction was mutual. Hooker was a constant correspondent and frequent visitor to Down House. The two men took long walks together, recounting tales of the sea and discussing the books and papers they were reading, Darwin's health, their families, and each other's scientific work. Given that many considered Darwin's theories to be sacrilegious, their conversations required absolute trust and confidentiality. In 1844, for example, while explaining the distribution of various species on the Gálapagos Islands, Darwin quietly confessed to Hooker that his sea voyage left him "almost convinced" that species were not "immutable."[32] Hooker's reply was, for Charles, the equivalent of manna from heaven: "I shall be delighted to hear how you think this [gradual] change [of species] may have taken place as no presently conceived opinions satisfy me on the subject."[33]

Unlike the natural theologists—who shoehorned the "facts" they discovered into awkward explanations of the Holy Scriptures—Darwin and Hooker searched for new ways of observing and thinking about nature. Both were fearless in letting the data they collected carry them to logical, fact-based conclusions. Darwin had many experts he could consult, but few colleagues were treasured more than Hooker. As Leonard Huxley articulated in his two-volume biography of Hooker, "The making of the *Origin* is not only the history of science—it is the history of a great friendship. In its fabric the two strands are indissolubly interwoven." Darwin described this intellectual partnership even more effusively to Hooker: "Talk of fame, honor, pleasure, wealth, all are dirt compared with affection; & this is a doctrine with which, I know from your letter, that you will agree from the bottom of your heart."[34]

ſ

For Darwin, the words, sentences, and paragraphs of Wallace's 1858 essay on natural selection represented his personal Pompeii. Far less stressful situations often precipitated a painful constriction of his chest, shortness of breath, and a rapid and bounding pulse. Then came a narrowing of his visual field—from the periphery to the center of his line of view—before he nearly fainted. This single post from an obscure journeyman naturalist, however, represented a watershed moment in Darwin's intellectual life. Everything in his carefully controlled world spun off in all directions: a weird version of life imitating the art of science or, at least, the second law of thermodynamics. He had good reason to be so upset. Wallace's essay was a near-perfect summary of the same complexities of thought that Darwin had pondered for more than two decades.

There is one chronological quibble with this version of events. Darwin *had* to know that Wallace was fast on his tail. In 1855, Wallace published an article in the *Annals and Magazine of Natural History* proposing, "Every species has come into existence coincident both in space and time with a pre-existing closely allied species." In so many words, Wallace was positing that new species evolved from existing ones as opposed to simply popping up in nature by means of a divine creation. Wallace called this concept the "Sarawak Law," for the northwest coast of the island of Borneo, where he first hit upon the idea and began searching for the data to support it.[35] Lyell showed Wallace's Sarawak paper to Darwin when visiting Down House in April 1856 and encouraged his friend to buckle down and write his book even though Darwin protested that his ideas were still not ready for print.[36]

One year later, on May 1, 1857, Darwin wrote to Wallace, "By your letter & even still more by your paper in *Annals*, a year of more ago I can plainly see that we have thought much alike & to a certain extent have come to similar conclusions. . . . I agree to the truth of almost every word of your paper & I daresay that you will agree with me that it is very rare to find oneself agreeing pretty closely with any theoretical paper; for it is lamentable how each man draws his own conclusions from the very same fact."

Even then, Darwin was savvy enough to plant his flag so there was no confusion as to who thought what first: "This summer will make the 20th year (!) since I opened my first-note-book, on the question how & in what way do species & varieties differ from each other.—I am now preparing my work for publication, but I find the subject so very large, that though I have written many chapters, I do not suppose I shall go to press for two years."[37]

ʃ

DARWIN WAS FORCED to set aside what he called his "trumpery feelings" over being scooped by Wallace to attend to a far more immediate crisis.[38] In June 1858, a deadly epidemic swept through the village of Down. Darwin's fifteen-year-old daughter Henrietta, or "Etty," was one of those stricken. Without knowing the cause but understanding its contagious nature, Charles and Emma Darwin agonized over the health of *all* their children. Epidemics were terrible and common events in England during this period. In Kent, the annual infant mortality rate—one of the best indicators of a community's health—was over 160 babies per thousand born, dying between birth and one year of age. Another 60 to 120 per thousand children between the ages of one and five years died each year there, too. The infant and child mortality rates were even higher in large, poorly sanitized cities like London. The most common causes of many of these early deaths were the once common childhood infections—diphtheria, whooping cough, bacterial meningitis, smallpox, scarlet fever, and measles—that we so rarely encounter today thanks to modern vaccines and antibiotics. Among the Victorian poor, the infant and childhood mortality rates were exacerbated by inadequate nutrition, overcrowding, and squalid living conditions.[39] Not even the wealthy were immune to these risks, and the loss of a child or two was a common experience in most British families. The Darwin clan was no exception. Emma and Charles had ten children; two of them died as infants and another died at age ten.[40]

On June 18, Henrietta awoke and complained of an almost burning sensation in the back of her throat. That night, the pain progressively got worse. Darwin wrote Hooker on June 23, explaining how the local doctor

Henrietta Darwin
as a young girl.

optimistically reported that "there was no choking, but immense discharge & much pain & inability to speak or swallow & very weak & rapid pulse, with a fearful tongue."[41] One can almost feel Darwin's relief when reading this letter in the archives. A youngster's sore throat during this era was often the harbinger of the childhood strangler known as diphtheria. Infection with the etiological bacteria *Corynebacterium diphtheriae* often causes suffocation. The microbe so inflames the throat that a thick, leathery patch of mucus and dead cells forms, completely closing off the airway. In most cases, this so-called "pseudomembrane" needs to be pierced and kept from resealing, thus allowing the child to breathe. For a modern-day reader, this may sound like a simple task, but the procedure known as intubation—the insertion of a straw-like tube through the obstruction and into the airway—did not exist in Darwin's day. And performing a tracheostomy—making a hole in the trachea by cutting through the neck—was a dire event fraught with complications, and many a parent watched their child choke or bleed to death. Even if a child escaped diphtheritic suffocation, there is a second phase to the infection that baffled physicians and parents for centuries. Just as the child appears to be recovering over several weeks, a cardiotoxin is silently produced by the diphtheria bacteria. This poison circulates through the body until it stops the heart from beating.[42]

Always a sickly child, this was hardly the first or last time Henrietta Darwin was the center of the Darwin family's attention and alarm.[43] Like her father, "Etty" suffered from poor digestion, although retrospective diagnosticians suggest her ill health may have been inspired more by attention seeking than organic disease. Between bouts of hypochondria, invalidism, and actual illness, she was at least healthy enough to help her father edit the proofs of *The Descent of Man*[44] and his *Autobiography*.[45] She also wrote and edited a two-volume biography of her mother, Emma.[46] Henrietta Darwin died at the age of eighty-four.

On June 24, Darwin "thanked God" that "after much suffering [Etty] is recovering." The evening before, however, they encountered another medical emergency. "Last night," Darwin wrote his cousin William Fox, "our Baby commenced with Fever of some kind."[47] That "Baby" was Charles Waring Darwin. The last of ten children, he was eighteen months old and severely developmentally delayed—a condition that has inspired a slew of diagnoses, especially when one considers the consanguinity of the marriage between two first cousins as well as Emma's age; she was forty-seven when she delivered him in 1857, under the sleepy spell of an anesthetic called chloroform, which had been recently developed as an obstetric tool by the cholera hunter, John Snow. On April 7, 1853, Dr. Snow famously administered chloroform to Queen Victoria when she delivered Prince Leopold.[48]

One common cause of cognitive, social, or developmental delays seen in a child with Charles Waring Darwin's medical and family history would be Down syndrome. This entity, however, was not described in the medical literature until 1866 by a British doctor named James Langdon Down, whose surname was only coincidental to both Down House and Down village. Dr. Down, the medical superintendent of the Royal Earlswood Asylum for Idiots in Surrey, called the syndrome "mongolism" because these children exhibited almond- or slanted-shaped eyes. It must be noted that terms such as "idiot," "moron," and "imbecile" were clinical jargon long before they became objectionable schoolyard insults. Hence, in an era where racism by whites against nonwhites was rampant, the phrase "mongoloid idiot" was long and widely used to describe the syndrome even if the cause remained unknown.[49] Nearly a century later, in 1959, geneticists ascertained the syndrome's molecular cause—an extra chromosome 21 in each cell of the body; hence the scientific name of trisomy 21.[50] It was not until the 1960s, after several physicians, nurses, and parents pointed out the offensive nature of the term "mongoloid," that pediatricians began referring to it by the far more neutral Down syndrome.

The physical attributes of Down children, however, are so striking that it is difficult to believe an astute observer like Darwin would miss describing a child with upward-slanting eyes or palpebral fissures (a skin fold that comes out from the upper eyelid and covers the inner corner of the eye), short neck, flattened facial profile and nose, small head, ears, and mouth, and a single crease across the center of the palms. What we do know for certain is that little Charles was developmentally challenged in an era when there were pitifully few ways to care for such children. These youths were often so devalued that many parents committed them to specialized, and unpleasant, institutions.

Charles Waring Darwin died on June 29, not from diphtheria but from scarlet fever. Heralded by a bright red rash that covers a child's torso and arms, scarlet fever raises a flock of goosebumps, causing the affected skin to feel like sandpaper.[51] The "British Hippocrates," Thomas Sydenham, characterized the rash as branny, "as if powdered by meal."[52] Other symptoms include swollen neck glands, a sore throat, scorching fevers, distinctive markings on the inner creases of the elbows, and brightly reddened taste buds, giving the tongue the appearance of an angry strawberry. A few weeks after the acute phase of scarlet fever subsides, the survivor endures an itchy flaking and peeling of the red sandpapery epidermis, just as a snake sheds its skin. Scarlet fever is highly contagious and among vulnerable children can spread with the speed of Mercury. Three days before his son's demise, Darwin confided to Lyell that "the Baby has much fever, but we hope not S. Fever.—what has frightened us so much is, that 3 children in village have died of Scarlet Fever, & others have been at death's door, with terrible suffering."[53]

Today we know that scarlet fever is caused by streptococcal bacteria, which secrete a powerful poison that quickly circulates through the body. Prior to the advent of antibiotics, this toxin often caused a repetitive, involuntary jerking motion of the face and extremities—quaintly referred to as St. Vitus's dance, after the patron saint of those afflicted with epilepsy. It can also yield rheumatic fever, kidney failure, heart disease, and, for some, an agonizing death.[54]

The obituary Darwin wrote for his son on July 2 is heartbreaking. Memorials, funeral announcements, condolence letters, and similar correspondence in Darwin's day were usually written on special stationery and envelopes edged with black ink, known as Victorian mourning paper. Although Emma and Charles Darwin used such paper for previous losses in the family, they may have been caught off guard by the toddler's sudden fever and death—a process that took only a few days—and had no such grieving materials in the house. Instead, Darwin pulled a sheaf of his favored foolscap and scratched out several pages of memories. There was only one trivial phrase scratched out and replaced. Contained within are a series of details on the little boy's antics:

> He was small for his age & backward in walking & talking, but intelligent and observant. . . . When crawling naked on the floor he looked very elegant. He had never been ill & cried less than any of our babies. He was of a remarkably sweet, placid, & joyful disposition; but had not high spirits, & did not laugh much. He often made strange grimaces & shivered, when excited; but did so, also, for a joke & his little eyes used to glisten, after pouting out or stretching widely his little lips. He used sometimes to move his mouth as if talking loudly, but making no noise, & this he did when very happy.

Darwin reminisced how the little boy loved to stand "on one of my hands" as he "tossed" him up in the air. Little Charles had a ready and "wicked little smile," loved kissing his father on the mouth, resting calmly in his lap, staring at Darwin's gentle face, and poking his fingers in his mouth. Darwin did, however, note the importance of teaching his son not to scratch or strike his siblings. Completing his memorial, Darwin wrote, "Our poor little darling's short life has been placid, innocent & joyful. I think & trust he did not suffer so much at last, as he appeared to do; but the last 36 [Emma later amended this time frame to be 24] hours were beyond expression. In the sleep of Death, he resumed his placid looks."[55] Henrietta Darwin later eulogized her youngest brother as having been born "without its full share of intelligence. Both my father and mother were infinitely tender towards him, but, when he died in the summer of 1858, after their first sorrow, they could only feel thankful. He had never learnt to walk or talk."[56]

When looking at the large marble and bronze statues of the great man—replete with massive forehead and brow, flowing beard, lumpy nose, and pensive face—it is hard to imagine Darwin dancing with his girls or playing soldiers with his boys.[57] But that he did and often. Darwin was a dot-

Darwin with his eldest son William.

Emma Darwin with son Leonard as a child.

ing, affectionate father, and great fun for any child. He encouraged his children to read books and think for themselves. He made up all sorts of games. He played croquet and billiards with the boys and caressed the faces of his girls. He gave them all endearing nicknames and took them on nature walks and through the Backs of his beloved Cambridge University, trips to the seashore, and excursions to London's great museums, gardens, and the Crystal Palace—which housed the Great Exhibition of 1851 and welcomed more than six million visitors that year. His sons assisted him in botanical experiments and breeding pigeons; he enlisted his wife and daughters in copying his letters and work into a more legible manner. The girls were educated at home by their mother, as was the custom of the day. The boys attended the finest preparatory schools for entry to and graduation from Cambridge University, except for his fourth son, Leonard, who attended the Royal Military Academy at Sandhurst.[58]

As with many Victorian men, Darwin was no stranger to the role of grieving father when little Charles died. His third child, Mary Eleanor, died twenty-three days after her birth in 1842, most likely from septic shock. His second child, Anne, or "Annie," died at age ten, in 1851, of what was then called bilious fever, the nineteenth-century humoral mélange of jaundice, nausea, vomiting, diarrhea, dehydration, and overwhelming sepsis. On April 23, 1851, Darwin and his sister-in-law, Fanny Wedgwood Allen, watched the little girl die in bed at the Malvern Hydropathic Institute, where Darwin had desperately taken her in search of a cure. Emma, who was eight months pregnant with their fifth son, Horace, stayed behind at Down. Darwin later told Emma that "when I gave her some water she said, 'I quite thank you;' and these, I believe were the last precious words ever addressed by her dear lips to me."[59] At noon, Fanny recalled, "we heard her breathe for the last time while the peals of thunder were sounding." Emma's diary for that date simply records, "12 o'clock."[60]

During her short life, the affection between Annie and her father was sweet to watch and, for Darwin, a joy to behold.[61] Six days after Annie's death, Darwins wrote his cousin William Fox, "Thank God she suffered hardly at all, & expired as tranquilly as a little angel.—Our only consolation is, that she passed a short, though joyous life.—She was my favor-

Annie Darwin.

ite child; her cordiality, openness, buoyant joyousness & strong affection made her most loveable. Poor dear little soul. Well, it is all over."[62]

Darwin wrote a memoir of Annie so that in "after years, if we live, the impressions now put down will recall more vividly her chief characteristics." It is lovingly archived and filed next to the one he wrote for little Charles. His closing thoughts are definitive and desolate: "We have lost the joy of the household, and the solace of our old age:—she must have known how we loved her; oh, that she could now know how deeply, how tenderly we do still & shall ever love her dear joyous face. Blessings on her."[63]

On the next page, he drew a map of the Great Malvern Priory cemetery, depicting the location of her grave, as if he would ever forget such information.[64] Her epitaph read, "A Dear and Good Child." The father, who could not bear to attend Annie's funeral, never ceased grieving her. The mere mention of Annie's name caused him physical distress. Henrietta Darwin recalled her father uttering her late sister's name only twice in the thirty-one years of his life that followed her death.[65]

Some have claimed that Annie's premature death was the reason Darwin turned away from his Christian faith. Many parents who lose children convert such tragedies into an angry atheism, but Darwin's religious fervor had been waning long before this tragic event.[66] The deaths of his first and second daughters, taken in concert with little Charles Waring's short life and developmental delays, have been interpreted by others as inspiration for the father's theory of natural selection and, to use a phrase Darwin did not originate, "the survival of the fittest." This explanation, too, holds little water for anyone familiar with the grief endured when losing a child. No parent—even one as biologically astute as Charles Darwin—views the death of his children as an example of nature's war at its most brutal. In the depths of such anguish, there is only love, regret, sadness, tears, and self-recrimination.

ʃ

Late in the evening of June 29, 1858, after little Charles exhaled his final breath, Darwin was deep in his grief. He wrote Hooker, "I hope to

God he did not suffer so much as he appeared. . . . Thank God he will never suffer more in this world." Darwin was so bereaved that he could not work, even as the clock was ticking on his priority issue with Wallace. He told Hooker, "I cannot think now on the subject, but soon will."[67] Not surprisingly, his digestive complaints of flatulence and stomach pain intensified—which Darwin duly recorded. In the months that followed, Emma worried whether her husband would hold up under the crushing weight of his grief, illnesses, and the massive amount of work he needed to complete so quickly. Fortunately, Darwin did endure. Aside from the love of his family and friends, it was his grand theory of all living things that sustained him.

# 2

# First

> I have been much gratified by a letter from Darwin, in which he says that he agrees with "almost every word" of my paper. He is now preparing his great work on "Species and Varieties," for which he has been collecting materials twenty years. He may save me the trouble of writing more on my hypothesis, by proving that there is no difference in nature between the origin of species and of varieties, or he may give me trouble by arriving at another conclusion; but, at all events, his facts will be given for me to work upon.
>
> —*Alfred Wallace, January 4, 1858*[1]

Most, if not all, scientists are obsessed with priority. In their realm, nothing compares to the claim, "I was there *first*." The victors are rewarded with far more than a footnote in a textbook or a tarnished plaque on the wall of a building eventually to be torn down and replaced with something else. Even the bountiful prizes billionaires bestow upon those making breakthroughs are but temporary glamour and gain. Discovering something important first remains a scientist's best shot at immortality. Yet history's behind-the-scenes perspective often reveals an unsettling characteristic of this high-stakes game. Since many minds make use of the same large figures and facts of their predecessors, there are often many actors trying to prove the same brilliant idea at the same time. The competition to be declared first can be brutal even when applied with a soft brush.[2] This was certainly the case with Alfred Wallace and Charles Darwin.

Why Darwin underestimated the progress Wallace reported between 1855 and 1858 remains a puzzle. In 1863, Charles Lyell recalled how he had "repeatedly urged [Darwin] to publish without delay, but in vain, as he was always unwilling to interrupt the course of his investigation; until at length Mr. Alfred R. Wallace, who had been engaged for years in collecting and studying the animals of the East Indian archipelago, thought out, independently for himself, one of the most novel and important of Mr. Darwin's theories."[3] Perhaps Darwin devalued Wallace as a mere prospector and purveyor of zoological specimens—a man without the proper breeding, family money, social connections, and advanced degrees. And he likely considered natural selection to be *his* brainchild and *his* alone. Indeed, a British upper-class sense of entitlement may have lulled Darwin into thinking he could take his time in publishing his work, willfully ignoring the possibility that others with more energy and ambition might supplant his ideas and claim to lasting fame.

Some have explained that Charles Darwin was a chronic procrastinator—cosseted and pampered at home, far away from the rough-and-tumble academic arena where one is constantly reminded of his competitors. The halls of scientific inquiry are littered with forgotten souls whose perfectionism prevented them from committing their ideas to the page—let alone publishing them for all to see and, possibly, denigrate or destroy. As with many nervous scholars, Darwin was fearful of crawling out on a limb only to have it sawed off by a competitor. A corollary to this explanation is that Darwin was stalling to gather more proof, more data, before committing his work to a scientific journal or as a scholarly tome. One might even posit that he was at a distinct disadvantage, since his illnesses and family obligations prevented him from conducting fieldwork in distant lands as he once did during his twenties.

By 1858, Darwin had already acquired most of the validation he needed to begin his book. He began this path during the 1830s, scribbling notes to himself in little leather-bound and brass-clasped notebooks, which he faithfully kept directly after his global expedition. Now preserved in the

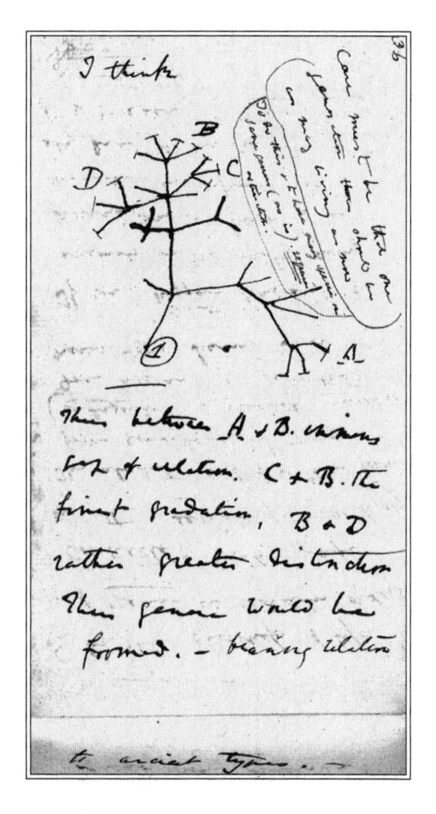


"Tree of Life," from Darwin's Notebook B on the transmutation of species.

Cambridge University Library Archives, they contain the seeds and sprouts of his theory of natural selection. Most famous is his diagram of the "Tree of Life," a stick figure he drew for his 1837 Notebook B depicting the transmutation of species. Darwin heads his schematic with two glorious words: *"I think."*[4] For more than two decades, he collected a mountain of corroborative evidence from his colleagues supporting and augmenting the theory that would soon make his name known around the world. His attempts to write it down includes a 35-page sketch on natural selection, scribbled in pencil in 1842, and a 189-page manuscript essay that he drafted in 1844. Shortly after, he hired an amanuensis to write up a 230-page copy.[5] Charles read portions of an incomplete manuscript on transmutation by natural selection "to Dr. Hooker as early as 1844 and some of the principal results were communicated to [Charles Lyell] on several occasions."[6] The proof of that pudding can be found in the tasting of Darwin's January 11, 1844, letter to Hooker:

> I have read heaps of agricultural & horticultural books, & have never ceased collecting facts—At last gleams of light have come, & I am almost convinced (quite contrary to opinion I started with) that species are not (it is like confessing a murder) immutable. Heaven forfend me from Lamarck nonsense of a "tendency to progression" "adaptations from the slow willing of animals" &c,—but the conclusions I am led to are not widely different from his—though the means of change are wholly so—I think I have found out (here's presumption!) the simple way by which species become exquisitely adapted to various ends. You will now groan, & think to yourself "on what a man have I been wasting my time in writing to."—.[7]

Moreover, Darwin was not exactly loafing while he put aside his theory on natural selection for so long. Even with all his chronic health issues, he composed important studies on his research aboard the *Beagle* (1839, and revised in 1845), coral reefs (1842), volcanic islands (1844), the geology of South America (1846), cirripedes, or barnacles (1851 and 1854), and several smaller projects. Nevertheless, it was only *after* Darwin read Wallace's 1856 article on "a natural arrangement of birds" that he returned to the divergence of various species lines by means of natural selection.[8]

Perhaps Darwin struggled so because he could only envision his *Origin* as a massive, intimidating, mutivolume study describing the nature of, well, nature. After receiving Wallace's letter, however, he saw that he no longer had the time to create an encyclopedic account. The book that resulted eighteen months later, *Origin of Species*, is actually an abbreviated version—or as Darwin referred to it, an "abstract"—of the more ambitious work he fantasized about for so long but never completed.

ſ

Darwin was right to worry that his ideas would inspire harsh criticism, or worse, from those who held that God created each species perfectly, in His image, and without the need for change over time. Darwin correctly assumed there was a legion of disciples lying in wait and ready to attack him. Darwin was, after all, boldly proposing a new set of natural laws dictating how species (including humans) actively evolved across eons, often in response to changing environmental, biological, or human-made conditions—a theory that ran counter to the greatest authority of his world—the Holy Scriptures.

In a June 1860 letter to Lyell, seven months after *Origin* was published, Darwin groused over the ongoing problem of the " 'Deification' of Natural Selection." Even though natural selection was central to his theory, he insisted, it "does not exclude still more general laws i.e., the ordering of the whole universe." For Darwin, natural selection was "to the structure of organized beings, what the human architect is to a building. The very existence of the human architect," he argued, "shows the existence of more general laws; but no one in giving credit for a building to the human architect, thinks it necessary to refer to the laws by which man has appeared." What perturbed Darwin most was that he had to keep explaining this point to his colleagues who should have known better. He complained about the double standard that then existed between practitioners of mathematics-based sciences, such as physics, and the still observation-bound life sciences: "No astronomer in showing how movements of Planets are due to gravity, thinks it necessary to say that the law of gravity was designed that the planets should pursue the courses which they pursue.—I cannot believe that there is a bit more interference by the Creator in the construction of each species, than in the course of the planets." The conundrum, Darwin knew, was a direct result of the wide allegiance to the Book of Genesis, Paley's *Natural Theology*, and too many Sunday sermons declaring that "living bodies" required an intelligent Creator.[9]

Fortunately, Emma Darwin, a devout Christian who regularly attended church services, supported her husband's intellectual thought experiments. Soon after they married, in February 1838, Emma wrote him a "beautiful letter," which he "safely preserved" among his papers. Her note was the scientific equivalent of a carte blanche, albeit phrased in a manner modern readers would find a bit too submissive and a whole lot racist. Emma lovingly wrote to Charles, "Everything that concerns you concerns me & I should be most unhappy if I thought we did not belong to each other forever." The next sentence, however, mars any beauty the first contained: "I am rather afraid my own dear Nigger will think I have forgotten my promise not to bother him, but I am sure he loves me & I cannot tell him how happy he makes me & how dearly I love him & thank him for all his affection which makes the happiness of my life more & more every day."[10]

It is, at the very least, upsetting to recall Emma's references to Charles with the "N-word—" supposedly to describe his habit of "working like a slave" or that he was "her slave."[11] In 1884, Francis Darwin recalled that his father "often said in fun that the woman was the real master in a house" and that he called her "Mammy," while his mother's "pet name" for his father "was 'Younigger'—(pronounced in one word)—or 'my nigger.' "[12] In other letters, Emma sometimes shortened the salutation to "My dearest N." and Charles occasionally signed communiques to Emma as "your old Nigger, C.D."[13] Charles and Emma did oppose the institution of slavery. Nonetheless, as with so many of their social class, the Darwins still considered people of color to be their inferiors and even a source of nasty ridicule.

In the late eighteenth and early-to-mid nineteenth century, several men of science had broached the origin of species and evolution, but none displayed the specificity or brilliance of Charles Darwin.[14] One of the most intriguing proponents was Charles's grandfather, Erasmus, who speculated on evolution in his 1794 scientific treatise *Zoonomia*: "As air and water are supplied to animals in sufficient profusion, the three great objects of desire, which have changed the forms of many animals by their exertions to gratify them, are those of lust, hunger, and security."[15] Erasmus also wrote poems on evolution, such as *The Botanic Garden* (1791), a discourse on "the economy of vegetation," and his posthumously published *Temple of Nature* (1803), which described the evolution of microscopic creatures to civilized human beings.[16]

In 1809, the French naturalist Jean-Baptiste Lamarck (1744–1829) famously outlined his view of evolution in *Philosophie Zoologique* (*Zoological Philosophy, or Exposition with Regard to the Natural History of Animals*).[17] He hypothesized that species changed by acquiring and passing down inheritable traits, as in his famous example of giraffes. These animals acquired their long necks after years of reaching for vegetation to eat that was formerly beyond their reach and then, Lamarck theorized, passing on the acquired trait of long necks to their offspring.

One of the most popular speculations on the origin of life on earth was an anonymously written book, *Vestiges of the Natural History of Creation*. Published in 1844, *Vestiges* sold more than one hundred thousand copies, appeared in several successive and revised editions, and "sparked

Erasmus Darwin.

one of the greatest sensations of the Victorian era"—it was so controversial and yet compelling. James Secord, the editor in chief of the Cambridge University Darwin Correspondence Project, wrote, "the book was banned, it was damned, it was hailed as the gospel for the new age."[18] Its author, Robert Chambers, was a talented publisher and writer from Edinburgh. Applying a breezy writing style and a talent for organizing his arguments into an understandable, narrative arc, Chambers offered cosmic notions of how the solar system was formed and the idea that life on earth emerged from a gaseous admixture. He suggested theories on the transmutation of species, spontaneous generation, and how to interpret the fossil record.[19]

Darwin "never paid down hard cash for *Vestiges.*" Instead, he read the book "with fear and trembling [in] the bustling, flea-infested British Museum" while visiting London on November 20, 1844, for a Geological Society council meeting. In Victorian popular culture, *Vestiges* preempted Darwin because it "advocated a natural origin for species in a framework of material causation and universal law," even though it erroneously suggested that new species simply emerged from old forms without the intermediate states that Darwin hypothesized.[20] Darwin may have hoped that

Jean-Baptiste Lamarck.

the book's popularity heralded the public's interest in what would become his far more scientific account of the topic.[21] But he also labored to distance himself from the book so as not to let its many inaccuracies seep into his own explorations.[22] As he told Hooker on January 7, 1845, "the writing & arrangement are certainly admirable, but his geology strikes me as bad & his zoology far worse."[23]

Despite its huge readership, *Vestiges* was criticized by both elite British scientists and the Anglican church. Just as Darwin's book became the talk of the 1860 BAAS annual meeting at Oxford, so, too, was *Vestiges* the cause célèbre of the 1845 meeting in Cambridge. For example, John Herschel, the astronomer, natural historian, mathematician, and one of the groundbreaking developers of photography, alluded to Chambers's "dangerous" volume while delivering his presidential address. Herschel was already one of Cambridge University's leading lights when Darwin was an undergraduate. From 1833 to 1838, he sailed to and settled in South Africa, with the express purpose of charting stars and other celestial objects in the Southern Hemisphere and, with his wife, describing the flora of the Cape.[24] When the *Beagle* dropped its anchor in Cape Town's harbor on June 3, 1836, Dar-

Robert Chambers.

John Herschel.

win and his captain, Robert FitzRoy, made a visit to see the great scientist and discuss their work.

On the day he was scheduled to speak, Herschel had a terrible cold, which was not helped by his self-confessed weak delivery and the "exceedingly wet" weather.[25] Without mentioning *Vestiges* by its title, he made the book's dangers clear by declaring "there was never a period in the history of science" when the moral influence of clerics and scientists was "more needed" to correct "the propensity to crude and over-hasty generalizations." For Herschel, Chambers's book was not only poorly written; it was based on "pure speculation" and, worse, faulty mathematics.[26]

A few days later, the *Cambridge Chronicle* reported how Samuel Wilberforce, then the forty-year-old dean of Westminster Cathedral, rose to denounce the author of *Vestiges* and bridge the relationship of science and Christian faith. Any man "who would dare to stand up in Cambridge and attempt to prove the connection of irreligion with science on the spot where the mighty Newton walked," he told the cheering crowd, was rash, if not foolish. After all, he preached, "a philosopher was the fittest person to receive Christianity, and if he became irreligious, such irreligion was not in consequence but in spite of his philosophy—(loud cheers)."[27]

Two years later, when the 1847 BAAS meeting was held at Oxford, the newly appointed bishop of that diose—the same Samuel Wilberforce—was invited to deliver a special sermon to the association's members. From the pulpit of Oxford University's Church of St. Mary the Virgin, he assailed evolutionists in general and Chambers in particular as he preached on "pride, a hindrance to true knowledge." Wilberforce warned that the theorist "grows to deal boldly with nature, instead of reverently following her guidance. He seals his heart against her secret influences. He has a theory to maintain, a solution which must not be disproved . . . and once possessed of this false cypher, he reads amiss all the golden letters around him."[28] The thunderous applause from the BAAS members confirmed Wilberforce's belief that he, too, was a bona fide member of that elite group of thinkers and doers—men of science—an assumption that would prove disastrously wrong only thirteen years later.

There was, of course, a major difference between these earlier speculative notions and Darwin's work, even if his critics did not always grasp the subtleties of his arguments. Unlike previous theories on the origin of species and evolution, Darwin introduced five main premises of natural selection. These foundational principles can be encapsulated with an acronym many of us learned in high school biology—**VISTA**:

**V**ariation: Offspring in a given species are not carbon copies of their parents; they vary depending on which traits they inherit.

**I**nheritance: Offspring *do* inherit many traits from their parents; some are positive, some are negative, others can be neutral or acquired and deleterious mutations.

**S**election: The offspring with positive traits that give them an advantage in survival tend to be selected out and reproduce at a greater rate than those with so-called negative traits that do not confer an advantage in survival.

**T**ime: Over eons of time, more of a species' offspring will contain the positive trait and they will thrive and reproduce at the expense of those carrying negative traits.

**A**daptation: Those better adapted to a particular environment and other factors can evolve to become a new species even though they share a common ancestor with the offspring that did not adapt.[29]

To a well-educated, modern-day reader, these points seem obvious. But when Darwin first explained them, such concepts were world-shattering. Natural selection divorced the scientist's gaze from the monotheistic deity's creative powers. Darwin looked not to the heavens but instead directly at nature—in the field and by way of experimentation—to describe the struggle for survival in a hostile world.

REMINISCING as an elderly man, Darwin insisted that "the most important event" in his life was the "voyage of the *Beagle*." This expedition, he wrote, "determined my whole career."[30] Others date the zenith of his career to November 1859, when *Origin* first appeared in print. In terms of his moral character, however, Darwin's most significant moments occurred between June 18 and July 1, 1858, as he found himself in a competitive race for primacy while contending with a major epidemic in his town, which caused serious illness in one child and the death of another. The easiest, and most sinister, path for Darwin to have taken after receiving Wallace's letter would have been to burn the manuscript and, given the long and watery distance the letter had traveled, pretend he never received it. Who would know? The answer is obvious. Charles Darwin would know. A firm subscriber to the Victorian gentleman's code of decency, honor, and honesty, he could not bear the base act of cheating to erase a competitor's work. Unlike so many scientists—indeed, unlike most people, period—Darwin conquered the temptation for singular glory and found a third, albeit imperfect, path forward.

On June 18, 1858, Darwin launched what appears, at first glance, one of his most honorable acts. This deed is contained in his letter to Charles

Lyell that evening, introducing Wallace and forwarding his essay. When reading these yellowed pages, filled with black ink that has turned brown with age, one can feel Darwin's nettled emotions: "Your words have come true with a vengeance that I should be forestalled. . . . I never saw a more striking coincidence. If Wallace had my M.S. sketch written out in 1842, he could not have made a better short abstract! Even his terms now stand as Heads of my Chapters." Darwin added, "I shall of course at once write & offer to send to any Journal." To Lyell, he privately mourned more than his dead son: "So all my originality, whatever it may amount to, will be smashed." To his credit, Darwin closed with a wish that Lyell would "approve of Wallace's sketch" so that he could quickly report the outcome to Wallace.[31]

One week later, on June 25, Darwin wrote Lyell a letter capturing his angst over the intellectual—if not existential—crisis:

> I am very, very sorry to trouble you, busy as you are, in so merely personal an affair. . . . I should not have sent off your letter without further reflection, for I am at present quite upset, but write now to get subject for time out of mind. . . . But as I had not intended to publish any sketch, can I do so honorably because Wallace has sent me an outline of his doctrine?—I would far rather burn my whole book than he or any man should think that I had behaved in a paltry spirit. . . . This is a trumpery letter influenced by trumpery feelings. . . . I will never trouble you or Hooker on this subject again.—"[32]

Darwin broke his vow of silence the very next day—June 26. He beseeched his "Lord Chancellor" Lyell to forgive him for writing still another letter and repeated his desire to handle the matter in an honorable and yet still favorable manner: "It seems hard on me that I should be thus compelled to lose my priority of many years standing, but I cannot feel at all sure that this alters the justice of the case."[33]

If such anguish was not enough for him to bear, he wrote to his friend Hooker three days later, on June 29, to tell him that "poor Baby died yesterday evening. . . . I have received your letters. I cannot think now on the subject [of Wallace' report], but soon will." What remains astounding was that in the depths of his grief over his son's death, he was able to summon up enough strength to remind Hooker, "I can easily get my letter to Asa Gray copied, but it is too short." Here Darwin was referring to an 1857 missive he wrote to Asa Gray detailing his theory of natural selection. Gray was the most distinguished American botanist of his generation. As a Harvard professor, he developed a central taxonomy of North American plants, developed relationships with all the leading scientists of his era, and wrote many important books and manuals on botany. In the next sentence of Darwin's letter to Hooker, however, he reverted to the role of grieving father and reported how "poor Emma behaved nobly & how she stood it all I cannot conceive. It was wonderful relief, when she could let her feelings break forth—God Bless you.—you shall hear soon as I can think."[34]

Who could blame Darwin for being unable to concentrate on anything other than the death of little Charles? In his state of mourning, however, Darwin *did* remain focused on his claims of scientific priority for natural selection. That same evening—hours after he told Joseph Hooker he could

Asa Gray,
circa 1870–1888.

not "think on the subject"—he did exactly the opposite and thought long and hard on it. Retiring to his study after dinner, he wrote a second letter to Hooker, who had already asked him for his documenting materials and "papers at once." In it, Darwin asked Hooker not to "waste much time. It is miserable in me to care at all about priority." He also made certain to send along Wallace's paper and a hastily composed "abstract of an abstract" of his letter to Asa Gray, noting that it "gives most imperfectly **only** *the means of change & does not touch* on reasons for believing that species do change." Perhaps to regain an ounce of dignity over Wallace's beating him to the post, he moped through the next few words he scribbled: "I hardly care about it."[35]

Whatever he thought or wrote that night, Charles Darwin *did* "care about it." His self-preservation, desperation, and pride of ownership all kicked into high gear as he thanked Hooker for his support and promised to send him the "sketch of 1844 **solely** that you may see by your own handwriting that you did read it.—I really cannot bear to look at it." Back and forth he went, up and down the emotional register, with each successive sentence he composed. A shaky Darwin finally closed his letter to Hooker: "The table of contents will show what it is. I would make a similar, but shorter & more accurate sketch for *Linnean Journal*.—I will do anything. God Bless you my dear kind friend. I can write no more. I send this by servant to Kew." That very night, one of Darwin's servants rode 20 miles in pitch-black darkness, from Down to Kew. That man—probably Darwin's butler, Joseph Parslow—and his horse risked their necks so that Hooker would have a few full days' time to compose the appeal to the Linnean Society.[36]

# 3

# Survival of the Fittest

> We were driven from home by Scarlet Fever, which caused the death of our poor dear little youngest child & was very bad in the village. We had other & bad illness in the House. As yet the sea has not done much for us. I the more regret that we shall not see you at Down at the time proposed, (but I hope at some other time) as I should be extremely glad (& grateful) to hear your objections to my species speculations. The difficulties which I can see are many & grave. I am now writing a pretty full abstract of all my notions on this subject.
>
> —*Charles Darwin to John Stevens Henslow, August 4, 1858*[1]

Lyell and Hooker went to work orchestrating the promotion of Darwin's work and the subtle devaluation of Wallace's essay. Darwin—whether he acknowledged it or not—was hardly an innocent naïf in the academic gamesmanship. When he enlisted Lyell and Hooker to settle the matter on his behalf, he knew he could not have placed his work in more agile hands. Perhaps most potent, by appointing his friends to manage the matter in the public arena, Darwin could feign plausible deniability if things went awry or if Wallace put up a fight.

Lyell and Hooker's machinations were so slick that most of their peers missed or ignored them. On reflection, however, the successful scheme they hatched remains difficult, if not impossible, to bless. Their task was enabled by a tightly knit web of British scientists, where almost all the professorships, fellowships, learned society memberships, curatorial positions,

Carl Linnaeus.

and funding streams were in the hands of an exclusive group of well-heeled men who studied the physical and natural sciences, often from the comfort of their book-lined studies. This self-selected, self-perpetuating group could be relied upon to favor Darwin and shout down the claims of an uppity outsider who dared break their ranks.

Lyell and Hooker began their tasks by assigning Darwin to write an essay describing the ideas he developed over the past twenty years—drawn from his notes, letters, and manuscripts. They would, in turn, package these materials and present them in a formal venue alongside Wallace's paper. There were two main thrusts to their design of attack. First, Lyell and Hooker framed all their communications with tact, generosity, and propriety mixed into a soufflé of mock objectivity and moral authority. The second, far enduring move, was to purposely order Darwin's name ahead of Wallace's at every instance and in each correspondence.

Unfortunately, most of their academic colleagues had already left the steamy, summer heat of London, Oxford, and Cambridge for cooler climes. As a result, there were not many high-level scientific meetings scheduled until the autumn, or Michaelmas, term. Yet to risk delay meant that Wallace might simply publish his essay elsewhere and win the scientific sweepstakes. Fortunately, Hooker and Lyell were both elected fellows of the Linnean Society, where they enjoyed the privilege of submitting and presenting papers at their meetings. Founded in 1788, the Linnean Society is the world oldest learned society devoted to the study of nature. It was, and remains, the institutional flame keeper of the Swedish botanist, physician, zoologist, and taxonomist Carl Linnaeus. Darwin, too, was a fellow of the society, having been admitted in 1854; Wallace would not achieve such lofty status until 1878.

On June 30, 1858, Lyell and Hooker wrote a memorial letter to the Linnean Society introducing two papers on the "same subject, viz. the Laws which affect the Production of Varieties, Races, and Species," containing the results "of two indefatigable naturalists, Mr. Charles Darwin and Mr. Alfred Wallace."[2] They described how both men "independently and unknown to one another conceived the same very ingenious theory to account for the appearance and perpetuation of varieties and of specific forms on our planet [and] may both fairly claim the merit of being original thinkers in this line of inquiry," even though neither had yet published their results.

Here resides the procedural rub: In almost every modern case of scientific priority, the gold standard was, and remains, being the first one to publish a definitive paper in a scientific journal. In Wallace's case, his manuscript was an unpublished stack of pages that he had sent to a colleague just before formal submission for publication. But in 1855 he had published his theoretical paper in the *Annals and Magazine of Natural History*, which vaguely hit upon the idea of natural selection and species differentiation. This was the same paper that Darwin had read and about which he wrote a congratulatory note to Wallace in May 1857.[3]

At the time he received Wallace's 1858 letter and manuscript, Darwin had only unpublished drafts of manuscripts, albeit many more pages written over a longer period.[4] Yet granting Darwin priority based on such dating is hardly conclusive because the rule of thumb is not *write* or perish; it is *publish* or perish. Thus, neither Wallace nor Darwin could claim priority in the tried-and-true, established manner of scientific reporting.

For their brief to the Linnean Society, Lyell and Hooker introduced the first section of Darwin's 1844 manuscript, which was "devoted to 'The Variation of Organic Beings under Domestication and in their Natural State;' and the second chapter of that Part . . . 'On the Variation of Organic Beings in a State of Nature; on the Natural Means of Selection; on the Comparison of Domestic Races and True Species.' "[5] To buttress Darwin's priority, Hooker and Lyell made certain to enclose the "abstract of a private letter addressed to Professor Asa Gray, of Boston, U.S., in October 1857, by Mr. Darwin, in which he repeats his views, and which shows that these remained unaltered from 1839 to 1857."[6]

Lyell and Hooker next pressed their fingers firmly on the scale by quoting from Wallace's February 1858 letter to Darwin—the same one that included his essay on natural selection—asking "that it should be forwarded to Sir Charles Lyell, if Mr. Darwin thought it sufficiently novel and interesting." Darwin, in Lyell and Hooker's carefully constructed narrative, was so impressed by Wallace's work that he sent it to Lyell for "the Essay to be published as soon as possible." To make their noble man look even nobler, Hooker and Lyell reported that they only agreed to do so "provided Mr. Darwin did not withhold from the public, as he was strongly inclined to do (in favor of Mr. Wallace), the memoir which he had himself written on the same subject, and which, as before stated, one of us had perused in 1844, and the contents of which we had both of us been privy to for many years."[7] Hooker and Lyell proposed that since neither "published his views, though Mr. Darwin has for many years past been repeatedly urged by us to do so, and both authors having now unreservedly placed their papers in our hands, we think it would best promote the interests of science that a selection from them should be laid before the Linnean Society."[8]

It was, however, the prolix, conditional, and final sentence of Lyell and Hooker's petition that sealed Darwin's place in history:

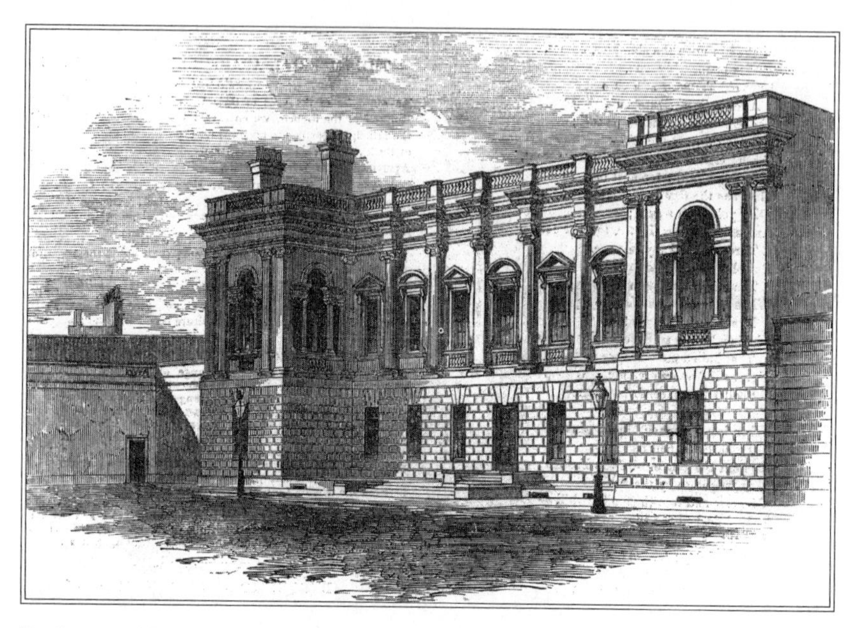

Burlington House.

> On representing this to Mr. Darwin, he gave us permission to make what use we thought proper of his memoir, &c.; and in adopting our present course, of presenting it to the Linnean Society, we have explained to him that we are not solely considering the relative claims to priority of himself and his friend, but the interests of science generally; for we feel it to be desirable that views founded on a wide deduction from facts, and matured by years of reflection, should constitute at once a goal from which others may start, and that, while the scientific world is waiting for the appearance of Mr. Darwin's complete work, some of the leading results of his labors, as well as those of his able correspondent, should together be laid before the public.[9]

THE LINNEAN SOCIETY was tucked into the east corner of the first floor of the stately Burlington House—just down the road from Piccadilly

Circus and across the street from Messrs. William Fortnum and Hugh Mason's fine food and grocery shop. The society had adjourned its business until the autumn owing to the death of its president, Robert Brown, on June 10. Brown was the Keeper of the Botanical Department at the British Museum and is credited with naming the headquarters of all living cells—the nucleus. He is also known for Brownian motion—the random movement of particles suspended in a liquid or gas.[10] Respect for the dead notwithstanding, the issues raised in the Hooker-Lyell appeal were so important that the society's officers decided to hold an extraordinary summer session on the evening of July 1, 1858. Both Darwin and Wallace were invited to present papers—along with several other agenda items that required attention but had been delayed because of Brown's death.[11]

The members typically met in a square chamber known as the Council Room, in what is now the Tennant Gallery of the Royal Academy of Arts.[12] The audience for the Darwin-Wallace meeting consisted of about thirty members all dressed in heavy frock coats and sweating profusely. Most of them were well known to Darwin even if the passage of time—with its attendant eating, drinking, and sitting—had rendered their faces corpulent, red, and wrinkled. As the evening wore on, some of them struggled to stay awake. They were helped by the benches and low-back chairs they were sitting on; such furnishings prevented any serious attempt at slumber, provided no lumbar spine support, and soon proved hard on all but the beefiest of rumps. Some of these men might have kept themselves alert by noticing the room's ornate architecture and furnishings. The heavy ceiling beams were enriched by crown plaster molding ornamented by dentils and acanthus-scrolled modillions, interspersed with flowers. The six-paneled mahogany wood doors were topped by ornate triangular architraves. Along the walls were a series of gas-lit brass sconces—each one blanketing the room's cream-colored paint with a coat of soot. Below them was a birchwood planked floor, sanded and smoothed to shiny perfection before being covered with a thick Persian rug.[13]

No transcript of the proceedings exists, but it is likely that the society's secretary, John Joseph Bennett, read Darwin's and Wallace's papers aloud to the members. Bennett was a botanist who had toiled for over thirty years as assistant Keeper of the British Museum's Herbarium and Library. That is, until Robert Brown's unexpected death, when Bennett was promoted as the museum's full-fledged, botanical Keeper. Hooker and Lyell were seated in front, ready to spring into action and address any queries their colleagues might have.

Darwin and Wallace were conspicuously absent. Wallace, who did not yet belong to the Linnean Society, was 8,500 miles away in Dorey, a village in northwest New Guinea. Darwin had a more distressing excuse for not coming into London—beyond his ailing gut—specifically, the village's epidemic, Etty's illness, and little Charles's death. Writing to Hooker on July 5, Darwin reported that he and Emma were "more happy & less panic-struck, now that we have sent out of House every child & shall remove Etty, as soon as she can move." By now, Etty's diagnosis had been revised from diphtheria to scarlet fever. In early July, one of the nurses Emma hired to take care of Etty—their butler's daughter Jane Parslow—developed "an ulcerated throat and quincy [sic] & the second [nurse] is now ill with the Scarlet Fever but thank God recovering. You can imagine how frightened we have been. It has been a most miserable fortnight."[14]

THREE DAYS AFTER Lyell and Hooker made their formal request for an extemporaneous meeting of the Linnean Society, July 4, Darwin covered his tracks by writing to Asa Gray. He asked, on the "unlikely chance," if Gray still had the original version of the "little sketch of my notions of 'natural Selection,'" so that he might check the date of his letter for the Linnean Society members. Explaining his recent receipt of Wallace's paper, Darwin described how Lyell and Hooker "urged me with much kindness not to let myself to be quite forestalled & allow them to publish with Wallace's paper an abstract of mine." He reminded Gray that "the only very brief thing which I had written out was a copy of my letter to you." Applying a modicum of grace, Darwin asked Gray not to "hunt for" the original letter but in case he did, Darwin was "sure it was written in September, October, or November of last year." He signed off with a revealing valediction: "I have troubled you with a long story on this head; so pray forgive me & believe me. Yours very sincerely, Charles Darwin."[15]

Only a day later, Darwin was already skittish as he equivocated to Hooker: "I can easily prepare an abstract of my whole work, but I can hardly see how it can be made scientific for a Journal, without giving facts, which would be impossible. Indeed, a mere abstract cannot be very short.—Could you give me any idea how many pages of Journal, could probably be spared me?" Darwin went on to decline Hooker's invitation to visit Kew to discuss matters fully. Referring to the immediacy of his family's threat from the scarlet fever epidemic in Down, he wrote: "Our children are too delicate for us to leave; and I should be mere living lumber." Emma and Charles sent their children—with the exception of the recuperating Etty—from Down to the home of Emma's sister, Sarah Elizabeth Wedgwood, who lived in nearby Tunbridge Wells. Darwin promised more work on the "abstract" once things quieted a bit: "Directly after my return home, I would begin & cut my cloth to my measure.—If the Referees were to reject it as not strictly scientific I would, perhaps publish it as pamphlet."[16] He also asked Hooker's help in explaining the Linnean Society proceedings to Wallace in a manner that ensured that he did not look like a scientific predator: "I should like this, as it would quite exonerate me. If you would send me his note, sealed up, I should forward it with my own, as I know address &c."[17]

The 1858 scarlet fever epidemic was to ever remain imprinted in Darwin's mind. Writing Hooker three years later—after one of the botanist's children contracted the dreaded microbe—Darwin commiserated, "I can sympathize with you about the fear of Scarlet Fever: To the day of my death, I shall never forget the sickening fear about the other children after our poor little baby died."[18] This was no exaggeration. At this point in human history, there were only two ways to combat this deadly, infectious scourge; going mano a mano, so to speak, by contracting the streptococcal microbe and cheering on one's immune system to victory; or, running away to a remote place before ever being infected, thus protectively sequestering from those who were contagious.[19]

By July 13, Charles, Emma, and Etty had reunited with the rest of their brood in Tunbridge Wells to prepare for their communal escape to the Isle of Wight. Darwin must have sighed in relief as he wrote Hooker:

> I always thought it very possible that I might be forestalled, but I fancied that I had grand enough soul not to care; but I found myself mistaken & punished; I had, however, quite resigned myself & had written half a letter to Wallace to give up all priority to him & should certainly not have changed had it not been for Lyell & yours quite extraordinary kindness. I assure you I feel it, & shall not forget it. I am **much** *more* than satisfied at what took place at Linn. Society—I had thought that your letter & mine to Asa Gray were to be only an appendix to Wallace's paper.[20]

Glowing with satisfaction, Darwin confessed, "You cannot imagine how pleased I am that the notion of Natural Selection has acted as a purgative on your bowels of immutability. Whenever naturalists can look at species changing as certain, what a magnificent field will be open,—on all the laws of variation,—on the genealogy of all living beings,—on their lines of migration, &c." Darwin's authorial angst, however, remained undiminished: "How on earth I shall make anything of an abstract in 30 pages of Journal, I know not, but will try my best."[21] Hooker assured Darwin that the Linnean Society would allow for up to 150 pages, perhaps more, to win the manuscript from any other publisher.[22] To complete the circle, that same day Darwin posted a package to Wallace containing both his and Hooker's explanations of their plan to publish their brief pieces that Lyell and Hooker presented to the Linnean Society members on July 1. It was sent to a general delivery address in Singapore, where Wallace would eventually pass through after completing his exploration of Borneo. In the meantime, Darwin distracted himself with the "fun here in watching a

Shanklin, Isle of Wight, circa 1900.

slave-making ant. . . . I have seen the migration from one nest to another of the slave-makers carrying their slaves (who are *house* & not field niggers) in their mouths!"[23]

NESTLED IN THE English Channel, the Isle of Wight is a seaside resort preferred by well-heeled Englishmen and their families for its easy access by ferry from either London or Portsmouth and for the striking coastal chalk and sand formations. Queen Victoria and Prince Albert made it the site of their summer home, Osborne House, which Albert designed to look like an Italian Renaissance palazzo. The Darwin family's destination was hardly as opulent as Victoria's but still required two full carriages for the family, their luggage, a maid, butler, and governess, a cat, and a wheelchair for Henrietta.

Once unpacked and settled, Darwin walked along the island's beaches and explored the local marine life, such as the thistle seeds and Coleoptera beetles washing up in the tide. He collected the beetles and brought them back to life for analysis. Even in crisis, Darwin was able to describe the Isle of Wight as "the nicest sea-side place" he had ever visited. "We like Shanklin better than other spots on S[outh] coast of the Island, though many are charming and prettier," he wrote Hooker.[24] Emma and the Darwin children—even the still infirm Henrietta—managed to regain their footing and frolicked and thrived on the seashore. By July 18, safely sequestered on the Isle of Wight, Darwin was comfortable enough to tell Lyell: "I am more than satisfied and do not think that Wallace can think my conduct unfair, in allowing you & Hooker to do whatever you thought fair. . . . I am going to prepare a longer abstract; but it is really impossible to do justice to the subject, except by giving the facts on which each conclusion is grounded & that will of course be absolutely impossible."[25] Tragedy struck the family again, that same day, when Darwin's older sister, Marianne Darwin Parker, died in Wales at the age of sixty.

ʃ

AT KEW, Joseph Hooker supervised the publication of the July 1 Linnean Society meeting presentations, from the editorial endorsements to reading the proof sheets. On July 21, after reading the proof sheets, Darwin moaned to Hooker, "I am disgusted with my bad writing. I could not improve it, without rewriting all, which would not be fair or worthwhile, as I have begun on better abstract for Linn. Soc. My excuse is that it *never* was intended for publication.—I have made only a few correction[s] in style; but I cannot make it decent, but I hope moderately intelligible. I suppose someone will correct the revise.—(Shall I?)"[26]

The final versions of the presentations read at the July 1 meeting appeared in the August 1858 issue of the *Journal of the Proceedings of the Linnean Society (Zoology)*.[27] The byline left little to chance: "Charles Darwin, Esq., F.R.S., F.L.S., & F.G.S. [Fellow of the Royal Society of London, the Linnean Society, and the Geologcial Society], Alfred Wallace, Esq." Hooker later claimed that the order of the paper's authors was "strictly alphabetical," but the notes of this sonata sound more flat than sharp. He must have understood the published version's implicit power for convincing readers that Darwin was listed first because he was the paper's "first" author—and hence creator of the novel theory. If only Wallace had sent his essay to a journal before querying Darwin, we might be celebrating Wallace in singular fashion today. If so, Darwin may have gone down in history as an odd chap who once traveled the world aboard the HMS *Beagle* and wrote an important book about barnacles. With the hindsight of more than a century and a half, one must wonder how Darwin would have felt had he been placed in the submissive, second-author's position. What might he have said if Wallace had been the man backed by well-placed advocates who thought it "fair" to accord *him* first billing? Alas, such speculation is entirely moot. The bell of discovery has long since pealed, and nothing can attenuate its loud reverberations.

Wallace would not receive the news of the July 1 Linnean Society meeting and the August publication of its minutes until October 6, 1858. Wallace's reply took a few more months to arrive to London, only in reverse. As with many chronic worriers, Darwin spent an inordinate amount of time agonizing over things that never came to pass. Alfred Wallace wrote Hooker a letter of appreciation on the same day he received Darwin's explanatory letter. After all, he told Hooker, "it would have caused me much pain & regret had Mr. Darwin's excess of generosity led him to make public my paper unaccompanied by his own much earlier & I doubt not much more complete views on the same subject, & I must again thank you for the course you have adopted, which while strictly just to both parties, is so favorable to myself."[28]

Wallace was not stupid—every author desires singular credit for his creations—but he also knew that only fellows of the Linnean Society could publish papers in their prestigious journal unless an elected member endorsed the paper of an outsider. The joint arrangement, then, represented a major coup for Wallace's career. Being so closely linked to Charles Darwin in an influential scientific journal was no minor accomplishment for the still obscure naturalist. It elevated Wallace's work and reputation in ways that the younger man had long dreamed of while, at the same time, subordinating his discovery to Darwin's.

For the rest of his career, Wallace remained in awe of Darwin, even when he disagreed with some of his hero's ideas. In 1889, Wallace coined the term *Darwinism* in support of "the overwhelming importance of natural selection over all other agencies in the production of new species." He adopted the role "of being the advocate for pure Darwinism."[29] Such deference did little to change the popular notion that he was the second banana in terms of who developed the theory of natural selection. Nevertheless, the exchange of letters, ideas, and theories between Wallace and Darwin—not to mention the subtle but effective machinations of Lyell and Hooker—represents a peaceful conclusion to what could have easily descended into an unholy war of words. For these powerful British men of science, it was a quiet and "proper conclusion" for the rather sticky problem of intellectual primogeniture.[30]

Some Linnean Society members were indifferent to the Darwin-Wallace presentations. In May 1859, for example, Thomas Hornsey Bell presented his presidential "state of the society" address. A dental surgeon, professor of zoology at King's College, London, and lecturer on anatomy at Guy's Hospital, Bell posited that the past year "has not, indeed, been marked by any of those striking discoveries which at once revolutionize, so to speak, the department of science on which they bear; it is only at remote intervals which we can reasonably expect any sudden and brilliant innovation which shall produce a marked and permanent impress on the character of any branch of knowledge, or confer a lasting and important service on mankind." Bell whined that there were no Bacons or Newtons, let alone "an Oersted or a Wheatstone, a Davy, or a Daguerre" gracing their meetings over the past twelve months. The advancement of science, he claimed, "is an occasional phenomenon whose existence and career seem to be especially appointed by Providence, for the purpose of effecting some great important change in the condition or pursuits of man."[31] The following year, 1860, Bell ignored Darwin's work in his presidential address to the society because he considered *Origin* to be too theoretical, controversial, and, according to his successor as Linnean Society president, the botanist George Bentham, "not accompanied by the disclosure of new facts or observations."[32] With the advantage of hindsight, it requires a hefty dose of self-control not to condemn Thomas Bell for so squarely situating himself on the wrong side of history.

In the late 1880s, in a letter to Darwin's son Francis, Hooker recalled the July 1, 1858, Linnean Society meeting far differently:

> The interest excited was intense, but the subject was too novel and too ominous for the old school to enter the lists, before armoring. After the meeting it was talked over with bated breath: Lyell's approval, and perhaps in a small way mine, as his lieutenant in the affair, rather overawed the Fellows, who would otherwise have flown out against the doctrine. We had, too, the vantage ground of being familiar with the authors and their theme.[33]

Less disputable was that by sending his paper to Darwin, Wallace did more than set the Linnean Society presentations into motion. He gave notice as to how close he was trailing on Darwin's coattails. Wallace confirmed Darwin's theory and freed the sickly naturalist from his epic writer's block. A Victorian man through to his marrow, Charles Darwin understood that the immortality he sought was best achieved between hard covers. Thanks to Alfred Wallace and his life-changing letter, the squire of Down finally settled down to work on a book that would change the world.[34]

# Part II

# THE BOOK

We have been reading Darwin's Book on the "Origin of Species" just now. . . . The book is ill-written and sadly wanting in illustrative facts—of which he has collected a vast number, but reserves them for a future book of which this smaller one is the avant-courier. This will prevent the work from becoming popular . . . but it will have a great effect in the scientific world, causing a thorough and open discussion of a question about which people have hitherto felt timid. So the world gets on step by step towards brave clearness and honesty! But to me the . . . explanations of processes by which things came to be, produce a feeble impression compared with the mystery that lies under the processes.

*—George Eliot to Barbara Bodichon, December 5, 1859*[1]

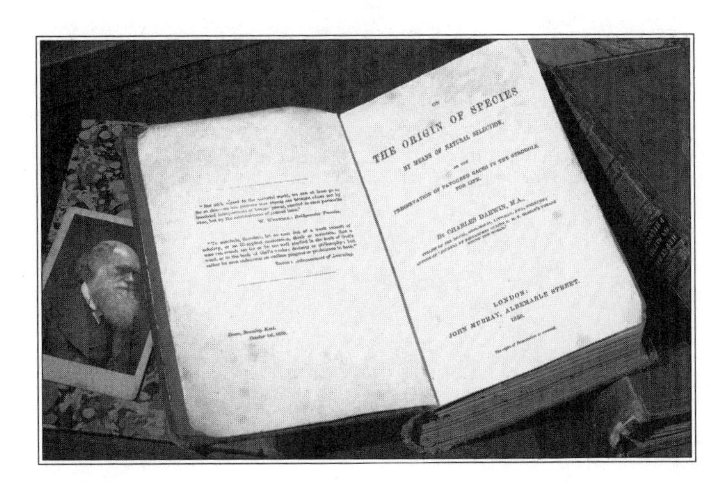

# 4

# The Devil's Chaplain

What a book a Devil's chaplain might write on the clumsy, wasteful, blundering low & horridly cruel works of nature!

—*Charles Darwin to J. D. Hooker, July 13, 1856*[1]

With Hooker's encouragement, Darwin set out to craft a far more detailed and longer "essay" on his theory of natural selection.[2] The initial goal was to publish it as a special supplement issue for the Linnean Society's journal. This, Hooker insisted, was the proper format for an important scientific study. At the time, Hooker was actively working to reorganize the society and inducing other "naturalists to concentrate their publications into well-established periodicals"—specifically the Linnean Society's journal—as opposed to chaotically scattering them about in provincial natural history publications that were "badly edited and badly supported."[3] Decorous expression and placement aside, a longer account of his natural selection "abstract" in the Linnean Society's journal would have been a far greater triumph for the Linnean Society than it would have been for Darwin.

Upon returning to Down in mid-August, Darwin arranged and rearranged the mounds of manuscripts, books, articles, and correspondence he had long accumulated in his study. Day by day, week by week, he composed his masterpiece. He never worked harder in his life, which is saying a great deal given his literary output over the previous three decades. To accommodate his long legs, Darwin sat on an upholstered "upright armchair," which was made even taller by "the addition of an ugly frame and castors." Instead of writing at his desk, he used a simple wooden board placed across his lap. On it, he clipped several pages of foolscap paper.[4] The maelstrom of the previous few months, along with his deep knowledge of the subject, unleashed a torrent of words. The pages seemed to fly off his desk as if his hand were attached to a mechanical printing press. Pausing only to insert a fresh, steel "William Mitchell's J-nib" into his dip pen or to take out his sterling silver snuffbox for a snort of tobacco, Darwin was no longer an invalid or grieving father.[5] He was a powerful creator—albeit, with a lowercase "c"—of what became *Origin*.

A great admirer of the leading lights of Victorian literature—the inimitable Charles Dickens, the pseudonymous George Eliot, and the industrious Anthony Trollope—Darwin understood the importance of creating a compelling narrative. His first book—*Journal of Researches*, which describes his work while aboard the HMS *Beagle*, 1832–1835—nicely demonstrates

Darwin's study.

this ability.[6] Over the course of several more monographs and thousands of letters, his powers of expression intensified. As with the aging of a cask of good whisky, by 1858, his writing style had deepened in flavor and color.[7] The eloquent sweep through nature that became *Origin* accomplishes far more than the mere recitation of facts, theories, and findings. Darwin gives voice, characterization, and descriptors to the parade of organisms he introduces on each successive page. He deftly applies analogies and metaphors to spin a complex yarn, one filled with contradictions and conflict—or, as he might say, the "struggle for life."

Most impressive was Darwin's mastery of his literary quarry. Virtually every point he makes is backed up by his own observations in the field or the measured findings of the other noted scientists he cites. One might not curl up to read *Origin* as easily as Dickens's *Bleak House,* Eliot's *Adam Bede*, or Trollope's *Barchester Towers*, but the sheer joy, attention to detail, and technical knowledge Darwin poured into his book proved essential to its persuasive argument.[8] This, he admitted, was no simple task:

> Nothing is easier than to admit in words the truth of the universal struggle for life, or more difficult—at least I have found it so—than constantly to bear this conclusion in mind. Yet unless it be thoroughly engrained in the mind, I am convinced that the whole economy of nature, with every fact on distribution, rarity, abundance, extinction, and variation, will be dimly seen or quite misunderstood.[9]

In early October, Darwin sent Hooker another progress report: "I am working most steadily at my Abstract; but it grows to an inordinate length; yet fully to make my view clear, (& never giving briefly more than a fact or two & slurring over difficulties) I cannot make it any shorter." Darwin predicted that he needed another "three of four months; so slow do I work, though never idle." A sentence later, he thanked Hooker, once again: "You cannot imagine what a service you have done me in making me make this abstract for though I thought I had got all clear, it has clarified my brains so much."[10]

For a sense of the speed with which he wrote, by Christmas—less than six months into the project—Darwin had filled 330 folio pages. At this point, he estimated he would need a minimum of 150, perhaps 200, more such pages, to "make a printed volume of 400 pages." His manuscript had grown so long, he wrote Hooker, that 'the subject really seems to me too large for discussion at any Society, & I believe Religion would be brought in by men, whom I know.—I am thinking of a 12mo [duodecimo] volume, like Lyell's 4th or 5th Edition of *Principles*."[11]

A month later, on January 25, 1859, Darwin told Alfred Wallace he was working on the penultimate chapter of what would be "a small volume of 400 or 500 pages."[12] Although he applauded Wallace's gracious acknowledgment of the publication of their Linnean Society papers, Darwin's letter was stained with a splotch of misdirection—something Wallace would not detect until years later: "Though I had absolutely nothing to do in leading Lyell & Hooker to what they thought a fair course of action, yet I naturally could not but feel but anxious to hear what your impression would be. I owe indirectly much to you & them." Wishing Wallace success, Darwin closed with mawkish apprehension: "I look at my own career as nearly run out: if I can publish my abstract & perhaps my greater work on the same subject, I shall look at my course as done."[13]

THE JOHN MURRAY PUBLISHING HOUSE, at 50 Albemarle Street in London's tony Mayfair district, was established in 1768 by John MacMurray. The founder's first major business decision was to drop the first three letters of his surname—what one friend called "the Wild Highland Mac"—to avoid the British prejudice against the Scots.[14] Initially, the firm was little more than a bookseller's shop. But Murray grew restless and expanded into the printing, publishing, and binding of new works. Over the succeeding two centuries, the Murray list of authors included Jane Austen, Lord Byron—whose literary executors burned his then-scandalous memoirs in the fireplace of the publishing house's drawing room—Samuel Taylor Coleridge, Arthur Conan Doyle, William Gladstone, Johann Wolfgang von Goethe, Washington Irving, David Livingstone (of "Dr. Livingstone, I presume" fame), Charles Lyell, Thomas Malthus, Herman Melville, Felix Mendelssohn, Walter Scott, George Bernard Shaw, Mary Shelley, William Wordsworth, and Queen Victoria.[15]

Darwin's work was well known to the company he hoped to keep. In 1845, he placed the second edition of his *Journal of Researches* with the Murray publishing house, by then under the command of the third John (1809–1890) in a long line of John Murrays. There Darwin metamorphosed into that rarest of literary characters, an author satisfied with his publisher.[16] The *Beagle* book became a staple of Murray's successful "Colonial Home Library" series and went through multiple editions and printings over the years, stacking up significant royalties for Darwin—who needed the cash more to remind him of his status as a bona fide man of letters than for balancing his checkbook.

John Murray III—the Murray whom Darwin dealt with—was in a constant state of dishevelment. A formal portrait, which now hangs in the National Library of Scotland, depicts a stern man attending to the manuscripts, books, bills, and letters littering almost every inch of his rolltop desk. His bulging forehead, bald, bumpy skull, and the thick, wild hair

John Murray III.

creeping down the sides of his face frame the piercing gaze of a busy editor and publisher. The mountain of paper before him was no mere artist's prop. The scope of Murray's reading was astonishing—from the latest novels by Dickens and Thackery to the most important scientific tomes and government reports of his day. His son, John IV, claimed that Murray possessed a photographic memory of the books he perused and could quote their contents at the drop of a monocle.[17]

In addition to the literary lights on his list, Murray published the annual reports of the British Association for the Advancement of Science, a popular series on chemistry and geology for the general reader, an influential literary and political magazine called the *Quarterly Review*, and a best-selling string of travel "handbooks," a genre Murray inaugurated long before Baedeker produced similar guides for befuddled travelers. Despite the many science books he published, however, Murray was not enamored with Darwin's concept of evolution nor Lyell's geology. In 1877, he wrote *Skepticism in Geology and the Reasons for It*, under the pen name "Verifier." Murray's book was in direct opposition to Lyell's theory that the planet was far older than the Bible's account and was always in a state of slow but constant change.[18]

Like Darwin, Murray walked for exercise. Every morning but Sunday—and until a few days before his death—the publisher marched 1½ miles to the railway station for a train that whisked him to his office and in reverse upon leaving London for the day. Murray's son quipped that John III "regarded a walk as the infallible cure for almost all bodily ailments, and I should hardly have been surprised to hear him recommend it for a broken leg."[19] Parenthetically, after Murray III's house was razed in 1922, the site became Centre Court of the All-England Lawn Tennis and Croquet Club at Wimbledon.

As with most successful publishers, Murray concerned himself with which books would sell and dismissed those that would not. One of his most successful authors was Paul Du Chaillu, a sensation-seeker who traipsed through the jungle in search of close encounters with the "savages" while hunting for leopards, elephants, crocodiles, and hippopotami. Du Chaillu's 1861 book *Explorations and Adventures in Equatorial Africa* was the type of adventure story that Victorian men, women, and children clamored for, especially for his descriptions of the "monstrous and ferocious ape, the gorilla . . . a hitherto, unknown animal." Some of his tales augured the 1933 motion picture *King Kong*; in one particuarly spicy chapter he told how "an immense gorilla stepped into the path [of two Mbondemo women], and, clutching one of [them], bore her off in spite of the screams and struggles of both." The British tabloid newspapers along Fleet Street thrilled to Du Chaillu's breathless account of how the woman returned to the village a few days later: "She related that the gorilla had misused her," even as she somehow managed to escape his evil clutch. " 'Yes,' said one of the men, 'that was a gorilla inhabited by a spirit.' Which explanation was received with a general grunt of approval."[20]

Du Chaillu knew precisely how to stroke his publisher's ego. While on a tour of North America, he wrote Murray, "Everybody who is not a donkey knows the name John Murray here, for your father, and your publications have gone all over the world, where the English language is spoken."[21] But Du Chaillu could also be quite demanding. In 1861, he ordered Murray to quickly arrange for the skin of one of his captured gorillas to be stuffed and mounted—at the publisher's expense of £25—and, later, in 1864, the bleaching and preservation of several gorilla skulls. Upon receipt, he added to his income by selling the skins and skulls to the British Museum.[22] Unlike Darwin and Lyell, however, Du Chaillu was a serial exaggerator, the type of science popularizer that drove—and continues to drive—serious scientists mad.[23]

On March 28, 1859, Darwin asked Lyell to approach Murray about publishing *Origin*. Darwin was so concerned about the controversial implications of his work that he instructed Lyell to make it clear to Murray from the outset "that I do not bring in any discussions about *Genesis* &c. and only give facts, & such conclusions from them, as seem to me fair . . . [the unorthodoxy of the book] is in fact not more than any Geological Treatise, which runs slap counter to *Genesis*."[24] Yet Darwin understood that no matter how well he finessed or avoided the issue, many would interpret his scientific account as an attempt to bury the Bible.

Lyell promised to see Murray the following day—acting as Darwin's literary agent without commission. The speed of the deal remains breathtaking. By March 31, Darwin was writing directly to Murray: "I have heard with pleasure from Sir C. Lyell that you are inclined to publish my work on the Origin of Species; but that before deciding & offering any terms you require to see my M.S."[25] One day later, April 1, Murray made an offer based on the strength of Darwin's introductory letter and his previous publications: "I can have no hesitation in swerving from my usual routine & in stating at once even without seeing the MS. I shall be most happy to publish it for you in the same terms as those on which I publish for Sir Charles Lyell."[26] Unable to take yes for an answer, Darwin wrote Murray back on April 2, freeing the publisher of all obligations if, after reading the manuscript, he felt the book was not likely to sell well. Darwin did, however, add, "but you will see that it would be a stigma on my work for you to advertise it, & then not publish it. My volume cannot be mere light reading, & some parts must be dry & some rather abstruse; yet *as far I can judge perhaps very falsely*, it will be interesting to all (& they are many) who care for the curious problem of the origin of all animate forms.—"[27]

Hooker was initially unhappy with Darwin's choice to work with a popular, or trade, publisher and told him so. In response, on April 2, Darwin politely responded, "Very many thanks for your letter of caution about Murray." Yet he could not help but crow about how "eager" Murray was to publish the book and how he offered "handsome terms & agreeing to publish without seeing M.S.!" He noted the caution he took, "owing to [Hooker's] letter" and how he "told [Murray] most *explicitly*, that I accept his offer solely on condition, that after he has seen part or all M.S. he has **full** power of retracting." As with most authors who fall in love with their words, Darwin hoped his book would become a popular success as measured by the number of books sold: "You will think me presumptuous, but I still think my book will be popular to a certain extent, enough to ensure heavy loss amongst scientific & semi-scientific men: why I think so is because I have found in conversation so great & surprising interest amongst such men." Unable to stop his rationalizing, Darwin iterated that Murray, not he, was the best judge of what to publish, and he did do a nice job with his *Journal of Researches*. With the skill of a practicing neurotic, Darwin declared that "if he chooses to publish it, I think I may wash my hands of all responsibility." Only a few paragraphs later, Darwin was compelled to close his letter: "I cannot help rather doubting whether Lyell would take great pains to **induce** Murray to publish my book; this was not done at my request, & rather grates against my pride. I know that Lyell has been **infinitely** kind about my affair but your dashed **induce**, gives idea that L[yell] had unfairly urged Murray."[28]

The stress of arranging for a publisher had a harsh effect on Darwin's gastrointestinal system and he was soon overwhelmed with fatigue, frustration, and flatulence. To Hooker, on April 7, he crabbed, "My God how I long for my stomachs'[sic] sake to wash my hands of it—for at least one long spell."[29] Sometime between April 8 and April 11, Hooker tried to cheer his friend with a note recanting any negative comments he may have made about the future success of the book or John Murray's "perfect good faith & kindness." Hooker also reassured Darwin as to how Lyell and he smoothed things over with the Linnean Society, which, thanks to Hooker's promissory notes, had been expecting to publish Darwin's manuscript.[30]

UNBEKNOWNST TO DARWIN, Murray *was* initially skeptical about the book's prospects. This uncertainty led the publisher to send the manuscript to his father's former literary adviser, George Pollock, who thought the book would make a huge splash, which could very well translate into huge sales. The book's other reviewer was Whitwell Elwin, the forty-three-year-old editor of Murray's magazine *The Quarterly Review*. Elwin, a conservative-leaning writer, ordained priest, and vicar of a parish in Norfolk, was less than impressed. On May 3, he wrote Murray that he had a "very high opinion of Mr. Darwin founded on his Journal of a Naturalist,"

which "is, as you have often heard me say, one of the most charming books in the language." The proposed book, however, was too flimsy for the publishing house. Indeed, Elwin pressed Murray to tell the author "and his friends to re-consider the propriety of sending forth his treatise in its present form." Oddly, for a book that is chock-full of facts, Elwin found it to be all theory and little evidence, which "would do grievous injustice to his views, & to his twenty years of observation and experiment. At every page I was tantalized by the absence of the proofs. All kinds of objections & possibilities rose up in the mind & it was fretting to think that the author had a whole array of facts, & inferences from the facts, absolutely *essential* to the decision of the question which were not before the reader. It is to ask the jury for a verdict without putting the witnesses into the box."[31]

Elwin—a self-confessed "smatterer in these subjects"—cooed that Darwin should drop his odd ideas on the origin of species and stick to writing about pigeons: "Everybody is interested in pigeons. The book would be reviewed in every journal in the kingdom & would soon be on every table."[32] Ironically, a similar suggestion was made by Sir Charles Lyell! But pigeons alone, as Darwin insisted in his book's first chapter, would not sufficiently describe the slow transitional steps of transmutation:

> I have never met a pigeon, or poultry, or duck, or rabbit fancier, who was not fully convinced that each main breed was descended from a distinct species. . . . May not those naturalists who, knowing far less of the laws of inheritance than does the breeder, and knowing no more than he does of the intermediate links in the long lines of descent, yet admit that many of our domestic races have descended from the same parents—may they not learn a lesson of caution, when they deride the idea of species in a state of nature being lineal descendants of other species?[33]

Fortunately, Elwin was not granted the final word. Three long days later, on May 6, Darwin huffed and puffed to Murray that Elwin's suggestions were "impracticable. I have done my best. Others might, I have no doubt, done the job better, if they had my materials; but that is no help." As if out of a tale told by P. G. Wodehouse—had old "Plum" been born a half century earlier—Darwin instructed his butler, Joseph Parslow, to book a train ticket to London and hand-deliver the first six chapters of his manuscript, neatly tucked and twine-tied into a brown paper wrapper. He informed Murray, "If I do hear (it must be soon) that you would like to see the *whole* M.S. my servant, who will bring the M.S., shall wait 2 or 3 hours & then call & bring home to the *latter* chapters, which I have just to run my eye once over again."[34] On May 10, he sent Murray "the first six chapters for press."[35]

The following morning, May 11, Darwin was still struggling to smooth out the more convoluted paragraphs of his manuscript. To Hooker, Darwin revealed both his delicate pride and racial insensitivity: "Thank you for telling me about obscurity of style. But on my life no nigger with lash over him could have worked harder at clearness than I have."[36] By May 14, however, he was assuring Murray, "You may **rely** on it, that my extreme wish for my health [sic] sake to get the subject temporarily out of my head, will not make me slur over the proofs, *I will do my utmost to improve my style.*"[37] This intense effort, however, was more than Darwin's fragile body could handle, and on May 18, he wrote his publisher to tell him that "my health has suddenly quite failed."[38] Exhausted by his Herculean effort, his fits of flatulence grew intolerable; the self-proclaimed "invalid" put his manuscript aside and dashed off to Moor Park, "for a week of Hydropathy."[39] Once there, he recuperated from "great prostration of mind & body," thanks to a combination of "entire rest & the douche & [George Eliot's novel] *Adam Bede*, which have together done me a world of good."[40]

ʃ

Reading proof sheets during the summer and fall of 1859 was a burdensome but essential chore. Fortunately, Darwin did not work in isolation on these pursuits—no sensible author does. He enlisted objective and supportive readers to make sure what he wrote made sense, that each sentence flowed well into the next one, and to identify any inaccuracies or typographical errors. He first turned to Emma Darwin, who reviewed the manuscript with attention to form and style rather than content. By now, she was used to the nature of Darwin's revolutionary ideas. If she did not fully support his secular explanations of the origin of species, she remained his greatest source of support—as both a life partner and early reader of his work. Then again, it is amusing to imagine how ardently she must have prayed for her husband's soul each night before retiring. Lyell read the proofs while vacationing and crisscrossing the Continent. So, too, did Emma's close friend Georgina Tollet, who was visiting from Staffordshire and staying at Down House for the summer.[41]

Hooker and his wife Frances were most essential to the final shaping of the manuscript. Their careful reading, fact-checking, and untangling of Darwin's sentences was a generous gift to the author and his readers. A literary disaster almost struck after Hooker mistakenly placed Darwin's chapter on the geographical distribution of species in a bureau drawer containing his children's crayons and drawing paper. Can you imagine Hooker's horror to find nearly a quarter of Darwin's manuscript marked up with his children's doodles, many of them obscuring Darwin's spidery handwriting? Hat in hand—or more accurately, pen in hand—Hooker confessed all to his friend. Darwin reassured him, "I **have** the old M.S. Otherwise the loss would have killed me!"[42] Even so, Hooker felt "brutified, if not brutalized, for poor D. is so bad that he could hardly get steam up to finish what he did. How I wish he could stamp and fume at me—instead of taking it so good-humoredly as he will."[43] Thankfully, the crisis was not nearly as dire as one in 1835, when Thomas Carlyle lent John Stuart Mill the only copy of his manuscript for *The French Revolution*. In that instance, Mill's housemaid mistook the pages for trash and used them to start the philosopher's morning fire, forcing the tart historian to rewrite his book from scratch.[44]

Mired in anxiety, Darwin solicited Lyell's encouragement: "As I regard your verdict as far more important in my own eyes & I believe in eyes of the world than of any other dozen men, I am naturally very anxious about it." Beseeching Lyell to keep an open mind, Darwin continued, "I shall be deeply delighted if you do come round, especially if I have a fair share in the conversion. I shall then feel that my career is run, & care little whether I am ever good for anything again in this life."[45]

Lyell moved rock formations to help Darwin, beginning with an address he gave to the Geology section at the 29th Annual Meeting of the British Association for the Advancement of Science, held in Aberdeen in September of 1859.[46] In his lecture on "works of human art in post-Pliocene deposits," Lyell plugged his protégé's book to maximum effect. The geologist explained how Darwin systematically concluded "that those powers of nature which give rise to races and permanent varieties in animals and plants, are the same as those which, in much longer periods, produce species, and, in a still longer series of ages, give rise to differences of generic rank. He appears to me to have succeeded, by his investigations and reasonings, to have thrown a flood of light on many classes of phenomena connected with the affinities, geographical distribution, and geological succession of organic beings, for which no other hypothesis has been able, or has even attempted, to account."[47] For the unwell Darwin, Lyell's praise was curative, and he thanked the geologist with a sweet response: "You would laugh if you knew how often I have read your paragraph, & it has acted like a little dram."[48]

Queen Victoria's prince consort, Albert of Saxe-Coburg and Gotha—a lover of all things modern, practical, and scientific—was that year's incoming president of the BAAS. Lyell made sure to delay his own talk until Albert "entered the Section room."[49] The next day, Albert invited Lyell and some of his colleagues for tea at Balmoral Castle—a full 47 miles away from the meeting site. The queen somehow managed to keep her famously short fuse in check and behaved in a most regal manner even though neither Darwin's book nor Lyell's conversation seemed to impress her. Later that evening, after the guests had all left, Albert complimented Victoria for

Prince Albert and Queen Victoria reading together.

being so "kind, sociable, and unselfish." By evening, she felt free enough to complain in her journal about having to entertain "four weighty [horse-drawn] omnibuses laden with philosophers & savants."[50]

Eventually, Darwin completed his mountain of proofreading.[51] So many changes were made that the printer delivered a bill for £72 and 8 shillings, which Murray generously picked up instead of passing on the charges to his author, as was traditionally and contractually done. This was one more sign of the publisher's business acumen and high hopes for the financial future of *Origin*.[52] By September's end, it was clear that Darwin was far too close to his book. He complained to his cousin William D. Fox, and several other friends, "So much for my abominable volume, which has cost me so much labor that I almost hate it."[53]

There were, however, a few tasks remaining. Darwin had to scribble out a draft for the title page, which he would send to Lyell and Murray for their stamp of approval. On it, Darwin proudly introduces himself as an authoritative, upper-class, learned gentleman; an elected member of some of the most prestigious scientific societies in Great Britain; and a scholar who easily traversed the heavy, iron gates guarding the ancient universities:

*An Abstract of an Essay*
*On the*
*Origin*
*of*
*Species and Varieties*
*Through Natural Selection*
*By*
*Charles Darwin, M.A.*
*Fellow of Royal, Geological and Linn Soc.—*
*London*
*&c &c &c &c*
*1859*[54]

When Murray saw the proposed title page, he suggested adding another credit to Darwin's byline: "Author of 'Journal of Researches During H.M.S. Beagle's Voyage Round the World,'" which was published, of course, by John Murray. To this request, Darwin acceded without comment. The publisher then ran his pencil through the words "An Abstract of an Essay," because he insisted general readers would guard their wallets and wait for the fuller version of the book. Although Darwin kept insisting that a longer and more definitive volume was in the works—if only in his head—Murray won this battle and took editorial privilege in replacing the word "by" for "through" in Darwin's prolix title. Darwin did, however, stand his ground in refusing his publisher's first choice for a title: *The Origin: Natural Selection*.[55]

For the book's epigraph, Darwin chose two quotations. The first was from William Whewell's *Bridgewater Treatise* (1833). When explaining the material world, Whewell wrote, "We can at least go so far as this—we can perceive that events are brought about not by insulated interpositions of Divine power, exerted in each particular case, but by the establishment of general laws."[56] The second quote Darwin selected was from Francis Bacon's 1605 foundational essay on empiricism, *The Advancement of Learning*, attesting that no "man can search too far" in his quest to better understand "God's words or God's works." Like Whewell, Bacon was a devout

Anglican who saw scientific exploration as a means of glorifying God. Darwin may have hoped in vain that these two citations might insulate him from criticisms that he was attacking the faith. Finally, he datelined the book "Down, Bromley, Kent, October 1, 1859," and it was done.[57] Not counting the decades of travel, research, correspondence, reading, and gestation in his beautiful mind, the composition of *Origin* took him thirteen months and ten days to complete.

ʃ

DARWIN'S LITERARY LABORS landed him in the hospital for nine weeks. This time, he chose the Wells House Hydropathic Establishment and Hotel at Ilkley, North Yorkshire. He had avoided Malvern ever since his daughter Annie died there in 1851. Moor Park, too, had lost its charm because his last few visits there "did not do [him] so much good as usual."[58] Darwin called Ilkley "the Malvern of the North"; most others nicknamed it "the Hydro." There was even a literary cachet to the place because the establishment was frequented by George Eliot—but not at the same time as Darwin—and was a mere 15 miles from Haworth, the village where Emily and Charlotte Brontë first spun their tales of *Wuthering Heights* and *Jane Eyre*, respectively.

To get there, Darwin took the train from Down to London on October 3 and took the evening off to visit with his brother Erasmus, who lived at 6 Queen Anne Street. The following morning, he hailed a taxi that whisked him to the King's Cross Station and boarded a train bound for Leeds. That journey took anywhere from seven to eight hours, depending on the number of stops. After alighting from the train and walking through the bustling station, he climbed into a waiting horse-drawn carriage, which was provided by the Wells House stables to transport him for the remainder of the journey. This leg of transportation, some 14 miles, added another few hours to the trek.

The chartered coach made its way along a twisting road, affording a beautiful view of the woodsy Wharfe Valley. It proceeded through the village of Otley and onto a brand-new lane, specially constructed with a shallow gradient to provide a smooth ride to Branhope Village. The car-

The Valley of the Wharfe.

riage's horses clip-clopped along the banks of the River Wharfe into the textile town of Burley-in-Wharfedale, passing by heather-covered hills and emerald-green moors, before fading right onto the Otley-Ilkley Road. The final score of minutes was spent driving up a steep hill to the curved corner of Wells Road and Broderick Drive. Finally, the coach pulled up to the grand porch of Wells House, "where porters in their dark-green livery would have shown Darwin into the impressive entrance hall and followed on with his luggage."[59]

There is one major omission to this description. At the time, Darwin's flatulence was in full force—hence the emergent need of a visit to Ilkley.

During his first two weeks there, Darwin enjoyed walking through Ilkley's lush gardens teeming and twisting with trees and foliage. Every morning, in the crisp, fresh air, he climbed a steep hill for the cold-plunge

Ilkley, Wells House.

baths at White Wells, filled with icy cold water drawn from the springs at the nearby Rombalds Moor. When he wasn't water-curing or strolling the grounds, Darwin played at an entirely different type of pool—"the American game" of billiards.[60] On October 15, he wrote Hooker, "You cannot think how refreshing it is to idle away whole day & hardly ever think in the least about my confounded book, which half killed me." Initially, Ilkley "brought out a different Darwin—an extrovert Darwin who responded to the company of young women and felt comfortable with convivial men."[61]

Two days later, on October 17, Emma and their youngest children (Henrietta, age sixteen, Elizabeth, age twelve, Francis, age eleven, Leonard, age nine, and Horace, age eight), along with the requisite governess, maid, and butler, arrived at Ilkley, hoping to keep Darwin cheerful and relaxed as he awaited the publication of *Origin*. They were disappointed to find their suite of rooms at the nearby North House to be cold, poorly lit, and gloomy.[62]

By October's end, Darwin's condition sharply deteriorated. He wrote Hooker, "I have been very bad lately; having had an awful 'crisis,' one leg swelled like elephantiasis—eyes almost closed up—covered with a rash and fiery boils; but they tell me it will surely do me much good.—it was like

White Wells, Ilkley.

living in hell."[63] One of the goals of hydrotherapy was to induce a "crisis," typically in the form of a minor skin eruption—redness really—as a means of casting off a diseased organ's "morbid excitement."[64] Still, it is difficult to imagine how Darwin's debilitating constellation of symptoms could lead to anything close to recuperative. A few weeks later, Darwin reported to Alfred Wallace how unsuccessful the Ilkley stay was: "I have profited very little. God knows when I shall have strength for my bigger book."[65] From the distance of more than half a century, Henrietta recalled Ilkley as "bitterly cold, he [Darwin] was extremely ill and suffering, the lodgings were uncomfortable, and I look back upon it as a time of frozen misery."[66]

Emma and the children returned to Down on November 24. Darwin stayed another two weeks and left Ilkley on December 7—first to London for two days of work with John Murray and another visit with his brother, Erasmus. Weak and wobbly, Darwin finally came home on December 9. There, in the comfort of his study, he braced himself for the oncoming avalanche of *Origin* book reviews.

# 5

# Best Seller

> My book on the Origin of Species has been some time finished, & will be published, as I hear from Murray, on Nov. 22nd—I fear, if you ever read it, that the conclusions will be abominable . . . & I must stop, otherwise I shall be ruined.
>
> —*Charles Darwin, 1859*[1]

*Origin* made its debut at a publisher's trade sale on November 22, 1859, and two days later, November 24, on booksellers' shelves.[2] The book cost 14 shillings—the 2022 equivalent of £76.81, or $104.70. Today, the price tag on a copy of the first edition might easily fetch £300,000, or $365,000![3] It was a thick, doorstopper of a book, weighing in at nearly 2 pounds and crammed into 513 pages, including index. Topping it off was a green cloth cover, with gold gilt lettering for the title and byline on the spine. At more than 150,000 words, *Origin* is longer than Jane Austen's lengthiest novel, *Emma* (1818) and shorter than one of Dickens's leaner works, *Great Expectations* (1860–1861).

A few weeks earlier, while still at Ilkley, Darwin received his author's copy. Murray attached a note explaining, "By this day's post I send you a specimen copy of your book bound—I hope it may receive your approval. Please reply by return & not a moment shall be lost in getting ready the early copies—your instructions seem quite clear & shall be carefully followed."[4] The ritual of an author opening the first copy of his newest book, glancing lovingly at the title page, and leafing through its successive pages never gets old for those odd birds who prefer constructing sentences to

ON

THE ORIGIN OF SPECIES

BY MEANS OF NATURAL SELECTION,

OR THE

PRESERVATION OF FAVOURED RACES IN THE STRUGGLE FOR LIFE.

By CHARLES DARWIN, M.A.,

FELLOW OF THE ROYAL, GEOLOGICAL, LINNÆAN, ETC., SOCIETIES;
AUTHOR OF 'JOURNAL OF RESEARCHES DURING H. M. S. BEAGLE'S VOYAGE ROUND THE WORLD.'

LONDON:
JOHN MURRAY, ALBEMARLE STREET.
1859.

*The right of Translation is reserved.*

Title page of *Origin of Species.*

building fortunes—and Charles Darwin was certainly one of those rare finches. In a note posted the day after receiving it, November 3, he wrote Murray, "I have received your kind note & the copy: I am *infinitely* pleased & proud at the appearance of my child. I quite agree to all you propose about price." A lower cost, Murray warned him on November 2, would decrease the publisher's profit margin and thus Darwin's royalties.[5]

Complete with warnings of the author's fragile health and the book's imperfect nature, *Origin* opens with a charming invitation into Darwin's world:

> When on board H.M.S. 'Beagle,' as naturalist, I was much struck with certain facts in the distribution of the inhabitants of South America, and in the geological relations of the present to the past

> inhabitants of that continent. These facts seemed to me to throw some light on the origin of species—that mystery of mysteries, as it has been called by one of our greatest philosophers. On my return home, it occurred to me, in 1837, that something might perhaps be made out on this question by patiently accumulating and reflecting on all sorts of facts which could possibly have any bearing on it.[6]

What made *Origin* such a major breakthrough was Darwin's insistence that there existed "laws" for the life sciences—just as Sir Isaac Newton proposed for the physical world in his 1686 elaboration of the laws of motion and universal gravitation.[7] Like Wallace, Darwin was strongly influenced by the political economist Thomas Malthus, whose 1798 book, *An Essay on the Principle of Population*, explained how populations grow and die based on the supply of resources, famine, and death.[8] Darwin first read Malthus's book while aboard the *Beagle*. As he later told his German colleague Ernst Haeckel, "When I happened to read 'Malthus on population' the idea of Natural selection flashed on me."[9] Darwin remained a loyal Malthusian for the rest of his life. In 1875, for example, he wrote, "Nothing can be more convincing and clear than the conclusions of Malthus, and yet every now and then some foolish author tries to disprove them."[10] Yet neither Newton nor Malthus so starkly positioned their laws against a monotheistic Creator and Designer of all living things as did Darwin. "These laws, taken in the largest sense," the naturalist carefully explained, were "Growth with Reproduction; Inheritance, which is almost implied by reproduction; Variability from the indirect and direct action of the external conditions of life, and from use and disuse; a Ratio of Increase so high as to lead to a Struggle for Life, and as a consequence to Natural Selection, entailing Divergence of Character and the Extinction of less-improved forms."[11]

ORIGIN IS often referred to as a best seller, but the adjective "scientific" is a necessary qualifier, especially when comparing Darwin's sales numbers to those of the popular authors of his day. Coinciding with his book's publication that year were Charles Dickens's *A Tale of Two Cities*, serialized in his new weekly magazine, *All the Year Round*, George Eliot's *Adam Bede*, George Meredith's *The Ordeal of Richard Feverel*, William Thackery's *The Virginians*, Alfred Lord Tennyson's *Idylls of the Kings*, and John Stuart Mill's *On Liberty*. These titles were hardly in direct competition with Darwin's tome, which was aimed at a much narrower audience. The first printing of *Origin* was 1,250 copies. Discounting the 80 books presented, with compliments, to select colleagues and reviewers, the initial run is said to have sold out in one day. By the end of December, *Origin* had sold approximately 3,800 copies.[12] In contrast, the weekly serialized installments of *A Tale of Two Cities* each sold 100,000 to 120,000 copies; in the years since, Dickens's most popular novel has sold more than 200 million copies.[13]

For a technical treatise, however, *Origin* was remarkably successful. Between 1859 and 1872, it went through six editions. Before the British copyright expired in 1901, John Murray printed and sold more than 56,000 copies in "the original format" and another 48,000 copies in "the cheap edition"—the then common term for a secondary run of a book, using thinner paper, double columns of text, fewer pages, and inexpensive binding and sold at a more affordable price. These numbers do not begin to account for the massive sales of the American and foreign-language editions—including Danish, Dutch, French, German, Hungarian, Italian, Polish, Russian, Serbian, Spanish, and Swedish—which were published during Darwin's lifetime, not to mention the multiple editions, printings, and translations after his death—especially after the copyright expired.

Nor was each successive edition merely a reprint of the first. For example, in the second edition (1860), Darwin took on some of the clergy's criticisms by cheekily asserting, "I see no good reason why the views given in this volume should shock the religious feelings of anyone."[14] In the third (1861) edition, Darwin added a useful essay entitled "An Historical Sketch of the Recent Progress of Opinion on the Origin of Species. Previously to the Publication of the First Edition of this Work," in which he cited those who pondered evolution before him and addressed the most critical reviews of his work.[15] In fact, Darwin amended and revised the book's successive editions for much of the rest of his working life. The final one he worked on—the sixth edition—appeared in 1871 and, in a corrected form, 1876.[16]

As a proud author, Darwin burned for glory. He wrote solicitous letters introducing his work to many of the leading naturalists of his day in England, Europe, and the United States, as well as to friends and relatives. For example, he mailed the book to Alfred Wallace in Malay and understatedly wrote he "had not seen one naturalist for 6 to 9 months owing to the state of my health and therefore I really have no news to tell you."[17] Another copy went to his brother Erasmus, who read it in only a few days and proudly wrote Darwin, "I really think it is the most interesting book I have ever read, & can only compare it to the first knowledge of chemistry, getting into a new world or rather behind the scenes." Adding a hypochondriacal flourish of his own, Erasmus kvetched, "I am so weak in the head that I hardly know if I can write . . . my ague [malaria] has left me in such a state of torpidity that I wish I had gone thro' the process of natural selection."[18]

JOHN MURRAY actively promoted *Origin* in advertisements in his other publications and magazines. He sent out hundreds of crisp, new presentation copies of *Origin*, "with the author's compliments." Given that Darwin and Murray posted so many gratis copies, the cost was enormous. The books alone amounted to £56, a sum that would be the equivalent of, at least, £6,000, or roughly $7,000, in 2022—not counting the postage.[19] Unlike most authors, however, Darwin could easily afford such a plan without it seriously affecting his household finances.

One of Murray's best marketing decisions involved selling 500 copies of *Origin* to Mudie's Lending Library, an enterprising means of distributing quality nonfiction books and novels to working, middle-class men and women. Most of these readers were unable to afford to buy their own copies of *Origin*, but many could manage the far smaller fee to "borrow" the book. Charles Mudie charged his subscribers 1 guinea, or £1 and 1 shilling, per year (about £105, or $120, in 2022 currency) to borrow one exchangeable volume at a time—whereas other lending libraries charged 4 to 10 guineas for the same privilege. Long before Amazon.com's virtual river of books flooded the world, Charles Mudie arranged for special railway cars and shipping lines to send books to customers across Great Britain and abroad. Having a book selected into Mudie's Library was an endorsement far better than appearing in a peevish literary review because each copy fell into the hands of so many readers.[20] Emma Darwin, a loyal Mudie's subscriber, was so delighted by her husband's newfound popularity that she wrote her son William—by then an undergraduate at Christ's College, Cambridge University—"It is a wonderful thing the whole edition selling off at once & Mudie taking 500 copies. Your father says he shall never think small beer of himself again & candidly he does think it very well written."[21]

Charles Lyell, too, beamed with pride over Darwin's success. As soon as he finished reading *Origin*, Lyell teased Darwin, "Right glad I am that I did my best with Hooker to persuade you to publish it without waiting for a time which probably could never have arrived tho' you lived till the age of 100 when you had prepared all your facts on which you ground so many grand generalizations." The book was, Lyell wrote, "a splendid case of close reasoning & long sustained argument throughout so many pages, the condensation immense, too great perhaps for the uninitiated but an effective & important preliminary statement which will admit even before your detailed proofs appear of some occasional useful exemplifications such as your pigeons & cirripedes of which you make such excellent use." Lyell *did* have some issues with the book but preferred to discuss his criticisms with Darwin privately rather than join the scrum yearning to tackle the book in the public sphere. His points of disapproval, however, were more than mere quibbles. He was, of course, concerned over the issue of enough proof to back Darwin's theory of natural selection. But documenting the proof needed was difficult because the changes Darwin suggested occurred so slowly and over such long periods of time.[22]

Equally troubling for the geologist—even though he famously demonstrated a world far older than that described in the Bible—was the battle between faith and science still being waged in his mind. Lyell fervently believed that the most important human quality—one higher than any other in the living world—was the God-given ability to think, reason, and create. What Darwin's theory of natural selection did, Lyell mourned, was reduce humans to a mere life-form, subject to the same stresses, changes, and adaptations as any other living creature, no matter how great or small. How could God grant us such supremacy, he asked Darwin, and still link our family trees to nothing loftier than primordial ooze? After reading the proof sheets of *Origin*, on October 22, 1859, the geologist confidentially wrote to Darwin: "I care not for the term Creation, but I want something higher than 'selection', unless the latter divinity be supposed to have produced the primeval egg or seed or germ of the first great branches of the organic kingdom—in some ante-Cambrian epoch." In other words, he desired a greater power pulling the strings of natural selection, either actively or remotely, in order to create sentient, articulate human beings. Finding such proof, he admitted, "is a vast step if secured among those humble advances which we may be permitted to make.[23]

It took several more years, but Lyell eventually accepted Darwin's theory of natural selection.[24] On March 7, 1863, the *Saturday Review* described his new book, *The Geological Evidences of the Antiquity of Man*, as "a trilogy, the constituent elements of which should be headed respectively, Prehistoric Man, Ice, and Darwin."[25] A few days later, March 11, Lyell confessed to Darwin the agony he endured while reconciling his Christian faith with natural selection: "My feelings, however, more than any thought about policy or expediency, prevent me from dogmatizing as to the descent of man from the brutes, which, though I am prepared to accept it, takes away much of the charm from my speculations on the past relating to such matters. . . . I have spoken out to the utmost extent of my tether, so far as my reason goes, and farther than my imagination and sentiment can follow, which I suppose has caused occasional incongruities."[26]

Hooker, on the other hand, was completely in Darwin's corner. Writing to his friend on December 12, 1859, the botanist admitted he was only halfway through the book but congratulated the naturalist for a tome "so cram full of matter & reasoning . . . I am perfectly tired of marveling at the wonderful amount of facts you have brought to bear & your skill in marshalling them & throwing them on the enemy." Hooker's postscript to this glowing letter shined even brighter: "I expect to think I would rather be author of your book than of any other in Nat. Hist. Science."[27]

ſ

Darwin knew that in order to succeed, *Origin* required a wave of acceptance by his younger colleagues, who were weary of the fuddy-duddy amateurs who had too long dominated the field. He was also well aware that the elders of natural history and theology would use all their rhetorical powers to strike down his grand theory. Yet Darwin was hardly prepared for the onslaught of reviews that appeared in the popular press, which helped inform a readership hungry for the latest discovery or idea. Every author dreams of the critical success inspired by an evergreen hit like *Origin of Species*. That climb is especially precipitous for scientists because "the average shelf life" of their publications—to crib Calvin Trillin's bibliographic dirge—"is somewhere between milk and yogurt."[28] As a result, few scientists become a household name; fewer still help shape the scientific agenda beyond their lifetimes. Darwin, for all his infirmities and neuroses, did. *Origin* unequivocally changed the world of ideas, even if relatively few people have read its entire contents. By the close of 1859, Charles Darwin stood at the precipice of fame that would ripple, swell, and expand far beyond his small circle of scientific colleagues. He was about to become **Darwin**, a name gilded with gold and carved into the marble walls of museums and libraries around the world, alongside Galileo and Newton. More than a century and a half after it appeared, *Origin* continues to influence and advance the life sciences.

ſ

The book's first review ran in the November 19 issue of the *Athenaeum*. It was not good news. Book reviewing in Victorian England was, by tradition, anonymous to supposedly facilitate the most honest of opinions and conclusions. Too often this practice was a facade for bad behavior, and no matter how nasty the review might be, the author remained unsure as to who wrote it. The author of the *Athenaeum* review was John Leifchild, a retired Congregational minister and biblical scholar. Talk about the wrong man for the job! Under the cloak of obscurity, Leifchild railed, "If a monkey has become a man—what may not a man become?" The minister ended his hostile review tersely: "Having introduced the author and his work, we must leave them to the mercies of the Divinity Hall, the College, the Lecture Room, and the Museum."[29]

The review's claim that *Origin* detailed monkeys begetting men demands some clarification. As early as 1837–1838, Darwin was privately speculating that "if we choose to let conjecture run wild then animals, our fellow brethren in pain, disease, death & suffering, & famine; our slaves in the most laborious works, our companions in our amusements, they may partake from our origin in one common ancestor; we may be all netted together."[30] Yet for a book many assume makes the definitive claim that the human family descended from a bunch of apes and monkeys, there is only one sentence in *Origin* that directly nudged the dialogue to include *Homo sapiens*. That vague statement appears on page 488, near the end of the book: "Light will be thrown on the origin of man and his history."[31] Darwin did speculate a few pages earlier that all life came from the same primordial seed, which, whether overtly stated or quietly inferred, includes human beings.[32] Darwin would not explicitly throw light on "the descent of man" until 1871, in his book of the same name, where he famously wrote, "We thus learn that man is descended from a hairy quadruped, furnished with a tail and pointed ears, probably arboreal in its habits, and an inhabitant of the Old World."[33] It mattered little that nowhere in *Origin* did Darwin yet make such a bold statement. The cultural die had been cast. For Darwin's opponents, the dictum was almost as clear as the equation 1 + 1 = 2. Instead of humankind being created in God's image, Darwinism held that man evolved from apes, monkeys, and "brutes."

The venomous nature of the *Athenaeum* review drove Darwin mad enough to misidentify its author as Samuel Woodward, a self-educated geologist who wrote on the Roman ruins and antiquaries in Norwich. For years, Darwin carried on a poorly aimed resentment at Woodward. On November 22, Darwin complained about him bitterly to Hooker: "As advocate he might think himself justified in giving argument only one side. But the manner in which he drags in immortality & sets the Priests at me & leaves me to their mercies, is base. [Woodward] would on no account burn me; but he will get the wood ready & tell the black beasts how to catch me.—I will not say to [a] soul that he is the author."[34]

Forewarned being forearmed, Darwin plotted a counterattack with Hooker. "It would be unspeakably grand," he wrote his confidante in the same letter, "if Huxley were to lecture on the subject [at the Royal Institution], but I can see this is mere chance. Faraday might think it too unorthodox." Darwin closed in triumph over Lyell's endorsement and, stiffening his upper lip, told Hooker, "You have cockered me up to that extent, that I now feel I can face a score of savage Reviewers."[35]

The Huxley referenced above was Thomas Henry Huxley, brilliant zoologist, anatomist, lecturer, and writer and the enfant terrible of British science. Darwin came to refer to the younger man as his "General Agent."[36] Faraday, of course, was Michael Faraday, whose brilliance helped usher in the age of electricity. The electric dynamos and batteries used to power our computers, appliances, lights, televisions, and even our automobiles are all a direct result of Faraday's research. For those who measure such things, his surname denotes a familiar unit of electric charge.[37] He was also a devout member of the Sandemanian Church in London, where he practiced "primitive Christianity," took Communion every Sunday, enjoyed communal meals with his brethren, and rejected personal wealth.[38] A brilliant speaker and communicator, Faraday was the most influential scientist of his day. From 1825 to 1866, he was the director of the laboratory at the Royal Institution of Great Britain, the grand dowager of science museums, founded in 1799 for the "promotion, diffusion and extension of science and useful knowledge."

In a few sentences, Darwin was angling for Huxley to lecture at the Royal Institution's famous Friday evening lecture series—the popular, standing room events where brilliant scientists explained their work to the working public in plain, everyday terms.[39] But here, we are getting ahead of ourselves, and Professor Huxley, so critical to this tale, demands a proper introduction.

# Part III

# FRIENDS AND FOES, 1859–1860

All I have seen of ["*your book*"] awes me; both with the heap of facts, & the prestige of your name, & also with the clear intuition, that if you be right, I must give up much that I have believed & written. In that I care little.

—*Reverend Charles Kingsley to Charles Darwin, December 18, 1859.*[1]

Most of Mr. Darwin's statements elude, by their vagueness and incompleteness, the test of Natural History facts.

—*Richard Owen,* Edinburgh Review, *April 1860*[2]

From left to right: Wilberforce, by Ape; Huxley, by Ape; Darwin, by Spy; Owen, by Spy.

# 6

# Darwin's Bulldog

Nothing I think can be better than the tone of the book—it impresses those who know nothing about the subject. . . . I trust you will not allow yourself to be in any way disgusted or annoyed by the considerable abuse & misrepresentation which unless I greatly mistake is in store for you—Depend upon it you have earned the lasting gratitude of all thoughtful men—And as to the curs which will bark & yelp—you must recollect that some of your friends at any rate are endowed with an amount of combativeness which (though you have often & justly rebuked it) may stand you in good stead—I am sharpening up my claws & beak in readiness.

*—Thomas Huxley to Charles Darwin, November 23, 1859*[1]

If Thomas Henry Huxley did not exist—to corrupt Voltaire's famous axiom—it would be necessary to invent him. In 1860, Huxley was thirty-five years of age and making a name for himself as an expert in comparative anatomy, natural history, and evolution. Unlike the shy and modest Darwin, Huxley was a voluble rabble-rouser. Whereas the former was rich, the latter came from humble beginnings where money was always a concern. The reclusive Darwin moved and shook British science from his wellborn perch in Down. The hard-driving and internally driven Huxley employed his wits at the Royal School of Mines on Jermyn Street in London—a narrow thoroughfare "as crowded and busy as Professor Huxley's own intellect."[2] Taken in full measure, these two men—who are so closely linked together in the annals of science—could not have been more different. The two major exceptions to this statement were their abandonment

Thomas H. Huxley in 1857.

of medical practice for the life sciences and a mutual passion for each other's work.[3]

Huxley was universally known as "Darwin's Bulldog." In 1895, Henry Fairfield Osborn, the paleontologist, cofounder of the American Eugenics Society, and longtime president of the American Museum of Natural History in New York City, noted the nickname while eulogizing Huxley's death.[4] During the winter of 1878–79, Osborn was a twenty-two-year-old student in London who "had the privilege of listening to [Huxley's] great course of lectures on comparative anatomy." One day, Darwin visited Huxley's laboratory and the young Osborn was gobsmacked by the appearance of the odd couple: "Darwin was instantly recognized by the class as he entered, and a thrill of curiosity passed down the room, for no one present had ever seen him before. . . . [His] grayish-white hair and bushy eyebrows overshadowed the pair of deeply set blue eyes, which seemed to image his wonderfully calm and deep vision of nature and at the same time to emit benevolence." Professor Huxley's "piercing black eyes and determined and resolute face were full of admiration and at the same time protection of his older friend." Once Darwin left the laboratory, Huxley confessed to his students, "You know I have to take care of him—in fact, I have always been Darwin's Bulldog."[5]

It takes a specific personality to happily adopt "bulldog" as one's nickname, and it fit Huxley as smartly as a Savile Row suit. Bulldogs give the appearance of pugnacity, no matter how friendly their owners claim them to be. In cartoons, they are often portrayed as drooling, sharp-toothed attack animals held back by a strained leash attached to their spiked collars. Huxley had to know, too, that "bulldog" was the colloquial term Oxford and Cambridge students used to describe their college proctors' attendants, or "sheriffs." These derby-wearing, able-bodied, hard-boiled, and dark-caped men maintained order among the students, collected their monthly rents and fees, and disciplined those caught violating the college's rules.[6] On evenings and weekends, the "bulldogs" patrolled the cobblestone streets, making certain the students wore their caps and gowns when strolling about after dinner and rounding up the rogues discovered in the local pubs and houses of ill repute.

Huxley was a bulldog of sorts long before he met Darwin. Throughout his career, Huxley *always* needed a battle. He *always* needed to prove himself to the posh types with fancy degrees; he was *always* fighting for position and legitimacy; *always* burying the fossilized men whose stubborn adherence to antiquated theories held back scientific progress; and *always* scrapping with the parvenus who perverted the facts of a matter for their own ideological purposes. Huxley lectured in a vigorous, firm, if not defiant manner. Students adored and admired his pugnacious style. Decades later, Huxley's name became a popular cultural symbol of academic combat. It was no accident that the 1932 Marx Brothers anarchic comedy of college life, *Horse Feathers*, takes place at the fictional "Huxley College." The college's president, played by Groucho Marx, introduces the hilarious patter song "Whatever It is, I'm Against It." The lyrics could easily have served as Huxley's credo. In case movie fans missed these cues, the rival school Huxley plays its climactic football game against is "Darwin College."[7]

Most self-made men reinvent themselves as they run away from the origins of their own species, and Huxley was no exception.[8]

Born in 1825, at "about 8 o'clock in the morning," in Ealing, Middlesex, a "little country village . . . within half-a-dozen miles of Hyde Park Corner," he was the second youngest of eight children. The Huxleys were a middle-class family who lived in a flat above a butcher's shop.[9] Huxley most identified with his mother, Rachel who, despite having "no more education than other women of the middle classes in her day, . . . had an excellent mental capacity . . . [and] rapidity of thought. If one ventured to suggest that she had not taken much time to arrive at any conclusion, she would say 'I cannot help it; things flash across me.' "[10]

Thomas's father, George, was a mathematics teacher at the Great Ealing School, a large semipublic school of high social standing, which once "claimed to be the finest private school in England."[11] Among its most successful pupils were W. S. Gilbert, who would someday write sparkling comic operettas with Arthur Sullivan; William Makepeace Thackery, who learned English grammar at Ealing decades before creating English literature; and Cardinal John Henry Newman, when he was still a young Anglican.[12] Huxley recalled his father as obstinate and hot-tempered. Disaster struck in 1835 when George either resigned or was fired from his post. He developed seizures and mental health problems—both were then considered signs of moral degeneracy demanding institutionalization. When George was well enough to warrant his freedom, he found odd jobs and eventually relocated to Coventry, where he "obtained the modest post of manager" at the local savings bank. The Huxley daughters "eked out the slender family resources by keeping school."[13] After Rachel's death in 1852, George descended into "worse than [a] childish imbecility of mind."[14] He died in his bed in 1855 at the Kent Lunatic Asylum, where his eldest son, James, was the medical superintendent and chief alienist (the nineteenth-century term for psychiatrist, since their patients were so "ailienated" from the so-called normal society).[15]

George Huxley's poor mental health forced Thomas to quit Ealing at age ten—after only two years of formal education. Leaving school turned out to be a blessing because "the society [he] fell into at school was the worst" he had "ever known." Ignored by their masters, smaller boys, like Huxley, were subjected to daily bullying or worse. Eventually, Huxley beat up a larger pupil who tormented him "until [he] could stand it no longer." The battle ended with Huxley giving his "adversary" a black eye. On their own accord, the two boys "made it up, and thereafter [Huxley] was unmolested." Twelve years later, Huxley was reacquainted with his "quondam antagonist" at a stable in Sydney, Australia, only to learn that "the unfortunate young man had not only been 'sent out' but had undergone more than one colonial conviction."[16]

Although the Huxley family tree produced many distinguished men and women, debilitating mood and mental health disorders ran deep within its trunk. Such maladies affected not only Thomas Huxley's father but also two of Thomas's brothers, George and James, his daughter, Marion, and two of his grandsons—Julian, a major scientist in his own right, who suffered a "nervous breakdown" in 1913, and Noel Trevenen, who committed suicide in 1914. Julian and Noel's brother, the novelist Aldous Huxley, was a longtime consumer of psychotherapy and, beginning in the 1950s, psychedelic drugs.[17]

Thomas Huxley, too, was prone to the highs and lows of depression and manic productivity. These episodes began at age thirteen, when he was apprenticed to train under a general medical practitioner. Huxley insisted that his melancholy was the result of attending his first autopsy. After helping to butcher the corpse, he developed "dissection-poisoning," a far too real risk in which the dissector nicks his hands with the scalpel. The bacteria dripping off its blade can yield an infection and—if severe enough to flourish in the bloodstream—sepsis and death.[18] Neither bacteriology nor the wearing of protective rubber gloves had yet been conceptualized when Huxley's injury occurred, but chronic depression is not likely to be caused by such wounds.

As with Abraham Lincoln and Charles Dickens—who fought their demons as they constructed and climbed ladders of success—Huxley papered over his family's squalor by means of reading all of his father's books and then some. Hungry for knowledge, he read by night in virtual darkness with only a candle to illuminate the pages and a blanket around his shoulders for warmth long after the evening's fire died. The breadth of his reading was as impressive as his ambition. One book he recalled best was James Hutton's two-volume *Theory of the Earth* (1795), which many, including Charles Lyell, credit as giving birth to the modern science of geology and the concept of uniformity. Huxley somehow taught himself enough German, Italian, Greek, and Latin, physiology, algebra, geology, chemistry, English history, and English grammar to pass the matriculation examination at the University of London.[19] Around this time, he developed a strategy "to make copious notes of all things" he read. Under a long list of subjects and disciplines, he urged himself on in his diary: "I must get on faster than this. I must adopt a fixed plan of studies, for unless this is done, I find time slips away without knowing it, and let me remember this, that it is better to read a little and thoroughly, than cram a crude, undigested mass into my head, though it be great in quantity. (This is about the only resolution I have!)"[20]

Huxley's love of natural history led him to audit courses on botany at Sydenham College in Chelsea. One May morning in 1842, he spied a flyer tacked onto a bulletin board announcing an open examination for the Society of Apothecaries' Gold Medal set to take place on August 1, 1842. As he looked "longingly at the notice," a classmate urged him to "go in and try for it." At first, Huxley "laughed at the idea, for [he] was very young, and [his] knowledge was somewhat of the vaguest."[21] Still, Huxley was nothing if not ambitious. He studied all summer, from "eight or nine in the morning until twelve at night. . . . A great part of the time [he] worked until sunrise." He put in so much time with his books that he developed an inflammation of the eyes, hindering his ability to read "for months afterwards."[22] In retrospect, Huxley wrote, "I should distinctly warn ingenuous youth to avoid imitating my example. I worked extremely hard when it pleased me, and when it did not, was extremely idle."[23]

The examination began at 11 o'clock on a hot summer's morning. Even in old age, Huxley could recall the industry of that day when he sat alongside "five other beings all older than myself at a long table." For hours, "nothing was heard but the scratching of the pens and the occasional crackle of the examiner's *Times* as he quietly looked over the news of the day." Although there was a short lunch break at 2:00 p.m., Huxley was only "half-done" with the examination by 4:00 p.m. and would not turn in his paper until 8:00 p.m. In a post-examination haze, he left with no idea of his performance until several weeks later when "an official-looking letter had arrived" in the mail. His sister Lizzie grew impatient and "could not restrain herself from opening" it. The news was terrific. Huxley came in second place and was to receive a silver medal at a banquet of "pudding and praise," held by the Society of Apothecaries on November 9, 1842. The self-trained swot beat all but one of the senior "University College men."[24]

Huxley went on to win a free scholarship for students "whose parents were unable to pay for their education."[25] Although he once aspired to become a mechanical engineer, he read chemistry, anatomy, physiology, and *Materia medica* at the Charing Cross Hospital and Medical College—an affiliate of the University of London—with the financially stable goal of becoming a doctor. One of Huxley's classmates was his brother James. In 1845, while still a medical student, Huxley published his first scientific paper, describing an anatomical structure at the base of a human hair shaft that became known as Huxley's layer.[26] The following year, he won a gold medal for anatomy and physiology and graduated second in his class. When Huxley came up for his MB (bachelor's in medicine) degree in 1846, he was too young to qualify at the College of Surgeons. As a result, Huxley did not sit for his second MB examination and was not granted a formal university degree. As with too many impoverished students, his scholarship ran out and he was left with serious debt. Later in life he admitted, "I am now occasionally horrified to think how very little I ever knew or cared about medicine as the art of healing. The only part of my professional course which really and deeply interested me was physiology, which is the mechanical engineering of living machines."[27]

ʃ

WITH THE AVENUE OF clinical practice closed to him, Thomas Huxley found the perfect job by joining the Royal Navy. From 1846 to 1850, he served as an assistant surgeon aboard the HMS *Rattlesnake*. The "28-gun frigate" was being outfitted for an exploration of New Guinea, Australia, and the South Sea, Huxley wrote to his sister Lizzie: "Our object is to bring back a full account of its Geography, Geology, and Natural History. In the latter department with which I shall have (in addition to my medical functions) somewhat to do, we shall form one grand collection of specimens and deposit it in the British Museum or some other public place, and this being the main object always being kept in view, we are at liberty to collect and work as we please."[28] It remains a noteworthy fact that three of the best English biologists of this era—Darwin, Hooker, and Huxley—all took some formal medical training. The nineteenth-century physician's emphasis on the symptoms, diagnosis, physiology, chemistry, natural history, and classification of disease could not have been wasted on their steel-trap minds. Thanks to the imperial explorations of Her Majesty's Royal Navy, these three men abandoned medical practice, faced dangerous risks on the high seas, studied nature in far-flung, virgin territories, made scientific sense of their findings, and—out of their labors—transformed the life sciences.

Huxley returned to England in October 1850. His voyage was cut short after the ship's captain died, and he used the occasion to request a "special leave" from the Admiralty to conduct research.[29] Adept at dissection and observation, he published several zoological studies in the prestigious *Philosophical Transactions of the Royal Society* and in the *Annals and Magazine of Natural History*, the latter being the same magazine where Alfred Wallace's work first appeared.[30] In May of 1851, Huxley was elected to the Royal Society. He was only twenty-six and the youngest of his class. He proudly wrote his sister Elizabeth of his great achievement: "I did not expect to come in till next year, but I find I am one of the selected. I fancy I shall be the junior Fellow by some years."[31] The following year, 1852, Queen Victoria presented him with the Royal Medal for the most important contribution to the advancement of natural knowledge. Huxley's award was for his "papers on the anatomy and the affinities of the family of *Medusae*," which had appeared in the *Philosophical Transactions* two years earlier.[32]

Even with these remarkable honors affixed to his curriculum vitae, Huxley was a young man in a hurry. He resisted authority figures, arbitrary rules, and outmoded intellectual boundaries. The elder he most resented was the paleontologist Richard Owen—who would soon become a powerfully negative force in both his and Darwin's career. As the grand old man of British science, Owen presided over an old boys' club of seniority, logrolling, toadying, and patronage. He admonished Huxley to winnow down his too-broad research agenda: "Make up your mind to get something fairly within your reach, and you will have us all with you." Huxley rejected Owen's unsolicited advice. With a Shakespearian passion, he railed against the older man's condescension, narrow-mindedness, questionable research ethics, and quid quo pro policy of granting favors. Huxley's hatred for Owen was so palpable that Darwin once warned him, "For Heaven sake do not come the mild Hindoo to Owen (whatever he may be): your Father confessor trembles for you."[33] Huxley ignored Darwin's advice and, as late as 1874, boasted of rarely missing an opportunity to "pound" Owen into "a jelly."[34]

After settling in London, Huxley maintained his naval commission as long as possible, because he needed his government salary until he could secure a paying professorship. There were a few academic nibbles—including a university post in Toronto that ultimately went "to the relative of a Canadian minister"—but no bites. He was rejected for similar positions at King's College, London, the University of Aberdeen, and Cork University. In 1853, as a result of refusing to report for active duty three times, he was formally dismissed from the Royal Navy list and its payroll. Huxley's prospects were so bleak and "intolerable" that he considered moving to Australia—which he had visited while aboard the *Rattlesnake*

and where, in 1847, he had met his future wife, an English émigré named Henrietta Anne Heathorn. One shudders to think of the fate of modern biology had Huxley pursued his fallback plan to "establish a practice in Sydney; to try even squatting or storekeeping."[35]

In early 1853, Huxley wrote Henrietta that the world was "no better than an arena of gladiators and I, a stray savage, have been turned into it to fight my way with my rude club among the steel-clad fighters. Well, I have won my way into the front rank, and ought to be thankful and deem it only the natural order of things if I can get no further."[36] Six months later, on July 6, with no job in sight, Huxley defiantly told her, "My course in life is taken. I will not leave London—I will make myself a name and a position as well as income by some kind of pursuit connected with science, which is the thing for which nature has fitted me if she had ever fitted any one for anything."[37] On August 5, he confessed that his greatest fear was "a life spent in a routine employment, with no excitement and no occupation for the higher powers of the intellect, with its great aspirations stifled and all the great problems of existence set hopelessly in the background." The young man melodramatically begged for Henrietta's support in his search for "truth and science . . . for there is only one course in which there is either hope or peace for me."[38] Drawn from Huxley's middle name of Henry, she took to calling him her "beloved Hal."[39] Long before the world realized it, Henrietta understood her fiancé's destiny as a world-renowned scientific authority.

Penniless for nearly a year and "quite out of sorts, body and mind, more at war with himself," chance finally favored Huxley's prepared mind. On July 20, 1854, Huxley was called to the professorship of natural history at the Museum of Practical Geology of the Royal School of Mines. The former professor there, Edward Forbes, had just accepted a position at Edinburgh University and recommended Huxley for his old post. The salary was £200 per annum for teaching two courses per term. Huxley calculated that his popular writing would net another £250 each year. "Therefore," he later explained, "it would be absurd to go hunting for chemical birds in the bush when I have such in the hand."[40] It turned out to be a wise choice. Huxley enjoyed a long and illustrious career at the School of Mines, which eventually became incorporated into the Imperial College of Science and Technology.

Huxley published dozens of papers, mostly on the paleontological and fossil collections of the Royal School of Mines. His studies ranged from primitive fish to ancient crocodiles, which were previously thought to be a fish. He examined the "highly convoluted" teeth of fossil amphibians and determined the age of the Elgin sandstones in Scotland, a region where many fossils of fish and herbivores were discovered. He also conducted work on the nervous systems of marine invertebrates and, eventually, the more complex neuroanatomy of vertebrates.

On July 21, 1855, after an eight-year engagement, he was finally financially secure enough to marry Henrietta and begin raising a family. As Huxley joked to Hooker, "I terminate my Baccalaureate and take my degree of M.A.trimony."[41] A few months later, in September, Darwin offered a preternatural nuptial warning: "I hope your marriage will not make you idle; happiness, I fear is not good for work."[42] In a cruel twist of circumstances, Huxley's personal happiness crashed and burned on September 15, 1860, when his first son, Noel, died of "brain fever," or encephalitis, caused by scarlet fever. The four-year-old boy was "the apple of his father's eye and chief deity of his mother's pantheon."[43]

On learning of Noel's passing, Darwin tenderly wrote Huxley, referencing the loss of his daughter Annie but not little Charles: "I cannot resist writing, though there is nothing to be said. I know well how intolerable is the bitterness of such grief. Yet believe me, that time, & time alone, acts wonderfully. To this day, though so many years have passed away, I cannot think of one child without tears rising in my eyes; but the grief is become tenderer & I can even call up the smile of our lost darling, with something like pleasure. My wife & self deeply sympathize with Mrs. Huxley & yourself. Reflect that your poor little fellow cannot have had much suffering. God Bless you."[44] Darwin's empathy turned out to be even more touching considering that a few weeks later, he and Emma helplessly watched their daughter Henrietta's precarious health over "9 days of as much misery as man can endure." The situation was so dire that Darwin confessed to Huxley, "My poor daughter has suffered pitiably, & night & day required three persons to support her. . . . I almost got to wish to see her die."[45] For both men, work served as an anodyne for their woes.

ʃ

HUXLEY MAY HAVE BEEN born with few social connections, but he was ingenious at "checkmating the snobs" in British scientific circles and beyond.[46] In 1864, for example, he founded a monthly supper club where the members shared ideas and plans for a scientific revolution, free of the distractions from their daily responsibilities. Huxley named his band of brothers "the X Club." It included Joseph Hooker; the physicist John Tyndall; the mathematician and publisher William Spottiswoode; John Lubbock, a banker, gentleman-biologist, and friend of Darwin's; Edward Frankland, the professor of chemistry at the Royal Institution; George Busk, a zoologist and paleontologist; physicist Thomas Hirst; and Herbert Spencer, the writer and subeditor of *The Economist*. All but Spencer were members of the Royal Society, but Spencer's loyal attendance at X Club meetings must have helped him in writing his popular books on psychology, biology, and sociology.[47]

The X Club met for the first time on November 2, 1864, at the St. George's Hotel on Albemarle Street, just a few blocks from Huxley's laboratory at the School of Mines.[48] Until its final meeting in March 1893, the members sat down to dine promptly at 6:00 p.m. every month except July, August, and September. Among the distinguished guests invited to discuss their work were Darwin, Asa Gray, Louis Agassiz, Hermann Helmholtz, and Bishop John Colenso, one of the few high churchmen amenable to Darwin's ideas. During its first few years, there were even summer weekend excursions for "members and their wives (x's + yv's, as the formula ran) to some place like Burnham or Maidenhead, Oxford or Windsor."[49] As the American historian and philosopher John Fiske observed in 1873, the group became known as "the most exclusive club in England."[50] Some have even joked that Huxley's intellectual salon was an excellent example of natural selection for a constellation of scientific luminaries.

THE BIOLOGIST'S STAR shined brightest as a public speaker. Huxley was blessed with the ability to convince the most recalcitrant audience of the validity of his scientific arguments and points. Initially a poor orator, he mastered the craft by delivering ten or more lectures a week over many years. Huxley appreciated, as few professors do, that the best lecturers speak *to* their audiences rather than *at* them, whether these gatherings were composed of scientific peers, pupils, or public audiences of varying levels of social class and education.[51] With time and practice, he acquired "a mastery of clear expression for which he deliberately labored, saying exactly what he meant, neither too much nor too little, and without obscurity." His speaking style was the antithesis of the shifty politician or devious lawyer and, instead, melded accuracy, truthfulness, science, literature, lucidity, memorable phrases, and an earnest enthusiasm. He employed the range and register of his voice like an orchestra, and he conducted his lectures as if they were bright, boisterous symphonies—guaranteed to inform, excite, and inspire. Huxley advised others hoping to emulate his success with a simple set of rules: "Say that which has to be said in such language that you can stand cross-examination on each word. Be clear, though you may be convicted of error. If you are clearly wrong, you will run up against a fact some time and get set right. If you shuffle with your subject, and study chiefly to use language which will give you a loophole of escape either way, there is no hope for you."[52]

Twenty years into Huxley's tenure at the Royal School of Mines, in 1875, he met Daniel Coit Gilman, a "Skull and Bones" Yale man, professor of geography and founder of Yale University's Sheffield Scientific School, and the second president of the University of California at Berkeley. Gilman was visiting the great universities of Europe in preparation to assume the presidency of the newly endowed Johns Hopkins University, which was to open in Baltimore, Maryland, the following year. Gilman is often credited with inaugurating modern postgraduate education in the United States.

At a dinner thrown by the eminent London physician Thomas Lauder Brunton, Gilman found himself sitting next to Huxley. Before dessert was served, Gilman asked Huxley if he might attend one of Huxley's lectures in South Kensington, where the Royal School of Mines had moved to in 1871. "They were given very early in the morning,—at nine o'clock, unless I am mistaken," Gilman recorded in his diary, "the exact subject I cannot tell, but it involved a minute delineation of the differences of vegetable and animal life in the earliest stages." The American educator was stunned "by the grace of his delivery; there were no 'hems' nor 'haws,' no repetition, no corrections. Every word came into place with perfect fitness." Huxley spoke on complex and technical issues without reading from a padded manuscript and took the stage armed only with a small slip of paper, "about as large as the palm of his hand, and this contained all his notes."[53]

After the lecture, Gilman visited Huxley's study, where he praised the zoologist and asked how he "acquired this exactness of speech." Even for the least informed members of the audience, Gilman gushed, "you have made, apparently without effort, a perfectly clear and interesting statement,—but without any manuscript." There was no secret, Huxley explained, other than paying close attention to the sentences he delivered and how the audience received them. "I always go before an audience with a definite scheme of what I am to say, and I know just what illustrations I am to introduce and where," he said. But how, Gilman asked, did Huxley choose the exact right word or phrase while at the lectern, as if in ex tempore? Surely, Huxley memorized such felicitous phrases beforehand. "Not at all," Huxley replied, after admitting he usually wrote out all his passages before giving a lecture, "but having carefully written what I wish to say, I avoid errors or inaccuracies, on this side and the other. Often, better words and phrases occur to me in speaking with the stimulus of an audience than I have thought of at my desk." Gilman later remarked that "these hints of

Huxley's methods I have often given to young men, for he was the most felicitous of lecturers on science whom I ever heard."[54]

ʃ

A TRADEMARK EXAMPLE of Huxley's ability to clearly communicate complex ideas in a muscular prose was his first review of *Origin*, which appeared, on Boxing Day 1859, in the *Times of London*. John Murray sent the book, per Darwin's instruction, to Samuel Lucas, the paper's book editor. Lucas may have been an excellent journalist, Huxley later explained, but he was "as innocent of any knowledge of science as a babe and bewailed himself to an acquaintance on having to deal with such a book." That unnamed person recommended that Lucas consult Huxley for help, which he dutifully did, with the editor's proviso that the review must begin with "two or three paragraphs of his own."[55] Huxley's son Leonard described the assignment in even grander terms: "Fortune put into his hands the opportunity of striking a vigorous and telling blow for the newly published book. Never was windfall more eagerly accepted."[56]

Huxley later admitted to Darwin's son Francis of being "too anxious to seize upon the authority thus offered of giving the book a fair chance with the multitudinous readers of the *Times* to make any difficulty about conditions; and being then very full of the subject, I wrote the article faster, I think, than anything I ever wrote in my life."[57] Although the review carried no byline, Huxley crowed how "the secret leaked out in time, as all secrets will, but not by my aid; and then I used to derive a good deal of innocent amusement from the vehement assertions of some of my more acute friends, that they knew it was mine from the first paragraph!"[58] Huxley admitted his authorship to Joseph Hooker on December 31, only a few days after the essay appeared—with the tacit understanding that the botanist would report this fact directly to Darwin: "I have not the least objection to my share in the *Times* article being known, only I should not like to have anything stated in my authority . . . as a scientific review the thing is worth nothing, but I earnestly hope it may have made some of the educated mob, who derive their ideas from the *Times*, reflect. And whatever they do, they *shall* respect Darwin."[59]

The review was an unqualified rave. "With a versatility which is among the rarest of gifts," Huxley wrote, Darwin turned "his attention to a most difficult question of zoology and minute anatomy; and no living naturalist and anatomist has published a better monograph than that which resulted from his labors." Explaining how *Origin* was more than twenty years in the making, Huxley argued, "Mr. Darwin abhors mere speculation as nature abhors a vacuum. He is as greedy of cases and precedents as any constitutional lawyer, and all the principles he lays down are capable of being brought to the test of observation and experiment." If that were not enough, Huxley explained how Darwin advanced the natural philosophy of Francis Bacon: "The path he bids us follow professes to be, not a mere airy track, fabricated of ideal cobwebs, but a solid and broad bridge of facts. If it be so, it will carry us over many a chasm in our knowledge and lead us to a region free from the snares of those fascinating but barren Virgins, the Final Causes, against whom a high authority has so justly warned us."[60]

Darwin hardly needed Hooker to alert him to the author of the anonymous, "splendid Essay and Review." Beaming with pride, he wrote Huxley on December 28 to compliment the "profound naturalist" who "writes & thinks with quite uncommon force & clearness; & what is even still rarer his writing is seasoned with most pleasant wit." Darwin recounted how his family laughed outloud at the most humorous sentences before stating the obvious: "Certainly I should have said that there was only one man in England who could have written this Essay & that *you* were the man." If his assumption was wrong, Darwin teased, he would like to meet this "hidden genius." Referring to the famously staid and authoritative *Times of London* by Anthony Trollope's nickname for it—*Jupiter Olympus*—Darwin asked how anyone convinced the editors to devote "3 and ½ columns to pure science? The old Fogies will think the world will come to an end. Well, whoever the man is, he has done great service to the cause, far more than by a dozen Reviews in common periodicals. . . . If you should happen to be *acquainted* with the author for Heaven-sake tell me who he is." In a rare exhibition of animus, Darwin could not help closing the letter with a swipe at what he presumed Richard Owen might say or do after reading the glowing review: "Upon my life I am sorry for Owen; he will be so d—d savage; for credit given to any other man, I strongly suspect is in his eyes so much credit robbed from him. Science is so narrow a field, it is clear there ought to be only one cock of the walk."[61] If such banter was not enough, Darwin rang in the new year by writing his ally, on January 1, 1860, "I will keep your secret."[62]

A FEW YEARS LATER, Huxley described how he worked through the more problematic aspects of Darwin's argument while planning a series of talks on "man's relations to the lower animals" aimed at working men: "Some experience of popular lecturing had convinced me that the necessity of making things clear to uninstructed people was one of the very best means of clearing up the obscure corners on one's own mind." On February 10, 1860, Huxley entered this battleground by delivering a Friday evening lecture at the Royal Institution. The topic was Darwin's theory as expressed in his *Origin of Species* and, always pushing the scientific envelope, how it might apply to humans.[63]

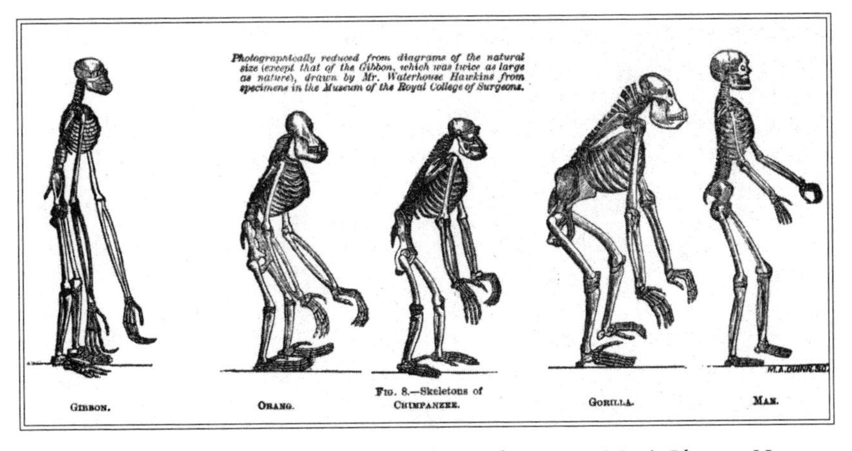


Frontispiece of Thomas H. Huxley's book, *Evidence as to Man's Place in Nature.*

This was a giant leap farther than Darwin was yet willing to go, even if it was the natural conclusion to his theory of evolution. The older man understood—just as the younger one eagerly ignored—there would be hell to pay for declaring such a blasphemous supposition aloud. In the concluding section of *Origin*, Darwin hemmed and hawed that "analogy would lead me one step further, namely, to the belief that all animals and plants have descended from some one prototype. But analogy may be a deceitful guide."[64] Huxley, however, had no such qualms, such as when he famously elaborated the human-ape link in his 1863 book *Evidence as to Man's Place in Nature*, which featured the now-familiar frontispiece of stooped ape skeletons evolving into a fully erect human.[65]

The Royal Institution's auditorium—a circular room that afforded direct eye contact between the speaker and his audience—was the perfect venue for a performer of Huxley's prowess. Well attended and publicized, his address was every bit the splendid show Huxley's friends and admirers had come to expect. He pointedly ignored "the heart-breaking . . . can't-learns and won't-learns and don't-learns" and, instead, spoke directly to "the do-learns."[66] To add some balance, the lecture was chaired by Sir Henry Holland—a second cousin to both Charles and Emma Darwin and a distinguished physician who attended to a long list of well-heeled patients, including members of the royal family. Holland was publicly skeptical of Darwin's assertion that all life descended from a single ancestral organism.

Huxley began by discussing, at great length, the comparative anatomy of horses, asses, and zebras before turning to Darwin's book and, in particular, the example of well-bred pigeons, "which the audience had an opportunity of examining." To help Huxley make his avian points, Darwin arranged for his poultry man to provide Huxley with "all the main kinds & some Dovecot & some pretty Toy Pigeons," in return for a few shillings and two tickets to the lecture.[67] At the climax of his talk, Huxley played the role of magician; but instead of pulling rabbits out of his top hat, he freed a kit of pigeons from the wicker baskets.[68]

He confidently concluded, "Mr. Darwin's must be regarded as a true theory of species, as well based as any other physical theory: they require, therefore, the most careful and searching criticism." More proof was needed to transform Darwin's theory into fact, he averred, but "I hardly know of a great physical truth, whose universal reception has not been preceded by an epoch in which most estimable persons have maintained that the phenomena investigated were directly dependent on the Divine Will, and that the attempt to investigate them was not only futile, but blasphemous." That said, Huxley predicted, Darwin's truth would ultimately triumph because science was a tenacious antidote to superstition. "Crushed and maimed in every battle," his voice soared with the force of an operatic tenor, "it yet seems never to be slain; and after a hundred defeats it is at this day as rampant, though happily not so mischievous, as in the time of Galileo. . . . *Origin* is not the first, and it will not be the [last] of the great questions born of science, which will demand settlement from this generation." Huxley closed his aria with phrases worthy of the best Mozart and Da Ponti collaborations: "Free discussion is the life of truth, and of true unity in a nation. Will England play this part?" he asked and, after theatrically pausing, answered, "That depends upon how you, the public, deal with science. Cherish her, venerate her, follow her methods faithfully and implicitly in their application to all branches of human thought; and the future of this people will be greater than the past." He nodded his head downward to cue the completion of his lecture. The applause was stupendous.[69]

It is difficult to imagine an author who would not be delighted by the dramatic conclusion to Huxley's talk. But the author in question was a sensitive fellow whose gut perked along when praised and quickly fluctuated when criticized. One week earlier, on February 2, Darwin had written a warning of sorts to Huxley: "My poor dear friend, you will curse the day when you took up the 'general agency' line."[70] By February 12, Darwin made his prediction come true as he complained to Lyell, "I think it was a great pity that Huxley wasted so much time in Lecture on preliminary remarks: he hardly gave idea of my notions."[71] Hooker, too, failed to appreciate the razzle-dazzle of Huxley's lecture. Writing to the Harvard botanist Asa Gray, he echoed Darwin: "Huxley made a failure of the R[oyal] Institution Lecture, which was a great pity, as he intended to have backed the book but unfortunately managed to damage it."[72] Two days later, February 14, Darwin grumbled, sotto voce, to Hooker:

> I succeeded in persuading myself for 24 hours that Huxley's lecture was a success. Parts were eloquent & good & all *very* bold, & I heard strangers say, "what a good lecture." I told Huxley so; but I demurred much to time wasted in introductory remarks; especially to his making it appear that sterility was a clear & manifest distinction of species, & to his not having even alluded to the more important part of subject. He said that he had **much** more written out, but time failed. After conversation with others & more reflection I must confess that as an Exposition of the doctrine the Lecture seems to me an entire failure.—I thank God I did not think so, when I saw Huxley, for

Thomas Huxley, as president of the Royal Society, 1883–1885.

he spoke so kindly & magnificently of me, that I could hardly have endured to say what I now think.[73]

ʃ

IN FACT, no one was more influential in promoting Darwin's cause than Thomas Huxley, even though he was compelled to admit that he did not endorse every single sentence of *Origin*. Huxley may have fashioned himself as Darwin's Bulldog, but he was no lap dog. On November 23, for example, he privately wrote Darwin immediately after finishing *Origin* for the first time: "The only objections that have occurred to me are 1st that you have loaded yourself with an unnecessary difficulty in adopting '*Natura non facit saltum*' ["Nature does not make jumps," or what today would be referred to as mutations] so unreservedly. I believe she does make small jumps—and 2nd, it is not clear to me why if external physical conditions are of so little moment as you suppose variation should occur at all—However, I must read the book two or three times more before I presume to begin picking holes."[74] Neither of Huxley's objections were entirely incorrect; and Darwin would later amend his thinking to consider these changes.[75]

In a brief but vigorous essay Huxley contributed to the December 1859 issue of *Macmillan's Magazine*, Huxley observed that Darwin's theories would "be settled only by the painstaking, truth-loving investigation of skilled naturalists. It is the duty of the general public to await the result in patience; and, above all things, to discourage, as they would any other crimes, the attempt to enlist the prejudices of the ignorant, or the uncharitableness of the bigoted, on either side of the controversy."[76]

Huxley next turned to writing a 57-page review of *Origin* for the April 1860 issue of the *Westminster Review*.[77] Overall, the review boomed with praise as he sliced the "pietists, whether lay or ecclesiastic, [who] decry it with the mild railing which sounds so charitable; bigots [who] denounce it with ignorant invective; old ladies of both sexes [who] consider it a decidedly dangerous book, and even savants who have no better mud to throw, quote antiquated writers to show that its author is no better than an ape himself." For Huxley, *Origin* was "a veritable Whitworth gun in the armory of liberalism" that every competent man had to admit was "a solid contribution to knowledge [that] inaugurates a new epoch in natural history."[78]

Unfortunately, it was, again, the occasional negative sentence embedded in Huxley's *Westminster Review* essay that Darwin (and his foes) remembered most. Upon closer contemplation of *Origin*, Huxley was compelled to note a few more issues with Darwin's argument. That they were loudly uttered by his chief advocate annoyed Darwin no end. But Huxley's bark and bite were so harsh because his points were so well made: "Mr. Darwin does not so much prove that natural selection does occur, as that it must occur; but, in fact, no other sort of demonstration is attainable." Huxley admitted that "Mr. Darwin is perfectly aware of this weak point and brings forward a multitude of ingenious and important arguments to diminish the force of the objection." Even though Huxley's "private ingenuity" did not find any fatal flaws in Darwin's tome, he had to confess "that, as the evidence stands, it is not absolutely proven that a group of animals, having all the characters exhibited by species in Nature has ever been originated by selection, whether artificial or natural." Unfortunately, he provided succor to Darwin's enemies by noting, "There is no positive evidence, at present, that a group of animals has, by variation and selective breeding, given rise to another group which was, even in the least degree, infertile with the first."[79] After reading the essay, Darwin agonized to Lyell on April 10: "There is a *brilliant* review by Huxley, with capital hits; but I do not know that he much advances subject. I *think* I have convinced him that he has hardly allowed weight enough to the cases of varieties of plants being in some degree sterile."[80] Here Darwin was being unfair and overly emotional. Huxley agreed with Darwin's grand theory and was simply asking for more proof of "the necessity of Natural Selection being shown to be a *vera causa* [true cause] always in action."[81] This Latin phrase, not incidentally, comes from Sir Isaac Newton's 1687 masterwork, *Philosophiæ Naturalis Principia Mathematica*, in which he cautioned, "We are to admit no more causes of natural things than such as are both true and sufficient to explain their appearances."[82]

Darwin was neither foolish nor vain enough to ignore Huxley's valid points. When it came to studying natural history, *vera causa* was not a matter of mathematical proofs and experimental data as it was for, say, Newton's physics or Boyle's chemistry. Only time and better research methods would bear out Darwin's empirical observations and descriptions. Equally difficult for Darwin was that his chapter on hybridism did not adequately demonstrate "that sterility is not a specially acquired or endowed quality but is incidental on other acquired differences."[83] For years, he labored to prove his concept on hybrid sterility, but was unable to accomplish such a task with the experimental methods available to him.

In early 1862, while recuperating from a "virulent influenza," Darwin acquiesced to Huxley, "No doubt you are right that here is great hiatus in argument; yet I think you overrate it."[84] Six years later, Wallace wrote Darwin to tell him that he still supported the notion that "natural selection *could* produce sterility of hybrids."[85] Darwin replied with a sigh that the cause was futile even if "no man could have more earnestly wished for the success of N[atural] selection in regard to sterility, than I did." For too long, Darwin struggled to prove the "terrible problem" that hybrid sterility was of some adaptive value, "but always failed in detail. The cause being, as I believe, that natural selection cannot effect what is not good for the individual. . . . It would take a volume to discuss all the points; & nothing is so humiliating to me as to agree with a man like you . . . on the premises & disagree about the result."[86]

# 7

# The Dinosaur

There is a natural and irrepressible tendency in the human mind to penetrate the mystery of the beginning of things, and above all that of the origin of living things, involving our own origin. But it is plainly denied to finite understandings to ascend to the very beginning, and to comprehend the nature of the operation of the First Cause of anything.

—*Richard Owen, 1842*[1]

Within weeks of the publication of *Origin*, Charles Darwin became a cause célèbre throughout England, Europe, and the United States.[2] Fueling the bonfire were a gaggle of critics, eager to attack or praise Darwin's book, which only spread the conflagration further. Throughout 1860, Darwin reckoned up dozens and dozens of reviews in both the popular and scientific press—spanning a full spectrum of positive to ferociously negative. John Murray made sure to send them all off to Down, where they were promptly digested. Darwin swooned over the good notices. He fretted over the bad ones. After a year of such notoriety, Darwin complained to Huxley, on December 2, 1860, "I have got fairly sick of hostile Reviews." To his credit, however, the next sentence Darwin wrote was: "Nevertheless they have been of use in showing me where to expatiate a little & to introduce a few new discussions."[3]

Many of the press appraisals were—to use a word Thomas Huxley coined nearly a decade later—*agnostic.* Such pieces read as if their authors wished neither to offend God's word nor the cause of modern science.[4]

This reporting style was best exemplified by the leading literary voice of Darwin's day, Charles Dickens—who was preoccupied with the composition of *A Tale of Two Cities* and a year later, *Great Expectations*, for his new weekly magazine, *All the Year Round* (2d, or two English pennies, weekly; 9d, or nine English pennies, monthly). The average weekly circulation of the periodical was well over one hundred thousand per issue, and for the special Christmas issue, as much as four times that many.

One might imagine the two men passing on the grand staircase of the Atheneum Club, but there is no proof that these two eminent Victorians ever met. That said, Dickens and Darwin were clearly aware of each other's work.[5] At the time of Dickens's death in 1870, both Lyell's *Principles of Geology* and Darwin's *Origin* were on the shelves of his library at Gad's Hill Place, in Rochester.[6] Dickens did know Richard Owen and wrote a charming essay about his work on gorillas for the May 28, 1859, issue of *All the Year Round.* And the Darwins were longtime fans of Dickens's big, thick novels—including *Oliver Twist*, *Nicholas Nickleby*, and *Martin Chuzzlewit*.[7] Emma often read the books aloud to the family, although she considered the "constant drinking" portrayed in *The Pickwick Papers*

Charles Dickens, circa 1860.

to be "a great blemish."[8] On November 9, 1861, one year after the publication of *Origin*, Emma and her sister Elizabeth attended one of the writer's famously popular public readings. For this performance, Dickens acted out both *A Christmas Carol* and the trial scene in *The Pickwick Papers*. Alas, Darwin's tender intestines forced him to stay at home, missing an opportunity for the author of natural selection to meet the great portrayer of human nature. Like Darwin, Dickens was hardly a picture of health. The following morning, Emma wrote her son William at Cambridge, "Dickens himself is very horrid looking with a light-colored ragged beard which waggles up and down. He looks ruined & a roue, which I don't believe he is, however."[9]

Between 1859 and 1861, Dickens ran three unsigned reviews of *Origin* in his magazine.[10] Late one evening, in the *All the Year Round* editorial office on No. 26 Wellington Street—a few blocks off the bustling Strand—a staff writer grappled with the early drafts of these reviews. Long after midnight, "the Inimitable" swept into the press room with his shockingly young mistress, the actress Ellen Ternan, on his arm. Before proceeding to the connected "private apartment" with his twenty-one-year-old consort, the forty-eight-year-old novelist spruced the essays up, giving each of them a bubble or two of his literary effervescence. All three articles were respectful synopses of the book but offered no hard opinions, pro or con, on its veracity.[11] As Darwin remarked to Lyell after the second review appeared: "There is a notice of me in the penultimate number of *All the Year Round*, but not worth consulting; chiefly a well-done hash of my own words."[12]

These reviews were far more important than Darwin's sour assessment of them. Dickens understood his huge readership better than just about any author in the English language. He knew, if Darwin did not, that they both identified with the beliefs of Christianity *and* were citizens of a modern world where new discoveries, technologies, and social changes were occurring at a dizzying pace.[13] By heavily quoting *Origin*—often without attribution—the magazine provided a tempting amuse-bouche of the book. The concluding paragraph of the second review nicely encapsulates the Dickensian touch and the need for more proof:

> If Mr. Darwin's theory be true, nothing can prevent its ultimate and general reception, however much it may pain and shock those to whom it is propounded for the first time. If it be merely a clever hypothesis, an ingenious hallucination, to which a very industrious and able man has devoted the greater and the best part of his life, its failure will be nothing new in the history of science. It will be a Penelope's web, which, though woven with great skill and art, will be ruthlessly unwoven, leaving to some more competent artist the task of putting together a more solid and enduring fabric.[14]

ʃ

THE RUSH OF negative notices bothered Darwin even though he claimed to dismiss them altogether. These essays tended to be written by skeptical or oppositional scientists, threatened theologians, and various defenders of the faith. The reasons behind their hostility ranged from intellectual and spiritual quibbles to pure jealousy and personal history. For Darwin, every hostile rebuke, no matter how relevant or nonsensical, represented another chip of marble struck off the theoretical sculpture he had so carefully worked on for over two decades.

One sarcastic slap came from his former captain aboard the *Beagle*, Robert FitzRoy. FitzRoy was far more than an accomplished naval officer and surveyor of vast stretches of oceanic coastlines. A former governor of New Zealand (1843–1845), he is best remembered for developing a telegraphic system of making weather predictions—or, to use the word he coined, forecasts. In 1854, he founded the British Meteorological (or Met) Office, which introduced the world's first "daily weather report" in September 1860.[15] Now a rear admiral and Tory member of Parliament (for Durham), FitzRoy interpreted Darwin's new book as an open decla-

Admiral Robert FitzRoy.

ration of war against the Bible's description of the Creation. In November 1858, FitzRoy wrote Darwin, "My dear old friend, I, at least, *cannot* find anything 'ennobling' in the thought of being the descendant of even the most Ancient *Ape*."[16] The squirrelly sea captain hardly stopped there. On December 1, 1859, he wrote a mean, pseudonymous letter to the editor of the *Times*, above the signature "Senex, wise old man." In a few paragraphs, FitzRoy took Darwin to task for his heretical views, arguing that natural selection was impossible because Man had only been on the planet for thirteen thousand or fourteen thousand years.[17]

Darwin, still soaking in the frigid baths at Ilkley, must have boiled as he read his morning *Times*. To Lyell, he carped, "[The letter] is, I am sure by FitzRoy; for he wrote me the other day. . . . It is a pity he did not add his theory of the extinction of the Mastodon &c from the door of [Noah's] Ark being made too small. What a mixture of conceit & folly, & the greatest newspaper in the world inserts it!"[18] A few paragraphs later, Darwin moved on to discuss his more formidable foe, Richard Owen: "How curi-

Richard Owen, circa 1892.

ous I shall be to know what line Owen will take,—dead against us, I fear; but he wrote me a *most* liberal note on the reception of my Book, & said he was quite prepared to consider fairly & without prejudice any line of argument."[19] Darwin had good reason to be concerned.

Portraits of Sir Richard Owen, KCB, FRMS, FRS, depict an embittered man who looks as if he just swallowed his own vomit. With cheekbones protruding from a gaunt face at acute angles, huge eyes, wild whiskers, and the black, scholar's skull cap he wore at work, he projected the appearance of a mad monk. He divided his colleagues into two categories: enemies and potential enemies. Too often, Owen directed his heavy-handed spitefulness at those who disagreed or competed with his research efforts and grandiose plans to build a deluxe museum of natural history in London. Owen's two biggest detractors in the latter goal, incidentally, were Huxley and Hooker, whose objection was that one man—especially one as imperious and ruthless as Owen—should not control such a vast collection. Instead, they argued, science would be better served by sharing the artifacts among several other institutions for further study.

While a student at the Lancaster Grammar School, Owen was regularly beaten for "impudent" behavior. The one advantage to attending this harsh institution was its most famous alumnus, William Whewell, the Cambridge natural philosopher, who often came to visit. Owen met Whewell while still a boy, and the meeting "ripened into" a beneficial, lifelong friendship. In 1820, at age sixteen, Owen "showed a taste for the study of medicine" and was apprenticed to a surgeon. He matriculated into Edinburgh University in 1824 but left Scotland a year later to attend London's St. Bartholomew's Hospital Medical College. In addition to his studies, he served as the prosector—an assistant who prepares anatomical material used in medical lectures—for the famed surgeon John Abernethy. Owen received his diploma from the Royal College of Surgeons in 1826.[20]

That same year, he established a medical practice in the heart of London's legal profession—on Serle Street, Lincoln Inn Fields—but soon grew bored attending "to the ailments of young barristers or solicitors' clerks." Well-read, "with musical abilities of no mean order, he was soon the center of a select social set, many of whom in after life remained his constant friends." Owen developed a well-honed knack for currying favor with those who could best advance his career.[21] In 1827, for example, he became engaged to Caroline Clift, the daughter of William Clift, the conservator for the Museum of the College of Surgeons. They married, after an eight-year-long engagement, in 1835.

While rummaging through the Royal College's vast collections, Owen discovered that the anatomical specimens preserved by the distinguished eighteenth-century surgeon John Hunter were gathering dust in the basement. In 1828, he asked Abernethy, who was then president of the College of Surgeons, to recommend him to the post of curating, cataloging, and illustrating its contents. Among Owen's selling points, Abernethy wrote, were that "the collection was located near his private residence; he could devote his leisure hours to the work; [and] there was no one else equally qualified to do so."[22] The letter worked like a charm, and Owen was given the appointment.

Upon completing this massive task of identification and cataloging in 1830, Owen met the French paleontologist Georges Cuvier during the great man's visit to London. The ambitious young Owen spent the following summer studying fossil vertebrates under Cuvier at the National Museum of Natural History in Paris.[23] This sojourn to Paris changed his life's work, even though it was temporarily put on hold in 1832 by a family tragedy that wound up benefiting Owen. His brother-in-law, William Clift Jr., was tapped to take over his father's job as curator for the Royal College of Surgeons but was thrown from a galloping cab on Fleet Street. Junior died of severe head wounds at the scene of the accident. Owen promised his grieving father-in-law that "he would take the son's place and never desert him or the museum."[24] Owen was so adept at combining research and self-promotion that by age thirty, in 1834, he was elected to the Royal Society, and at age thirty-seven, in 1837, appointed the Hunterian Professor of Anatomy and Physiology of the Royal College of Surgeons.

Family and professional connections aside, Owen built himself into the doyen of British science. Widely known as a comparative anatomist with an expertise in the nascent field of paleontology, it was Owen to whom Darwin sent the fossils he found in South America.[25] For most of the 1840s and 1850s, Owen focused his attention on the remains of extinct reptiles discovered across England, including the carnivorous *Megalosaurus*, the herbivorous *Iguanodon*, and the armored *Hylaeosaurus.* He noted how all three dinosaurs shared a fused vertebral body at the end of their long spines, suggesting a common ancestor—a finding that predated the publication of *Origin* by eighteen years.[26] Describing these creatures' anatomy was a major chore in taxonomy because the extinct reptiles were so different from anything currently alive. In fact, Owen coined the word *dinosaur* while speaking for over two hours at the 1841 annual meeting of the British Association for the Advancement of Science.[27] He compounded two Greek words to form the new one: δεινός (deinós), or "terrible," and *σαῦρος* (saûros), or "lizard."[28]

George H. Lewes, circa 1878.

Owen was remarkably clever at exploiting the public's fascination with his ancient "terrible lizards." For example, the paleontologist's acquaintance with Charles Dickens inspired the novelist's description of a *Megalosaurus* waddling its way up Holborn Hill in the opening paragraph of his 1852 masterpiece, *Bleak House*.[29] Perhaps Owen's wackiest publicity stunt occurred outside the Crystal Palace as it prepared to move from Hyde Park to Bromley, in southeast London. The sculptor Benjamin Waterhouse Hawkins was commissioned to create a pack of life-sized dinosaurs for a "Dinosaur Court," scheduled to open in early 1854, under Owen's direction. On New Year's Eve 1853, the sculptor invited eleven guests to have dinner inside the *Iguanodon*, one of the newly constructed clay and cement molded "dinosaurs." Seated at the head of the table, which corresponded with the dinosaur's head, was the guest of honor, Richard Owen.[30] In 1856, he was named superintendent of the natural history department of the British Museum, "a post that was specially made" for him.

One dark December afternoon in 1859, Owen granted journalist George H. Lewes an interview. They met on Owen's turf in the Royal College's Hunterian Museum. While they were talking, a gentleman approached them and asked if the professor might identify "the nature of a curious fossil," which one of the workmen on his property had recently dug up. Before the man could fully draw the fossil out of his bag and hand it over to Owen for inspection, the wizened professor softly said, "That is the third molar of the under-jaw of an extinct species of rhinoceros." Of the three men present, the only "under-jaw" that did not drop from this confident display of expertise belonged to Owen.[31] In the same magazine article, however, Lewes warned against Owen's derision of Darwinian theory as he implored his readers not to tax "the followers of . . . Mr. Darwin with absurdities [he has] not advocated; but rather endeavor to see what solid argument [he has] for the basis of [his] hypothesis."[32]

ʃ

ON NOVEMBER 11, 1859, Darwin sent Owen a fawning note, promising a presentation copy of *Origin* and warning, "I fear that it will be abominable in your eyes; but I assure you that it is the result of far more labor than is apparent on its face." Darwin asked Owen to read his book straight through as he openly admitted "that my meaning will not be clear to anyone, without a considerable amount of reflection. Whether I be in main part right, as I naturally think myself to be, or wholly wrong, the old saying of *magna est veritas et prevalebit* ["truth is great and shall prevail"] is a grand conclusion to all doubts."[33] Four weeks later, on December 10, Darwin wrote Lyell with crushing news:

> I have [had a] very long interview with Owen, which perhaps you would like to hear about, **but please repeat nothing**. Under garb of great civility, he was inclined to be most bitter & sneering against me. Yet I infer from several expressions, that *at bottom* he goes immense way with us.—He was quite savage & crimson at my having put his name with defenders of immutability. When I said that was my impression & that of others, for several had remarked to me, that he would be dead against me: he then spoke of his own position in science & that of all the naturalists in London, "with your Huxleys," with a degree of arrogance I never saw approached. He said to effect

> that my explanation was best ever published of manner of formation of species. I said I was very glad to hear it. He took me up short, "you must not at all suppose that I agree with in all respects".—I said I thought it no more likely that I [should] be right on nearly all points, than that I [should] toss up a penny & get heads twenty times running.[34]

Upon asking Owen what his specific criticisms were, Owen claimed to have "no particular objection to any part." He then added, in a tone Darwin found offensive, "If I must criticize, I should say 'we do not want to know what Darwin believes & is convinced of, but what he can prove.'"[35] Darwin must have intuited that Owen was as hungry as the carnivorous dinosaurs he studied. Only now, *Darwin* was the prey! In fact, Owen was already telling others that Darwin's theory of natural selection and transmutation was biologically boneheaded and not backed up by sufficient evidence. Darwin would have to wait another four months before Owen's extensive critique was set into type, published, and widely quoted.

OWEN'S "ANONYMOUS" ESSAY appeared in the April 1860 issue of *Edinburgh Review*—one of the most influential British magazines of the nineteenth century. The piece was a relentless weapon of words that viciously plundered and pillaged the book in hand.[36] The basis of his epic animosity toward Darwin is complicated, as are most human conflicts. Some of it can be blamed on his naturally suspicious and nasty demeanor. Others have ascribed the venom of the review to professional jealousy and that as the leading light in paleontology, *he*—not Darwin—had scientific priority in explaining the origin, definition, and extinction of species. This was more than mere ego alone. For example, his 1841 study of dinosaurs suggested a common ancestry of those beasts. Eight years later, Owen published an important and often ignored book, *On the Nature of Limbs.* In this book, he defined homology, a word he coined in 1843, as "the same organ in different animals under every variety of form and function." This work, through a careful dissection and description of multiple species, demonstrated how all vertebrates, from the simplest to the most complex, shared similar skeletal systems—an observation that became an essential fact of evolutionary biology.[37]

It is also important to recall that Owen was a well-connected, God-fearing, conservative Briton. He served Queen Victoria, the Defender of the Faith, as her children's natural history tutor. With this royal pat on the back, he assumed the role of the nation's "high priest" of science. In 1849, he published a long and tedious essay, "On Parthenogenesis," ascribing the origin of all creatures to "an all-wise and powerful First Cause of all things."[38] His conceit dictated that he alone was best suited to resolve the discrepancies between the Bible's account of Creation and those discovered or theorized by scientists.[39] As such, he conjured an explanation that whatever the external forces changing a species—natural, sexual, or even the breeder's push toward artificial selection—the resulting variations were superficial, at best. If species *did* show variation over time, he vamped, it was because "some pre-ordained law or secondary cause is operative in bringing about the change."[40] Several years later, Huxley brutally derided Owen's thesis as one that "may be read backwards, or forwards or sideways, with exactly the same amount of signification."[41]

Unfortunately, Owen's *Edinburgh Review* essay was a review of several books—a time-honored trick used by hostile reviewers to bait and switch the reader into belittling a particular author's work by praising the work of others. In 45 pages—and more than twenty thousand words—Owen misrepresented Darwin's arguments to advance his own ideas. As with most reviews designed to create havoc, only the most knowledgeable and careful of readers could have detected the essay's false tone. Hardly anonymous, the review glowingly referred to the work of the "eminent paleontologist" Richard Owen more than twenty times—or about once every other page. He even quoted from his 1858 presidential address to the British Association for the Advancement of Science, specifically citing his work on the parthenogenesis of a small ocean predator known as *Hydrozoa*: "The first acquaintance with these marvels excited the hope that we were about to penetrate the mystery of the origin of species; but, as far as observation has yet extended, the cycle of changes is definitely closed."[42]

From the review's opening sentence to its last, Owen could not temper the war within. He attacked Darwin's prose. He claimed that the author misused scientific terminology or, worse, created it out of thin air—an especially odd complaint by the man who concocted the word *dinosaur*. Owen even questioned Darwin's competence as a naturalist, alluding to him as a wealthy country squire, an amateur who played at science, as opposed to those who labored in the university professoriate. Splashing more iodine into a gaping wound, Owen took Darwin to task, along with Hooker, Huxley, and the Oxford mathematician Baden Powell, for allegedly trashing natural theologians and God-fearing men and women as one-dimensional, backward, and naive.[43]

Perhaps most devastating was the query that echoed Huxley's reviews: "Is there any one instance proved by observed facts of such transmutation? We have searched the volume in vain for such. When we see the intervals that divide most species from their nearest congeners, in the recent and especially the fossil series, we either doubt the fact of progressive conversion or, as Mr. Darwin remarks in a letter to Asa Gray, 'one's imagination must fill up very wide blanks.'"[44] Owen went in for the kill by noting the absence of fossilized transitional species. In so doing, he amplified Huxley's quibble of how Darwin failed to fulfill Isaac Newton's first rule of reasoning of *vera causa*, or true cause. Newton made an excellent and lasting point. Proof—not theory alone—is what science demands of all its practitioners. Yet in the hands of a malicious opponent, the cry for *vera causa* can be used to strike down almost any nascent theory that another scientist might posit. One bit of irony to this reasoning is that Owen did not place the same criterion for Genesis as an explanation, beyond faith alone. Unfortunately, Darwin did not and would never have definitive proof of natural selection; the "wide blanks" from the incomplete fossil record were, of course, gradually, albeit partially, filled—thanks to far more advanced methods and technologies—but not until long after his death.

Owen's critique of *Origin* was so relentlessly negative that several other prominent dons lined up to agree. One of them, Sir John Herschel, the Cambridge astronomer and proponent of delineating laws of nature and physical science, called Darwin's theory of natural selection "the law of higgledy-pigglety." There was no finessing this comment, and Darwin correctly interpreted it as "evidently very contemptuous."[45] Herschel's disdain must have been both excruciating and perplexing for Darwin, who first met the astronomer while sailing on the *Beagle*. A few years earlier, while at Cambridge, Darwin was enthralled by Herschel's book, *Introduction to the Study of Natural Philosophy*, which held that nature was ruled by a set of scientific laws. The soundest laws were proven by means of mathematical formulas, such as Newton's laws of gravitation. But the "derived" laws of nature, Herschel explained, remained a challenge for natural philosophers to discern because instead of mathematics, they required observation and inductive reasoning to reveal unified explanations.[46] Years later, Darwin recalled how Alexander von Humboldt's *Personal Narrative* and Herschel's book "stirred up in me a burning zeal to add even the humblest contribution to the noble structure of Natural Science. No one or a dozen other books influenced me nearly so much as these two."[47]

THE DELIVERY OF the *Edinburgh Review*'s April issue to Down House initiated a long and painful weekend for the author, his family, and their two guests—Huxley and Hooker. The following Monday, April 9, he wrote to his publisher, John Murray, in anguish: "I am thrashed in every possible way to the full content of my bitterest opposers. The article is very venomous, & is manifestly by Owen. I wish for auld lang syne's sake he had been a little less bitter."[48] On Tuesday, April 10, Darwin wrote Lyell: "It is extremely malignant, clever, & I fear will be very damaging. He is atrociously severe on Huxley's lecture & very bitter against Hooker. So, we three *enjoyed* it together; not that I really enjoyed it, for it made me uncomfortable for one night, but I have quite got over it today." Nevertheless, Darwin added, one had to go over the essay with a fine-tooth tweezer "to appreciate all the bitter spite of the many remarks against me; indeed, I did not discover all myself—It scandalously misrepresents many parts. He misquotes some passages altering words within inverted commas. . . . It is painful to be hated in the intense degree with which Owen hates me.[49]

No author so stabbed ever forgets a sharp review. Yet to complain about it by writing a point-by-point rebuttal only makes matters worse, especially since most of its readers have forgotten it and moved on to the next unscrupulous review assailing some other poor soul's hard work. Nevertheless, Owen's vitriol—which appeared in a widely read and influential periodical—unleashed a powerful and lasting narrative dismissing Darwin's idea as nonsensical, if not heretically dangerous.

On May 8, Darwin groused to his former Cambridge professor, John Stevens Henslow, "Owen is indeed very spiteful. He misrepresents & alters what I say very unfairly. . . . The Londoners says he is mad with envy because my book has been talked about: what a strange man to be envious of a naturalist like myself, immeasurably his inferior! From one conversation with him I really suspect he goes at the bottom of his hidden soul as far as I do!"[50] A month later, he wrote Asa Gray, "No one fact tells so strongly against Owen, considering his former position at Coll[ege] of Surgeons, as that he has never reared one pupil or follower."[51] Darwin's uncharacteristic bitterness never abated. Fourteen years after Owen's review appeared, in 1874, Darwin wrote Hooker, "What a demon on earth Owen is. I do hate him."[52] In his brief 1882 autobiography, Darwin went a few steps further by repeating Hugh Falconer's assessment of Owen as "not only ambitious, very envious and arrogant, but untruthful and dishonest. His power of hatred was certainly unsurpassed. When in former days I used to defend Owen, Falconer often said, 'You will find him out someday,' and so it has proved."[53]

Presentation of the Darwin statue at the British Museum of Natural History, 1885.

ʃ

THIS TALE OF scientific acrimony does, however, have a happy, if belated, Darwinian victory. In 1885—three years after his death—Darwin's admirers induced the British Museum of Natural History in London to honor the naturalist by erecting a massive, snowy-white marble statue of him on the grand staircase of its magnificent Central Hall. The statue commanded a vast room where, today, the skeletal remains of various creatures hang from a colorful tiled ceiling adorned with 162 panels representing different plants from across the British Empire.

Owen, who outlived Darwin by a decade, had to walk by the marble tribute every day at work, until a stroke in 1890 paralyzed and confined him to his drawing room and bedroom. Toward the end of November 1892, Owen stopped eating and, a few weeks later, on December 18, he died. His grandson concluded two long volumes of the paleontologist's life with a lyrical and decidedly exaggerated summation: Sir Richard "passed peacefully away, without a struggle, leaving the

Richard Owen statue at Museum of Natural History, London.

world poorer by the loss of an untiring worker and of a most genial and kind-hearted man."[54]

In 1927, his heirs orchestrated a successful movement to replace the Darwin statue with a dark bronze figure of Owen. For the next eight decades, the Darwin piece was cornered away in a side room near the cafeteria, representing the museum's equivalent of buyer's remorse. In 2009—the bicentennial year of Darwin's birth and the 150th anniversary of the publication of *Origin*—his life-sized statue was rightfully restored to its original, majestic position. High above—in a dimly lit mezzanine gallery under the shadow of a secondary staircase—Richard Owen glowers and picks a dinosaur's bone. One likes to think he is in a permanently foul humor over having been displaced by Darwin. The bronze Owen's back is pointedly presented to the marble Darwin, as if the statues are still fighting their feud. Directly across from the paleontologist—neatly tucked in front of an arched window—is an imposing marble statue of Thomas Huxley. Darwin's Bulldog glares at the museum's founder; his

Owen's "Cathedral" of Natural History in the South Kensington section of London.

pose gives him the appearance of being ever ready to leap out of his chair and attack.

Richard Owen "invented" the word *dinosaur* and for years presided over a "splendid mansion in which to store the treasures of nature."[55] The imposing, Romanesque-style, terra-cotta-clad British Museum of Natural History remains one of England's architectural gems. Lines of cheerful tourists and chattering schoolchildren await the opening of its doors every morning. And yet it is Charles Darwin who revolutionized our understanding of the life sciences—which begs the question: Who's the dinosaur now?

# 8

# Soapy Sam

Great power of concentration was combined with an incessant readiness to turn aside to fasten upon any new object which came before him. . . . His observation was sleepless and made him an excellent naturalist. If you were driving him across a country that was new to him, the conversation would be again and again interrupted by some remark upon its geology or vegetation . . . everything interested Bishop Wilberforce, so he could not help bearing a hand in whatever interested him.

—*Arthur Rawson Ashwell, 1880.*[1]

Just as "Darwin's Bulldog" is irrevocably pasted onto Huxley, the moniker "Soapy Sam" has proven twice as gummy for Samuel Wilberforce. Legend has it that Prime Minister Benjamin Disraeli concocted the nickname because he found Wilberforce to be so "unctuous, oleaginous, saponaceous," but there exists no documentary evidence to support this claim.[2] Disraeli was born Jewish and baptized at the age of twelve. He and Wilberforce did, however, represent opposite sides on "Jew bills" with Disraeli calling to strike the phrase "on the true faith of a Christian" from the Parliamentary membership oath.[3] The two men were later aligned in their distaste for Darwin's ideas, but they never really liked one another.[4] In 1870, Disraeli based a less than savory character on Wilberforce in his best-selling novel *Lothair*.[5] Wilberforce failed to appreciate such inclusion, but as he wrote to a childhood friend that May, "My wrath against D[israeli] has burnt before this so fiercely that it seems to have burnt up all the

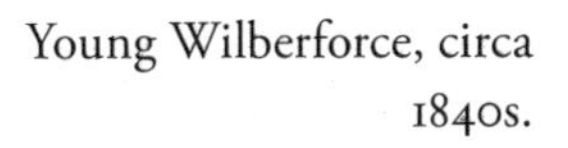

Young Wilberforce, circa 1840s.

materials for burning and to be like an exhausted prairie fire—full of black stumps, burnt grass, and all abominations."[6]

Another story centers on Wilberforce's brother, Robert, who was overheard gossiping in the Parliament's guest gallery, sometime in 1849 or 1850. He supposedly recounted how Sam, as a little boy, "was always washing his hands." Later, Robert amended matters by claiming that the name originated from the four letters—S.O.A.P.—that are carved into the entrance gate of Cuddesdon College, the Church of England training school for Oxbridge men that Wilberforce established in 1853 and opened for students the following year. This explanation proved specious, because "Soapy Sam" had been in use long before even "a stone of Cuddesdon College had been laid."[7] In fact, the letters S.O.A.P. commemorate "Samuel Oxon"—as Wilberforce then signed his name—and the college's first principal, Alfred Pott.

The *Oxford English Dictionary*, that venerable scorekeeper of the English language, records "Soapy Sam" as having first appeared in print in an 1854 diary entry (even though the book containing it was not published until 1928). The diary belonged to Ellen Dwight Twistleton, the wife of the

Cuddesdon Palace.

Right Honorable Edward Twistleton—a Poor Law Commissioner from 1845 to 1847. She wrote this particular entry after attending a breakfast that included Wilberforce: "The Bishop of Oxford I never do like. He has a very fine head and a very bright blue eye, and is considered, as you know, one of the most agreeable men in London society, but he has the most Jesuitical mouth and his manner, when Lords are in presence, richly merits his popular sobriquet of 'Soapy Sam.'"[8] Mrs. Twistleton's 1854 observation supports an earlier, if unclear, origin of the nickname.

The following year, 1855, "Soapy Sam" entered the annals of literature by way of Anthony Trollope's novel *The Warden*—the first of his *Chronicles of Barsetshire* series. In it, Trollope introduced a character named Samuel, or "dear little Soapy," who was clearly based on the bishop: "To speak the truth, Samuel was a cunning boy, and those even who loved him best could not but own that for one so young, he was too adroit in choosing his words, and too skilled in modulating his voice."[9]

In 1860, four months after *Origin* was published, a controversial book appeared with the generic title *Essays and Reviews*. Written by six Liberal Anglican churchmen and one layman, *Essays and Reviews* flew off the booksellers' shelves faster than Darwin's tome; within two years it sold over twenty-two thousand copies. The book included essays on recent biblical historical research; the evidence, or lack thereof, for Christian prophecies and miracles; religious thought in England; and the cosmology of Genesis. In early 1861, a vituperative book review of *Essays* appeared in John Murray's *Quarterly Review*. The anonymous notice was written by Samuel Wilberforce, which not only raised the book's profile but also sold out five printings of the *Quarterly Review*. Wilberforce ratcheted up the pressure by attempting to organize his clerical colleagues to condemn the book and pull it off library shelves.[10] Four years later, according to the *Dictionary of National Biography*, Wilberforce remained so incensed over the book's existence that he proposed a resolution in the House of Lords censuring the book.[11] On July 15, 1864, the resolution was soundly defeated. During the debate, the Liberal Lord Chancellor (Lord Westbury), Richard Bethell, placed the sobriquet "Soapy Sam" on parliamentary record. He not only deemed the motion to be illegal; he derided Wilberforce's opinions on religion and science as nothing more than "a well-lubricated set of words, a sentence so oily and saponaceous that no one can grasp it."[12]

The following year, October 1865, *The Pall Mall Gazette* aimed a dart directly at the bull's-eye, even if the score was credited to Wilberforce. Apparently, the Bishop of Oxford was overheard in a conversation he had with another passenger while traveling on an eastward train to Norwich for a church congress. The woman seated across from him "commented in flattering terms on the eloquence of the great Anglican divine," Samuel Wilberforce, without realizing she was speaking directly to him. "But why, sir," she asked, "do people call him Soapy Sam?" The bishop, who did not reveal his identity, replied, "Well, madam, I suppose it is because he is always in a good deal of hot water and always manages to come out with clean hands."[13]

Samuel Wilberforce's's father, William Wilberforce, was the evangelical Christian missionary, politician, and fiery abolitionist known as "the Great Emancipator." During the fall of 1823, a few days before Samuel matriculated into Oriel College, Oxford University, William counseled his third and favorite son that he was to "be tried by a different standard from that which is commonly referred to, and be judged by more rigorous rule. . . . Remember, my dear boy that you have *my* credit in your keeping as well as your own."[14]

While at Oxford, Wilberforce walked a narrow path of trying to please his chums, his tutors, and his father at every instance. A member of the United Debating Society (the forerunner to the famed Oxford Union Society), he condemned the character of Queen Elizabeth I and "fagging in our public schools"—the boyish tradition at the Eton, St. Paul's, Harrow, Rugby, and Winchester schools whereby older students conscripted the newest ones as their personal servants on a good day and with physical and, for some, sexual abuse on the worst ones.[15] Despite the many hours he spent studying in the library, Wilberforce loved the outdoors and often went hunting and riding on the estates of his wealthier friends. By 1826, he won his bachelor's degree with an admirable first honors degree in mathematics and a second in classics. That same year, however, he failed to win a prestigious fellowship at Balliol College, prompting him to take the wealthy, nineteenth-century British equivalent of a gap year by traveling across the Continent. Balliol's Master invited Wilberforce "to offer himself again when a vacancy should occur, but before that time his plans had changed . . . [and] marriage and his future profession occupied all his thoughts."[16]

Oriel College during the first half of the nineteenth century was a hotbed of Christian faith that veered perilously close to Catholicism, thanks to a charismatic and brilliant college fellow named John Henry Newman. In 1833, Newman founded what is now known as the "Oxford Movement" of Anglo-Catholicism. As a student, Wilberforce "expressed the keenest admiration for the intellect and powers" of Newman but kept his distance, perhaps fearing Newman's mysticism, love of ritual, and, as Wilberforce biographer Arthur Ashwell claimed, a "dread of surrendering his judgment into the keeping or guidance of another."[17]

By 1845, Newman had left the Church of England and Oxford Uni-

Oriel College, Oxford University.

versity for the Roman Catholic Church, ultimately becoming a priest in 1847 and a cardinal in 1879. Between 1850 and 1863, two of Wilberforce's three brothers, Robert and Henry—both graduates of Oriel College, Oxford—followed Newman into the Catholic Church, despite Wilberforce's entreaties for them to remain with the Church of England.[18] Decades later, in 1868, his daughter Ella and her husband—an Anglican priest named Henry Pye—were also received into the Catholic Church, much to Wilberforce's anguish.

John Henry Newman did not think highly of Samuel Wilberforce. In 1838, for example, Newman described Wilberforce as "so far from anything higher than a dish of skimmed milk that we can hope nothing from him." Three years later, in 1841, Newman took glee in recounting how he and his Oxford cronies "stigmatized [Wilberforce] as a humbug for many years."[19] In turn, Wilberforce had a "tense relationship" with Newman.[20] Nonetheless, as a bishop he used his influence to block those seeking to squash the Oxford Movement, one of the great church controversies of his tenure. For many years, Wilberforce served as a conduit between those wishing to reinstate the older, Catholic customs and liturgy—removed from the Anglican church by King Henry VIII during the Reformation—and those congregations who favored evangelicalism, modern English-language Bibles, and sermonizing.

ʃ

SAMUEL WILBERFORCE married the love of his life, Emily Sargent, on June 11, 1828. She was the daughter of a prominent evangelical clergyman named John Sargent and Mary Smith (who was William Wilberforce's first cousin). Upon their return from a three-month "wedding tour" of Europe, Wilberforce sat for his clerical examinations and passed with distinction. He was ordained as a deacon on December 21 and appointed curate of the church at Checkendon, near Henley-on-Thames. Within a year, he was already displaying his trademark flip-flopping depending on the occasion at hand. To one friend, he wrote, "As to politicks,[sic] Sir, I don't know what to say are my opinions. I was once, as you know, a Radical. I believe I am now, with some exceptions, a very high Tory."[21] One year later, December 20, 1829, he was ordained a priest and in 1830 became Parson of Brightstone on the Isle of Wight. In less than a decade, 1839, he was promoted to Archdeacon of Surrey and in 1840 was appointed to the prestigious post of Canon at Winchester Cathedral. He was neither a "great reader for reading's sake" nor a "profound thinker." Wilberforce's meteoric rise in the church was a function of the sheer force of his personality, eloquence, organizational skills, preaching at churches across the British Isles, and attendance at every important meeting that came his way.[22]

Wilberforce's occupational success was marred by the sudden death of his wife Emily on March 10, 1841, less than a month after delivering their son and fifth child, Basil. Wilberforce's grief was so great that he canceled his invitation to deliver the prestigious Bampton Lectures on Theology at Oxford. He asked his mother-in-law, Mary, to move in and run his household, which she did with an iron hand until her death in 1861. For the rest of his life, Wilberforce suffered from anniversary reactions on the dates marking Emily's birthday and death; the mere mention of her name was enough to evoke the depths of sadness. What Emily's death could not extinguish was his burning ambition.[23]

Six months after becoming a widower at age thirty-six, Wilberforce made a "striking public appearance" at an antislavery meeting. Prince Albert, the chairman of the gathering, was so impressed that he made Wilberforce his personal chaplain. In 1843, Albert promoted Wilberforce to subalmoner and, from 1847 to 1869, Lord High Almoner to Queen Victoria.[24] The year 1845 was a banner one for Wilberforce; he became Dean of Westminster Abbey, was elected to the Royal Society, took his Doctor of Divinity degree at Oxford, and made national news by speaking on the partnership of science and religion at the British Association for the Advancement of Science meeting held in Cambridge. In October of the same year, the Conservative prime minister Robert Peel appointed him Bishop of Oxford, where he remained for twenty-four years until the Liberal prime minister—and a friend from his Oxford days—William Ewart Gladstone, appointed him to the bishopric of Winchester.[25] When taking the oaths of priest and bishop, Wilberforce answered a hearty "Yes" to two requisite questions:

> Are you persuaded that the Holy Scriptures contain all Doctrine required of necessity for eternal salvation through faith in Jesus Christ?

> Are you determined, out of the said Scriptures, to instruct the people committed to your charge, and to teach nothing, as required of necessity to eternal salvation, but that which you shall be persuaded may be concluded and proved by the Scripture?[26]

With this oath, Wilberforce swore allegiance to the unswerving certainty of church doctrine. In a fascinating aside, only a few weeks before assuming his duties as Bishop of Oxford, he wrote down several additional pledges of propriety, including one he rarely succeeded at: "Beware of exaggerating, either in praise or blame:—guard my conversation more. Let me not be a slanderer."[27]

Wilberforce's love for "public business" became obvious soon after assuming his bishop's seat in the House of Lords in 1845.[28] At Westminster Palace, he argued for closer connections between church and state and against liberalizing the divorce laws. Predictably, he opposed the opening of museums or exhibitions on Sundays, the taxation of churches, and Jewish emancipation. In other matters, he proved sympathetic to the plight of the poor, improved sanitation, and prison reform. His insinuation into so many political matters was, however, at odds with the advice of the royal family, who desired their court priest to stay out of the political arena and focus on lending spiritual support to the government and Crown.[29]

John William Burgon, the rabidly anti-Darwinist dean of Chichester Cathedral and defender of Biblical infallibility, wrote that Wilberforce was "large-hearted, liberal, generous to a fault." Nonetheless, Burgon was compelled to add:

> He was *too* clever, *too* self-reliant, whereby he often put himself in a false position, and exposed himself to unfriendly criticism . . . he was *too* persuasive, *too* fascinating in his manner, *too* fertile in expedients; and thus, he furnished not a few with pleas for suspecting him of insincerity. Sure of himself and unsuspicious of others, he was habitually *too* confiding, *too* unguarded in his utterances. But above all, his besetting fault was that he was a vast deal *too facile* . . . [as a consequence] he was sometimes obliged to 'hark back,'—to revoke,—to unsay. This occasioned distrust.[30]

Wilberforce's first of many Waterloo moments occurred during a nasty fight with Renn Dickson Hampden, the Regius Professor of Divinity at Oxford. In 1847, Prime Minister John Russell recommended Hampden to the queen for appointment as Bishop of Hereford. The "effect of this announcement was electrical, and the excitement . . . was instantaneous."[31] The source of the charged political storm—according to the conservative factions of the Church of England—was Hampden's "distasteful" theology. In 1832, when Hampden presented his Bampton Lectures on Christianity at Oxford—a series entitled "The Scholastic Philosophy Considered in Its Relation to Christian Theology"—he used inductive reasoning to separate "the original truth" of Christianity from subsequent "rituals and rules" made by scholastic theologians during the Middle Ages.[32] Hampden dispassionately "surveyed the whole field of Christian theology with reference not to the doctrines taught but to the words and phrases used in expressing them."[33] He riled the powers that were by insinuating that the clergy was more interested in maintaining power than in personifying the teachings of Christ. In late 1834, for example, John Henry Newman criticized Hampden's *Observations on Religious Dissent*: "While I respect the tone of piety which the Pamphlet displays, I dare not trust myself to put on paper my feelings about the principles contained in it; tending as they do, in my opinion, altogether to make shipwreck of the Christian faith."[34]

After Prime Minister Russell's announcement, Wilberforce rebuked Hampden as unqualified to become a bishop and asked the premier for permission to quiz the professor on his knowledge of Christianity. Wilberforce further demanded that Hampden remove his published volume of his Bampton lectures—as well as *Observations*—from circulation because of their heterodoxy.[35] Formally charging the professor with immorality and heresy, Wilberforce went as far as to enjoin a tribunal to rule on Hampden's character—and ruin his reputation—before the Court of Arches, an ecclesiastical court of the Church of England.

None of the allegations stuck, and because he so coveted the prestige of becoming a bishop, Hampden agreed to pull his books from the shelves. After Wilberforce finally read them—instead of relying on Newman's hastily written précis of them and the swirl of clerical gossip—he admitted that while he disagreed with their arguments, there was no evidence of heresy.[36] Standing on rapidly shifting ground, Wilberforce withdrew his call for a tribunal, publicly admitted his error, and absolved Hampden. On December 28, 1847, he wrote a mealymouthed apology to the professor. Soapy Sam claimed he was simply following his conscience and that he bore Hampden no malice. After putting down his pen, Wilberforce promptly contracted influenza and took to his bed.[37] Apologies aside, the fiasco cost him dearly. He lost royal favor with both Prince Albert and, more importantly, Queen Victoria, who thereafter doubted the bishop's "sincerity and disinterestedness."[38] The episode, as Burgon acidly noted, "reached a singularly lame and impotent conclusion."[39]

The bishop's stoops to slander remained a problem for the rest of his career. Long after the Hampden debacle, Wilberforce employed the same dodgy tactics over and over again. Too often, he abused his position to act as judge, jury, and executioner—without doing the necessary work of reading the lawyer's brief. When demonstrated to be in the wrong, Soapy Sam's preening arrogance and self-importance led him to believe he could slither away with a cry of "Oops" or, at worst, a hollow apology. As he advanced and backpedaled on various issues of the day, Wilberforce's reputation eroded bit by bit until, for many, he became the burlesque of a bishop. By the time he was invited to lend his name to the list of honorary vice-presidents for the 30th Annual BAAS Meeting, in June 1860, one might assume he had absorbed enough wisdom to sail into a different direction. To his lasting discredit, he did not.[40]

ʃ

THE VIEWS Wilberforce unfurled at the BAAS meeting in Oxford were substantiated in print a few days later—thanks to his unsigned review of Darwin's book in the July 1860 issue of the *Quarterly Review*. The nearly eighteen-thousand-word essay was a raw hunk of rib eye steak marbled with the juicy fat of Richard Owen. In fact, Owen coached Wilberforce throughout the process of drafting the review, and Wilberforce quoted him extensively.[41]

The bishop opened his long piece with a few disarming compliments. Darwin's "scientific attainments, his insight and carefulness as an observer, blended with no scanty measure of imaginative sagacity, and his clear and lively style, make all his writings unusually attractive." Wilberforce predicted that this "most readable book, full in facts in natural history, old and new," would appeal not only "to men of science exclusively, but to everyone who is interested in the history of man and of the relations of nature around him to the history and plan of creation."[42]

The gentle prelude turned vicious as Wilberforce attacked Darwin's Christian faith and parodied the evolution poems written by Darwin's grandfather, Erasmus Darwin. The bishop, of course, objected to the notion that "mosses, grasses, turnips, oaks, worms, and flies, mites and elephants, infusoria and whales, tadpoles of to-day and venerable saurians, truffles and men, are all equally the lineal descendants of the same aboriginal common ancestor, perhaps of the nucleated cell of some primæval fungus, which alone possessed the distinguishing honor of being the 'one primordial form into which life was first breathed by the Creator.' "[43]

Throughout 38 pages of single-spaced print, Wilberforce adroitly pulled out the loose threads of *Origin* until there was little material left to spare. Constitutionally unwilling to abandon the teachings of the Scriptures, he misquoted or quoted Darwin out of context to support the teachings of natural theology. The bishop found so "much and grave fault" with the book that his text read like a fire-and-brimstone sermon:

> Mr. Darwin's 'argument' [is] . . . absolutely incompatible with the whole representation of that moral and spiritual condition of man which is its proper subject matter. Man's derived supremacy over the earth; man's power of articulate speech; man's gift of reason; man's free-will and responsibility; man's fall and man's redemption; the incarnation of the Eternal Son; the indwelling of the Eternal Spirit,— all are equally and utterly irreconcilable with the degrading notion of the brute origin of him who was created in the image of God and redeemed by the Eternal Son assuming to himself his nature.[44]

Many paragraphs later, Wilberforce pointed out how no one had seen or documented the changes Darwin proposed to have occurred by way of natural selection. Again, the cry of *vera causa* was shouted. Where was the proof—fossil or otherwise—of these transitional species Darwin hypothesized? Pigeons and dogs, Wilberforce noted, no matter how they were bred, remained pigeons and dogs.[45] Darwin's so-called "transmutationist" endeavors were, in the bishop's mind, an "utterly rotten fabric of guess and speculation."[46]

But it was probably the essay's final paragraph that was most hurtful to Darwin. Either on purpose or by chance, the bishop hit upon the naturalist's embarrassing flatulence. *Origin*, he asserted, was "*the frenzied inspiration of the inhaler of mephitic gas.*" Instead of looking at nature as it was, Wilberforce wrote, Darwin was "capable of believing anything . . . he is able, with a continually growing neglect of all the facts around him, with equal confidence and equal delusion, to look back to any past and to look on to any future."[47]

With each successive page of Wilberforce's review, Darwin labored to remain calm. On his copy of the magazine, he angrily scrawled, "Mere words!"[48] To Hooker, on July 2, Darwin noted how Wilberforce "picks out with skill all the most conjectural parts, & brings forward well all difficulties."[49] Darwin penned similar letters to many other friends, all of them communicating the same suspicions and stiff upper lips. On July 20, he told Huxley, "The *Quarterly* is uncommonly clever; & I chuckled much at the way my grandfather & self are quizzed. I could see here & there Owen's hand—By the way how comes it that you were not attacked? Does Owen begin to find it more prudent to leave you alone? I would give five shillings to know what tremendous blunder the Bishop made; for I see that a page has been cancelled & new page gummed in."[50] Two days later, July 22, he wrote Asa Gray, again claiming that the review made him "chuckle with laughter at myself."[51] He carped to Charles Lyell, on July 30, "Owen is really wonderfully clever in his malevolence."[52]

He did complain to John Murray, in early August, about the *Quarterly*'s poor choice of a book reviewer. With his tongue firmly embedded in his cheek, Darwin wrote, "I really believe that I enjoyed it as much as if I had not been the unfortunate butt. There is hardly any malice in it, which is wonderful considering the source whence many of the suggestions came. The bishop makes me say several things which I do not say, but these very clever men think they can write a review with a very slight knowledge of the Book reviewed or the subject in question."[53] Murray, on the other hand, may have purposely placed a hostile reviewer, knowing that controversy often translates into book sales. Such tactics failed to warm Darwin. As with Owen's hostile *Edinburgh Review* piece, Bishop Wilberforce's article slashed and burned its way through his theory of natural selection. No matter how hard he tried to laugh them away, there was nothing remotely funny about either of these horrible reviews.

# 9

# A Mysterious Malady

I hope that you will have a pleasant *[BAAS]* meeting—my health is too bad to come.

—*Charles Darwin to Oxford entomologist, John Obadiah Westwood, June 25, 1860*[1]

Famously unwell from the tip of his cranium to the bottom of his bowels, Darwin ached, pained, and erupted in all sorts of weird ways nearly every day of his adult life.[2] Darwin complained of tingling and numbness of the hands and fingers, extreme sensitivity to changes in the weather, shivering and trembling, severe headaches ("a swimming head"), eczema, boils, depression, crying spells, fainting, insomnia, joint pain, mouth ulcers, dental problems, frequent infections, exhaustion after a few hours of working at his desk, nausea, and frequent episodes of cyclical retching and vomiting of an "acid, slimy (sometimes bitter)" material that corroded his teeth.[3]

But it was his embarrassing "fits of flatulence" that remained his chief complaint and reason for keeping out of the public eye (and nose). As early as 1847, in preparation to visit Oxford University for the annual BAAS meeting, Darwin requested, "If I can have my meals to myself & a room to be by myself in, for, as you know, my odious stomach requires that." His condition was so severe, he begged off attending a dinner hosted by Charles Daubeny at Magdalen Hall. Always polite, Darwin worried, "I must explain to him how it was I refused him, & his party, would never have suited me."[4]

From January 1849 to mid-January 1855, he kept a diary of health—a work he sandwiched in between writing scientific monographs and letters to his far-flung colleagues, conducting a full menu of experiments, describing biological specimens, and contemplating his theories. In his daily diary, Darwin dutifully recorded the intense discomfort in his bloated abdomen and each "fit." He experienced these anywhere from ten to fifty times a month—sometimes as many as four or five per day. The collection of so much gas, stretching his viscera far more than they were designed to contain, was painful enough to incapacitate him. Darwin, the quintessential describer of nature, was so troubled by the vulgar words used to illustrate his problem that he often abbreviated them in his journal with the word "fit" or, if only occasional and mild, "fl," often accompanied by a battery of adjectives such as "bad," "baddish," "very vigorous," "longish," and, if both heavy and easily tolerated, "heazyish."[5] More than a few Darwinian wags have cracked that the frequency with which his wife, Emma, recorded "windy" or "ill winds" in her diary reflected her husband's gastrointestinal disorder rather than the weather.

In a familial version of folie à deux, the Darwins mirrored his ill health with a cacophony of complaints and illnesses. Emma delivered ten children over a decade and a half, replete with morning sickness, painful and sometimes dangerous deliveries, multiple feeding schedules, diaper changes, housekeeping, meal preparation, cleaning, supervision of servants, and child-rearing. She was alternatively sweet and stern in her duties as a loving wife and mama to her brood. Emma's primary mission, according to her son Francis, was to shield Darwin "from every possible annoyance, and omitted nothing that might save him trouble, or prevent him from becoming overtired, or that might alleviate the many discomforts of his ill-health."[6] In a charming memoir, Gwen Raverat—Charles and Emma's granddaughter—recalled how it "was a distinction and a mournful pleasure to be ill" at Down House. "Of one thing I am certain," Raverat explained, "that the attitude of the whole Darwin family to sickness was most unwholesome. At Down, ill health was considered normal."[7]

ON BOTH SIDES of the Atlantic Ocean, physicians of this era possessed a poor understanding of most diseases and their treatment. Many of these doctors characterized the malfunctioning gastrointestinal tract as victim to its own Four Horsemen of the Apocalypse: dyspepsia, constipation, autointoxication, and neurasthenia. The first two referred to severely upset stomachs and a paucity of regular, soft, painless bowel movements. The latter two—thought to be the result of so-called morbid poisons emanating from bits of rotting, undigestible foodstuff and stool stuck in the gut—yielded a constellation of chronic complaints, including nervous exhaustion, low energy, and depression.

Indigestion was rampant across nineteenth-century Europe and North America. In 1826, Dr. James Johnson, the physician extraordinary to King William IV, described how "[indigestion] knocks at the door of every gradation of society, from the cabinet minister planning the rise and fall of empires to the squalid inhabitants of [the slums] . . . the philosopher, the divine, the general, the judge, the merchant, the miser, and the spendthrift are all, and in no very unequal degree, a prey to the Protean enemy."[8] Others held that upset stomachs were the domain of creative, overworked, or troubled minds. For example, the British homeopathic physician John Henry Clarke claimed in his popular 1888 book *Indigestion: Its Causes and Cure* that "if Darwin's stomach had recovered from the effects of seasickness, he would doubtless have been a happier man, and his view of humanity might possibly have been a more generous and exalted one."[9] As a young man, Benjamin Disraeli, Great Britain's future prime minister, was said to have grumbled about the rumbling in his belly, "Dyspepsia always makes me wish for a civil war."[10] With due respect to such philosophical, effluvial, and emotional explanations, most cases of gastrointestinal distress were likely caused, or exacerbated, by the rich, heavy diets of the era: a menu dominated by fried and breaded foods, huge amounts of fatty meat, whole milk and cream, lard-based sauces or gravies, butter and cheese—and an excess of salt, pepper, spices, pickled condiments, and sugar—all washed down with whole milk, cream, overboiled cups of coffee and tea, and glasses of whisky, followed by snorts of snuff, unfiltered cigars, and pipes stuffed with tobacco.

Too many physicians during Darwin's day—whether they practiced in the distant provinces or on London's tony Harley Street—were too quick to bloodlet their patients. It was hardly coincidental that when Dr. Thomas Wakley founded what became the British Empire's most important medical journal in 1823, he named it *The Lancet*. Doctors also applied hot glass cups to affected areas of the body in an attempt to blister, scarify, and draw out foul or inflammatory humors. For other conditions, they overprescribed toxic cathartics and emetics, yielding a steady stream of vomit or stool, to rid the body of its supposed poisons. Darwin assiduously avoided lancets, hot cupping, and other barbaric methods employed by these so-called "heroic" physicians. Darwin did, however, sample a long list of prescriptions, including arsenic, bismuth, amyl nitrate, pepsin, quinine, morphine, cinnamon, logwood, antacids, bitters, aloe, iron-containing tonics, purgatives, and the mercury-containing calomel.[11]

In March 1849, a forty-year-old Darwin complained to a colleague, "Indeed all this winter I have been bad enough, with dreadful vomiting every week, & my nervous system began to be affected, so that my hands trembled & head was often swimming. I was not able to do anything one day out of three & was altogether too dispirited to write to you or to do anything but what I was compelled.—I thought I was rapidly going the way of all flesh."[12] Desperate for relief, he contacted a physician named James Manby Gully, who maintained a successful hydropathy, or "water cure," spa in Malvern near the Welsh border, where he treated more than six thousand patients annually. Charles and Emma read Gully's popular book, *The Water Cure in Chronic Disease*, with great interest because Gully prescribed treatments that, at least, caused little harm or untoward side effects.[13]

Hydrotherapy consisted of cold and hot baths with water drawn from local mineral springs, sheets doused in either hot, warm, or cold water wrapped around one's belly and limbs, ice packs, "Russian baths"—where wet towels were slapped and vigorously rubbed against the aching body—and sitting in freezing cold pools of water up to the hip. Such therapies were thought to cause a counterirritation to the skin, reducing inflammation of the affected inner organs, ridding the body of toxins, and numbing overexcited nerve endings. Other treatments included enemas, douche showers using a hose device of 1 to 2 inches in diameter and from a height of 20 feet, long walks and outdoor exercises, and a strict diet forbidding dairy products, sugar, salt, meat, spices, condiments, coffee, tea, alcohol, and tobacco. Instead, the menu emphasized fresh fruits, vegetables, and whole grains. Instead of milk, patients were ordered to drink large quantities of water—eight or more glasses per day.[14]

Darwin was hardly the only luminary of his generation to seek such soggy pursuits. The rich, famous, and worried-well all flocked to medical spas and grand hotels—from Baden-Baden in Germany's Black Forest and Vincenz Priessnitz's famous hydropathic institute in Grafenberg, Austria—where the now familiar prescription for eight glasses of cold, pure water per day is said to have originated—to Lourdes, France, and the Battle Creek Sanitarium in Michigan.[15] Eminent Victorians, including Charles Dickens, Florence Nightingale, Alfred Lord Tennyson, Thomas Carlyle, Herbert Spencer, and the Right Reverend Samuel Wilberforce, frequented such places for respite from their indigestion. One of Darwin's favorite novelists, Mary Anne Hawes—better known as George Eliot—was a chronic dyspeptic, as was her common-law husband, George Henry Lewes—the philosopher, literary critic, and popularizer of science. Lewes's best-selling *Physiology of Common Life*—which was published the same year as *Origin*—contains a thick chapter titled "Digestion and Indigestion." In sentences one imagines Darwin could have easily written, Lewes concludes, "Much of our happiness depends in health and health cannot continue without digestion . . . the misery of mankind, springing from many causes, is intensified by Indigestion, which lessens the fortitude to endure calamities, and increases the tendency to indulge in painful forebodings."[16]

Darwin made his first visit to Gully's hydropathic institute in the spring of 1849. The results were spectacular. Darwin told friends how "Dr. G[ully] feels pretty sure he can do me good, which most certainly the regular doctors could not. . . . I feel certain that the water cure is no quackery."[17] After returning home in early summer, Darwin extolled his aquatic adventure as a "grand discovery." On July 7, 1849, he wrote, "I go on with the Treatment here, just the same as at Malvern, though in a somewhat relaxed degree, so as to avoid bringing on a crisis.—I am building a douche, for Dr. Gully tells me I shall have to follow treatment for a year. I consider the sickness as absolutely cured." Following Gully's prescriptions, his vomiting and flatulence ceased within three weeks. "How sorry I am I did not hear of it," he added, "or rather I was not compelled to try it some five or six years ago."[18]

By mid-October, Darwin was back at Malvern, where he felt "certainly a little better every month; my nights mend much slower than my days. . . . My treatment now is lamp 5 times per week & shallow bath for 5 minutes afterwards; douche daily for 5 minutes & dripping sheet daily." The water was "wonderfully tonic," and he enjoyed "more better consecutive days this month than on any previous ones." As Darwin told a friend, "The cold-water cure, together with 3 short walks is curiously exhausting; & I am actually *forced* always to go to bed at 8 o'clock completely tired.—I steadily gain in weight & am never oppressed with my food. . . . Dr. Gully thinks he shall quite cure me in six or nine months more."[19]

For more than a decade, Darwin made visits to Malvern and the hydropathic institutes at Ilkley, in North Yorkshire, and Moor Park and Sudbrook Park, in Surrey—including the week the 1860 BAAS meeting was held in Oxford.[20] The Moor Park institute was one of his favorite spots to heal. It boasted a beautiful mansion for its guests, surrounded by moors, glades of trees, and the North Branch of the River Wey. Best of all, Moor Park was only 40 miles from Darwin's home. While there, on April 28,

1857, he wrote, "I can walk & eat like a hearty Christian; & even my nights are good.—I cannot in the least understand how hydropathy can act as it certainly does on me. It dulls one's brain splendidly, I have not thought about a single species of any kind, since leaving home."[21] A year later, he wrote Emma about a relaxing, ninety-minute stroll he took there: "At last, I fell fast asleep on the grass & awoke with a chorus of birds singing around me, & squirrels running up the trees & some Woodpeckers laughing, & it was as pleasant a rural scene as ever I saw, & I did not care one penny how any of the beasts or birds had been formed."[22]

Darwin eventually soured on hydropathy, but the "tyranny" of his vomiting and flatulence continued for the remainder of his life.[23] Desperate for some respite or cure, Darwin consulted a quiver of doctors. In May 1865, he sought out a physician named John Chapman, who was the editor and publisher of the *Westminster Review* and a friend to both George H. Lewes and George Eliot. Chapman wrote a popular book prescribing ice applied to the spine—placed in a bag he designed for the treatment of nausea, dyspepsia, and the stomach gyrations of seasickness, epilepsy, neuralgia, chol-

Moor Park hydropathic institute, Surrey, circa late 1850s.

era, paralysis, and menstrual problems. The idea behind this frigid therapy was that the ice supposedly acted as a "direct sedative" to the spinal cord by reducing the blood supply to that portion of the central nervous system and thus reducing its "automatic or excitomotor power."[24]

In preparation for the consultation—a house call, no less—the fifty-five-year-old Darwin called out each of his symptoms as Emma dutifully wrote them down in her neat, cursive handwriting. The list filled the front and back of a sheet of paper with Darwinian precision, especially its conclusion: "For 25 years extreme spasmodic daily & nightly flatulence, occasional vomiting; on two occasions prolonged during months. Extreme secretion of saliva with flatulence . . . I feel nearly sure that the air is generated somewhere . . . lower down than [the] stomach and as soon as it regurgitates into the stomach the discomfort comes on." The ice packs, like the water baths, failed to effect a lasting cure.[25]

ʃ

Darwin's odd amalgam of symptoms has long mystified and fascinated a hive of historians. For decades, prominent medical and scientific journals have published a slew of self-proclaimed solutions to the question, What made Darwin so ill during most of his adult life?[26] In fact, so many retrospective diagnoses have been offered by doctors—who never met, let alone examined the man—that if compiled into one volume, they would constitute a thick textbook of medical speculation.[27]

The diagnoses range from asthma to zoonotic infections. One of the most publicized medical opinions, Chagas disease, hails from the latter category. It is passed on to humans via what Darwin called the "great black bug of the Pampas"—*Triatoma infestans*. Many of these insects, but not all, carry *Trypanosoma cruzi*, the causative protozoan behind Chagas disease. The *insects* bite vertebrates, often around the mouth (hence the name "kissing bug" or "vampire bug"), deposit their feces into the victim's bloodstream, and if they themselves are infected with *T. cruzi*, spread the parasite. Thus begins a long cycle of infection, with swelling at the bite, inflammation, fever, and fatigue, eventually progressing to blood clots in the extremities, heart disease and failure, severe neurological problems, constipation, and other digestive disturbances.

Some have claimed, without definitive proof, that Darwin contracted this infection sometime between 1834 and 1835 during his world-famous voyage on the HMS *Beagle*. In one version, one of the ship's officers kept a *Triatoma* bug as a pet and allowed it to feed on his blood. This bug may have bitten and left excrement under Darwin's skin, by way of an eczema lesion, a boil, or an ordinary cut. Darwin also recorded encounters with insect bites during his many overland trips through Chile and Argentina and across the Andes. These, too, may have exposed him to the endemic but, at the time, undescribed infection.[28]

The diagnosis of Chagas disease first received global attention in 1959, after an Israeli physician named Saul Adler published a paper on it—and Darwin—in *Nature*.[29] The infection's chronic phase—which can take years to develop—does feature *some* of the symptoms experienced by Darwin. For example, the infection *can* cause enough damage to the small intestine to permanently alter one's digestion. But the parasite also causes inflammation of the heart and brain, with dire results such as brain abscesses and heart failure. Darwin suffered occasional heart palpitations—which began before the bug bites—but he endured no significant cardiac event until his death in 1881 from a garden-variety myocardial infarction, or "heart attack." Flatulence—so important to Darwin's clinical picture—is not one of the more common symptoms seen with Chagas disease. Given the specificity of Darwin's symptomatology and the absence of the illness's most serious complications during his long life, the diagnosis of Chagas disease seems more fanciful than factual.

Others have speculated, with even less evidence, that Darwin suffered from a complex of allergies, including one to pigeons; Crohn's disease; gout; cyclical vomiting syndrome; an inherited mitochondrial disorder; chronic seasickness; brucellosis; diaphragmatic hernia; narcolepsy; diabetes; arsenic poisoning; gallbladder disease, chronic appendicitis; amoeba infection; hepatitis; peptic ulcer; acute intermittent porphyria; Ménière's disease; rheumatic disorders such as systemic lupus erythematosus; along with many other organic explanations—all of which, upon closer examination, prove more far-fetched than Chagas disease.[30]

Psychiatrists, too, have entered the fray, testifying that Darwin suffered from hypochondria, panic disorder, psychogenic neurasthenia, and depression, compounded by his shy, gentle nature and a conflict-adverse personality.[31] Others have suggested that he suffered from post-traumatic stress disorder, caused by the loss of his mother when he was eight, witnessing surgical procedures without anesthetic while he was a medical student, and the death of three of his children. Not a few retrospective diagnosticians ascribe Darwin's nervous stomach to his heretical ideas about the origin of species. The theory, here, is that the controversial scientific dilemmas he tackled brought on waves of worry and indigestion. All or some of these nerve-racking concerns *might* have manifested as stomach pain, heart palpitations, and panic attacks, but they hardly explain the severity of his vomiting and flatulence.

As WITH MANY great minds marred by dysfunctional bodies and accumulated neuroses, Darwin found salvation in the rigid adherence to solitary, creative work. In 1846, he described his routine to Robert FitzRoy: "My life goes on like Clockwork and I am fixed on the spot where I shall end it."[32] On most days, he awoke at 7:00 a.m., went out for a morning walk, and breakfasted alone. Darwin then retreated to his study for a few hours of concentrated writing until about 12:15 p.m. After a brief walk, he enjoyed luncheon, the family's main, and largest, meal of the day. He rested until 3:00 p.m., took another walk, and rested again while Emma read to him or helped him attend to his correspondence. In the late afternoon, he dressed for dinner—although at table, he rarely took more than tea "with an egg or a small piece of meat." The family entertained themselves with nightly rounds of backgammon, word games, communal readings of the latest popular novel, and Emma's piano playing—especially if guests were present. Darwin usually excused himself to read in his study and by 10:00

Down House.

p.m. retired to bed—only to sleep badly. All these activities depended on his level of energy, exhaustion, and fits of flatulence.[33]

Darwin's principal source of exercise—walking—revolved about a group of trees arranged in a lopsided rectangle at the edge of his garden, where he had a path cleared and covered with gravel. He called it his "sand-walk." Strolling among the blossoming bluebells and orchids, wild grass and ivy, and the birch, dogwood, privet, and holly trees, Darwin inhaled fresh air and basked in the sunshine. The walks had another virtue in that they allowed him to think through the scientific problems awaiting him at his desk.[34] At each turn—about a quarter mile in distance—he threw down a stone or a pebble to mark the distance he had walked. On days when his work was not going well, he struggled with four- or five-pebble problems. As he told one friend in 1861, "Observing is much better sport than writing."[35]

On family holidays, the restless Darwin brought along a valise full of

The sand-walk.

work, proof sheets, and experiments to contemplate or conclude. His son Leonard recalled that "the truth was that he was never quite comfortable except when utterly absorbed in his writing. He evidently dreaded idleness as robbing him of his one anodyne, work."[36] Similarly, Darwin insisted, "I know well that my head would have failed years ago, had not my stomach always saved me from a minute's over work."[37] He explained the secondary gain of his illnesses more plainly in an 1876 draft of his *Autobiography*: "I have had ample leisure from not having to earn my own bread. Even ill health, though it has annihilated several years of my life, has saved me from the distractions of society and amusement."[38]

Darwin's unbending self-isolation seemed odd, if not comical, to those who did not understand the extent of his debilitating fatigue. The Irish poet William Allingham portrayed Darwin in August 1868 as "tall, yellow, sickly, very quiet. He has his meals at his own times, sees people or not as he chooses, has invalid's privileges in full, a great help to a studious man." The next day,

Allingham described Darwin's absence from a social gathering: "Charles Darwin expected but comes not. Has been himself called 'The Missing Link.' "[39]

The most striking aspect of Darwin's gastrointestinal symptoms was their predictability. As he described the problem in 1864: "It rarely comes on till 2–3 hours after eating, so that I seldom throw up food, only acid & morbid secretion; otherwise, I should have been dead, for during more than a month I vomited after every meal and several times most nights."[40]

Darwin's reclusiveness meant he took most of his meals in the comfort of his home. Fortunately, in terms of the historical record, the Darwin family saved virtually every piece of paper they scribbled on, and Emma Darwin's recipe book is no exception. It consists of a mélange of rich, dairy-filled foods. Every day she served up vegetables swimming in milk, cream, and butter; creamed soups; meat and fish entrées doused with "white gravy," or Béchamel sauce; pies and prime beef encased in butter and lard-based crusts; as well as dairy-rich puddings, custards, drinks, and desserts. Darwin's favorite dessert was crème brûlée—or as Emma called

Darwin, circa 1854.

it, "burnt cream." One serving alone contained half a pint of whole cream, 1 tablespoon of flour, 2 eggs, and 2 ounces plus 1 teaspoon of extra-fine sugar. For late-night snacks, Darwin often hankered for bread dunked in milk and custards of all kinds. He especially enjoyed a tonic concocted by Emma consisting of castor oil, sugar, and milk. More than half of the recipes in Emma Darwin's cookbook are desserts or appetizers laden with cheese, butter, cream, milk, and eggs. Ironically, Emma's cookery probably exacerbated, if not initiated, many of Darwin's digestive issues.[41]

Although he tried to eliminate rich desserts as early as 1844, Darwin rarely adhered to these restrictions for long periods of time. Writing that year to his sister Susan Elizabeth Darwin, he told her to "thank, also, my Father for his medical advice—I have been very well since Friday, nearly as well, as during the first fortnight & am in heart again about the non-sugar plan."[42] Five years later, in 1849, he complained to Susan about his diet at Gully's water cure institute: "At no time must I take any sugar, butter, spices, tea, bacon, or anything good. . . . I grieve to say that Dr. Gully gives me homeopathic medicines three times a day, which I take obediently without an atom of faith."[43] And in 1866, he told the famed London physician Henry Bence Jones how, for more than two decades, cheese and coffee—typically taken with milk or cream—violently disagreed with him.[44] Interestingly, his symptoms disappeared, or at least improved, when taking the water cure at the various hydropathic institutes—where he was prohibited from consuming the rich, creamy, fatty, sugary "excitatory" foods he so loved.

After reviewing his symptoms, *this* physician believes Darwin most likely suffered from systemic lactose intolerance (see Table 1). The diagnosis, however, must be credited to a team of biochemists from Cardiff and a gastrointestinal pathologist and a historian of science from Leeds.[45] Alas, lactose intolerance—a problem that can be largely avoided by not consuming lactose-containing dairy products—was not figured out, let alone described in the medical literature, until the 1960s, more than a century after Darwin published *Origin*.[46]

## Table 1: Symptoms of Systemic Lactose Intolerance Compared with Darwin's Symptoms

| Symptoms of Systemic Lactose Intolerance | % People with Lactose Intolerance with this Symptom* | Darwin's Description of His Symptoms | Occurrence of Darwin's Symptoms |
|---|---|---|---|
| Gastrointestinal symptoms (pain, bloating, diarrhea) | 100 | Stomachache | Common |
| Flatulence | 100 | Flatulence (belching) | Common |
| Headache | 86 | Headache | Common |
| Light-headedness and loss of concentration | 82 | "Swimming head" and difficulty concentrating | Common |
| Nausea and vomiting | 78 | Vomiting, retching | Very common |
| Muscle and joint pain | 71 | Rheumatic joint pain | Often |
| Tiredness and chronic fatigue | 63 | Chronic fatigue and exhaustion | Very common |
| Allergy (eczema, sinusitis, hay fever, rhinitis) | 40 | Skin rash, eczema, and boils | Often |
| Mouth ulcers | 30 | Mouth sores | Common |
| Heart palpitations | 24 | Palpitations in the chest | Common |
| Depression | Common, but not quantified | Depression | Frequent |

* Represents proportion of people diagnosed as lactose intolerant who have this symptom within 48 hours of ingesting lactose. Darwin's occurrence is based on his notes and letters during periods of the episodes.

Adapted from A. K. Campbell and S. B. Matthews, "Darwin's Illness Revealed," *Postgraduate Medical Journal* 81 (2005): 248–251.

Primary lactose intolerance typically emerges from an inherited trait.[47] This genetic variant causes an inability to break down and absorb lactose—the major sugar in milk and dairy products. Normally, the lining of the human small intestine produces an enzyme known as lactase, which breaks lactose into its two simple—and absorbable—sugar molecules: glucose and galactose. Without this enzyme, however, the relatively large lactose molecules can pass through to the large intestine, where they are fermented by the colonic bacteria residing there. Secondary lactose intolerance can also appear after a serious infection, injury, surgery, or other damage to the lining of the gut that impairs its ability to produce lactase—such as severe Chagas disease or typhoid fever. Finally, there is congenital lactose intolerance, in which a baby is born with an inability to produce lactase. This is caused by an extremely rare, autosomal recessive trait, and before its elucidation, these babies often died in infancy.

One result of lactose intolerance is the overproduction of methane, nitrogen, and carbon dioxide gas in the gut—which is often expelled in the form of angry bouts of bloating, diarrhea, and flatulence. Neither medical terminology nor crude slang expressions can fully describe what happens after an afflicted person consumes dairy products. The gas produced is especially noxious to the nose and hangs in the air with a vengeance. Equally distressing, lactose intolerance often results in ileus—the abnormal movement of the gut—resulting in a reflux of the food back into the stomach, acid in the esophagus, painful retching, and, ultimately, vomiting up the stomach's watery mix of hydrochloric acid and undigested sludge.

There is even a Darwinian twist to this dietary malady. A major factor involving the origin of mammals was the production of milk to feed their young. Human babies manufacture the lactase enzyme in their gut while they are nursing at their mother's breast. A few months into their lives, the gene coding for this enzymatic production turns off in affected individuals. It can take twenty or more years for one's lactase production to dip below 50 percent of what a normal person produces, leading to unpleasant symptoms whenever the affected individual quaffs down a glass of milk. For example, Darwin was about thirty when he began experiencing his severe flatulence and episodes of retching and reflux after meals. Contributing to this case study of natural selection was the advent of dairy farming, which began six thousand to eight thousand years ago and resulted in human beings consuming massive amounts of cow and goat milk, cheese, butter, cream, and other dairy products. During a single heartbeat of evolutionary history, humans went from human breast milk drinkers only during infancy to lifelong dairy consumers. In Darwin's time, anywhere from 80 to 100 percent of the European population consumed dairy foods daily.[48]

ADMITTEDLY, it is impossible to make any definitive conclusions on the diagnosis of lactose intolerance—or any other malady—especially without some type of tissue, DNA, or relic from Darwin's corporeal body. Who knows what other spectacular diagnoses lay in wait? Thankfully, those administering Westmister Abbey have no intention of taking the drastic step of exhuming Darwin's remains.

There may, however, be a less-invasive answer to this pathological puzzle. Carefully preserved in the Charles Darwin Collection at the Cambridge University Library is a 3-inch lock of brown hair. Clipped from the head of his daughter, Anne Elizabeth Darwin, the hair reposes in a sealed cellophane wrapper dated April 23, 1851. It was obtained as a memento, just after her premature death at age ten. I made multiple requests to analyze Annie's hair using the latest genetic tools, but they were all denied. As the Cambridge University Keeper of Manuscripts and Curator of Scientific Collections wrote me in late April 2022, "This is not something we would like to take forward at this time."[49] Even if the Darwin heirs and the Cambridge University Library *did* consent to my weird request, the 171-year-old hair specimen appears to have no follicles—where the DNA is most readily available—and there is no guarantee that Annie's tress carries genetic evidence of lactose intolerance. For now, a potential solution to her or her father's poor health may remain tightly guarded within that lock of hair.

Whether Darwin's symptoms were in his gut, in his head, or elsewhere in his tortured body—the aches, pain, depression, and gaseous nature of his maladies were *very* real to him. His awful illnesses framed almost every step he took and thought he conjured. Just like Marley's ghost in Dickens's *A Christmas Carol*—a Darwin family favorite—Darwin, too, found himself in shackles.[50] Every "fit of flatulence" represented another carbonated, clanking kink in his torturous bowels. Ebenezer Scrooge was obliged to see spirits, while Darwin was forced to ghost his friends and communicate from the seclusion of his study. "If the character of my father's working life is to be understood," his son Francis explained, "the conditions of ill-health, under which he worked, must be constantly borne in mind . . . for nearly 40 years he never knew one day of the health of ordinary men, and his life was one long struggle against the weariness of strain and sickness."[51]

It is painful to imagine the embarrassment that Darwin's delicate condition imposed—particularly when interacting with others outside his small circle of family and friends. It forced this proper English gentleman to stay far out of reach of the most important scientific meeting of his storied career. Instead of defending his controversial work to his colleagues at Oxford, the self-proclaimed invalid was at a water cure in Surrey.

# Part IV

# OXFORD

I hold that the function of science is the interpretation of nature—and the interpretation of the highest nature is the highest science. What is the highest nature? Man is the highest nature. But I must say that when I compare the interpretation of the highest nature by the most advanced, the most fashionable and modish, school of modern science with some other teachings with which we are familiar, I am not prepared to say that the lecture-room is more scientific than the Church. . . . The question is this—Is man an ape or an angel? My lord, I am on the side of the angels.

*—Benjamin Disraeli, November 25, 1864,*
*Speech given at the Sheldonian Theatre, Oxford University*[1]

British Association
for the Advancement
of Science
Thirtieth Meeting.
Oxford 1860.
Associate's Ticket.
Admit Mr Geo: Ward
No. 1444 John Taylor Treasurer

# 10

# The Association

If it would not be selfish, I think the plan which would cause least trouble would be for you, if not asking too great a favor, to reserve me rooms in a near college; & then if I am prevented coming, I will promise to write 2 or 3 days before the meeting *[of the BAAS]*, so that the rooms would be at your disposal for someone else.

—*Charles Darwin to George Rolleston, June 6, 1860*[1]

I have given up Oxford; for my stomach has utterly broken down & I am forced to go on Thursday for a little water-cure, to "Dr. Lanes Sudbrook Park, Richmond Surrey", where I shall stay a week, & should stay rather longer had it not been for Etty. Etty improves slightly, but so slightly that it takes weeks to perceive any difference; she cannot sit up in bed for more than few minutes.—Farewell—I hope that you will have pleasant time at Oxford; I much wished to have been there, but I could not stand it, or indeed anything.

—*Charles Darwin to Charles Lyell, June 25, 1860*[2]

The axis of the planet had just completed its maximum tilt toward the sun when over a thousand people descended upon Oxford's academic village. Visitors from the British Isles, Europe, and North America discovered what novelist Evelyn Waugh later described as "the rare glory of her summer days . . . when the chestnut was in flower and the bells rang out high and clear over her gables and cupolas, exhaled the soft vapors of a thousand years of learning."[3] It was not the university's "cloistral hush" at the close of the Trinity term that attracted the "intervening clamor" of "intruders"; instead, it was the 30th Annual Meeting of the British Association for the Advancement of Science.[4]

When the association was founded in 1831, it was just that—a gathering of men of varied backgrounds, classes, and wealth who shared mutual interests in the production and dissemination of scientific knowledge.[5] From its earliest days, the association promoted a profession of "scientists"—a word coined by its first president, William Whewell, to replace the more common term of "natural philosopher." The new designation described those committed to objectively advancing the field using facts, figures, careful observation, and reproducible evidence. Others fancied appellations such as "men of science" or "cultivators of science."[6] Charles Darwin, incidentally, described himself as a naturalist.

Unlike the elite Royal Society of London, which required election by one's pedigreed peers, the BAAS opened its rolls to anyone of sufficient

The 1847 British Association General Meeting, in the Sheldonian Theatre, Oxford University.

good character. Membership cost £2 during a member's first year and £1 annually thereafter; those with £10 to spare subscribed for life memberships. The dues helped defray the costs of meetings, occasional publications, and the annual report.[7] In 1869, the BAAS began publishing *Nature*, a weekly journal that eventually became so influential that scientists today clamor to see their work appear on its pages.

Every year, the association met in a different provincial city for "one week or longer." These conclaves were held during the summer or early fall to avoid the rainy, cold, and dark winters of Great Britain—conditions that made horse-drawn coach and buggy travel on muddied roads even more arduous and dangerous than it was in fair weather. Other than different standards of dress, customs, and expression, the BAAS meetings would be easily recognizable to modern-day attendees of scientific conventions. Both then and now, the attendees enjoyed the scientific deliberations on new discoveries, side excursions to popular tourist attractions, heavy meals, and gossip about those present or absent from that year's meeting.[8]

Sheldonian Theatre (exterior), Oxford University.

ſ

The "opening general meeting" of the 1860 conference began at 4:00 p.m. on June 27 in Oxford's cavernous Sheldonian Theatre. Designed by Christopher Wren, the auditorium boasts perfect acoustics and was built without load-bearing columns in the central space, so that its interior resembles an ancient Roman theater. The university's chancellor, vice-chancellor, and many of its department heads hung on every word as the outgoing BAAS president, Prince Consort Albert, spoke. His Royal Highness was in a good mood, having just visited his son Bertie, the Prince of Wales, who was then studying at Christ Church College, Oxford. With a mixture of modesty and humor, Albert told the audience, "I cannot but express a hope that the interests of science have not suffered during my year in office." The distinguished listeners, who would never agree to such a self-deprecating statement, cheered with gratitude for his service.[9]

As the prince consort handed off the gavel "in modest and appropriate terms" to his successor—the Right Honorable John, Lord Wrottesley—Albert grandly stated, "It gives me peculiar gratification to resign the office of president into such able keeping."[10] The prince was being polite. Although Wrottesley served as the president of both the Royal Astronomical Society (1841–1842) and the Royal Society (1855–1856), he is now an obscure if not completely forgotten figure. As described in the *Notes and Records of the Royal Society of London*, "the reasons for this neglect are understandable: he made no contributions to the conceptual structure of science and his experimental work in astronomy, whilst painstaking, was otherwise undistinguished."[11] When he took the gavel of the BAAS, Wrottesley was a wealthy, sixty-one-year-old, bloated fellow who rarely ventured out into the sun. He preferred the night air, when the stars were out in full force. In an 1857 carte de visite (a small card containing a photographic portrait), Wrottesley's right eyelid droops—no doubt the result of middle-aged ptosis—giving him the appearance of permanently squinting though his telescope in the observatory of his country estate in Wolverhampton.

In his interminably long presidential address, Wrottesley did not spe-

John Wrottesley, 1857.

cifically mention Darwin's new book. He did, however, offer a stilted but telling thought experiment on faith and science. "Let us assume," he suggested, "that to any of the classical writers of antiquity, sacred or profane, a sudden revelation has been made of all the wonders involved in Creation accessible to man." Wrottesley pondered what if it had been "disclosed not only what we now know, but what we are to know hereafter, in some future age of improved knowledge . . . would they not have been delighted to celebrate the marvels of the Creator's powers?" The astronomer expelled several more gusts of hot air before answering his own question: "*Marvels* indeed they are, but they are also mysteries, the unraveling of some of which tasks to the utmost the highest order of human intelligence." He completed his speech by bowing his head to pray: "Let us ever apply ourselves seriously to the task, feeling assured that the more we thus exercise, and by exercising improve our intellectual faculties, the more worthy shall we be, the better shall we be fitted to come nearer to our God."[12]

Lord Wrottesley's words received a smattering of "warm applause." The only recorded comment was made by Lord Derby, Edward Smith-Stanley, the thrice-term prime minister. Lord Derby did not, however, acknowledge Lord John's quasi religious drivel and instead praised Prince Albert, "who from the moment when he first set foot on the shores of England, had exhibited earnest and unvarying interest in all that tended to the advancement of the sciences and the arts." The audience clapped with gusto for a beloved prince who would be dead eighteen months later.[13] His Majesty and his retinue left as soon as the session adjourned and "shortly afterwards returned to London by special train."[14] One wonders if Albert promised Queen Victoria that he would be home in time for dinner.

ʃ

THE MEETING'S AGENDA was divided into seven sections, prosaically labeled A: Mathematics and Physics; B: Chemistry and Minerology, including their application to agriculture and the arts; C: Geology; D: Botany and Zoology, including Physiology; E: Geography and Ethnography; F: Economic Science and Statistics; and G: Mechanical Science. Nearly three hundred papers were presented.[15] The seven sections met concurrently, and the attendees divided themselves according to their research interests. Despite these partitions, there was plenty of section-hopping and overlap, especially when a paper presented in one section appealed to a member subscribed to another. The quality of the papers varied greatly, from such luminaries as social reformer Edwin Chadwick and African explorer David Livingstone to those who were neither distinguished nor accomplished. With respect to the latter category, George Rolleston—the Oxford naturalist, physiologist, and physician and Huxley's loyal protégé—derided the membership of section D (Botany and Zoology) as a "pack of loafers all trying to drink as much and work as little as possible."[16] And yet it was the presentation of two "unimportant" papers—both of which noted Darwin's work—to this very section of "loafers" that constituted the opening shot in the fight to come.[17]

Charles Daubeny, the professor of botany and chemistry at Oxford, was the first to introduce Darwin's theory to the official proceedings. He was a heavyset man with thin, combed-over hair and pursed lips. Daubeny had studied medicine at Edinburgh and later in London before taking his med-

George Rolleston, MD, circa 1881.

ical degree from Oxford in 1818 and passing the examination to become a fellow of the Royal College of Physicians. By 1822, he was Oxford's professor of chemistry; in 1834, Daubeny was elected to the Royal Society; promoted to the chair of Botany and Rural Economy; and in 1856, he served as president of the BAAS. For many years, his research interests centered on the volcanoes, earthquakes, and thermal springs of France, Germany, Hungary, and Italy, which he described, at length, in a turgid prose.[18] At Oxford, the doctor's kingdom was the university's botanical garden, where he grew all sorts of medicinal plants, including foxglove for congestive heart disease, peppermint to soothe upset stomachs, and "intoxicating" *Cannabis sativa* and *indica*—from which, as he wrote in his *Popular Guide to Physick*, "the demoralizing effects are well known."[19]

Daubeny's June 28 paper to the Botany and Zoology section would have been completely dispensable had it not been for its provocative title: "On the Final Causes of Sexuality of Plants with Particular Reference to Mr. Darwin's Work *On the Origin of Species by Natural Selection*." His primary explanation for the existence of sexual organs in plants, for example, was

Charles G. B. Daubeny, MD, circa late 1850s.

"to minister to the gratification of the senses of man by the beauty of their forms and colors."[20] At the conclusion of his talk, Dr. Daubeny timidly disclosed that "if we adopt in any degree the views of Mr. Darwin with respect to the origin of species by natural selection, the creation of sexual organs in plants might be regarded as intended to promote this specific object." Yet when pressed upon the question of whether he supported Darwin's ideas, he skittered away from anything close to a full-blooded sponsorship. The *BAAS Annual Meeting Report* noted that while he "gave his assent to the Darwinian hypothesis, as likely to aid us in reducing the number of species, he wished not to be considered as advocating it to the extent to which the author seems disposed to carry it." Instead, Daubeny punted and "recommended to Naturalists the necessity of further inquiries, in order to fix the limits within which the doctrine proposed by Darwin may assist us in distinguishing varieties from species."[21]

John Stevens Henslow, the session's president, was so befuddled by Daubeny's fuzzy conclusion that he called upon Thomas Huxley for explication. Huxley's primary goal that morning was the passage of his proposal to separate the burgeoning field of physiology from botany and zoology and give it its own dedicated section of the association's proceedings.[22] Before the meeting he considered that its enactment was "a matter of course," but the motion failed to carry. Huxley may well have still been cogitating on this "cruel" defeat when Henslow called upon him.[23] Regardless of his inner thoughts, however, Huxley promptly "deprecated any discussion of the general question of the truth of Mr. Darwin's theory [because] he felt that a general audience, in which sentiment would unduly interfere with intellect, was not the public before which such a discussion should be carried on." Daubeny must have boiled to the point of dehydration as Huxley took one step further by noting that the paper "had brought forth nothing new to demand or require remark."[24] The Oxford man closed his allotted time by inviting the section members to visit his experimental garden, where he had several ongoing projects on agricultural chemistry that he thought might be of interest, "the nature of which he proposed to explain on the spot."[25]

The audience was far more eager to hear the next set of papers listed on the program. That is, everyone except for Richard Owen, who refused to allow the subject of Darwin's new book to "drop." The *Athenaeum* quoted the rascally paleontologist as simply desiring "to approach this subject in the spirit of the philosopher." But Owen had far more mischievous intentions when he noted that "there were facts by which the public could come to some conclusion with regard to the probabilities of the truth of Mr. Darwin's theory. Whilst giving all praise to Mr. Darwin for the courage with which he had put forth his theory, it must be tested by facts."[26] Not coincidentally, Owen continued, those facts resided in the huge body of work *he* had published on "the structure of the highest Quadrumana [four-handed apes] as compared with man." Without much proof other than his authoritative say-so, Owen declared that a gorilla's brain "presented more differences as compared with the brain of man, than it did when compared with the brains of the very lowest and most problematical of the Quadrumana." These "immense" differences could be found—or not found—in "the posterior lobes of the cerebrum."[27] Specifically, Owen claimed to have identified a structure known as the "hippocampus minor" in human but *not* in gorilla brains—thus "*proving*" his contention that men and women were unique in their thinking abilities and, thus, human beings merited a special subclassification of the species, separate and apart from any other creature. What Owen did not know was that he was propagating an artifactual error; subsequent anatomical studies, with better visualizing techniques, *have* identified a hippocampus minor in the brain of apes, thus tossing most of his hippocampal blathering into the overflowing dustbin of history.[28]

Owen insisted that "the same remarkable differences of structure were seen in other parts of the body." For example, "the structure of the great toe in man, which was constructed to enable him to assume the upright position; whilst in the lower monkeys it was impossible, from the structure of their feet, that they should do so." This was no insignificant point; the big toe of the human foot, which enables balanced, bipedal, or two-legged, walking, *is* a major difference between man and apes. Nor was Owen incorrect to demand from Darwin "the necessity of experiment. The chemist, when in doubt, decided his questions by experiment; and this was what was needed by the physiologist."[29] The rub was that designing such experiments, as Owen knew all too well, was not yet possible by mere observation of nature alone.

The dispute over the brains of apes and men was a hot topic in 1860 for all the obvious reasons—each side finding what they wanted to see in their gross anatomical dissections. Soon enough Huxley could no longer tolerate his senior colleague's unfounded yammering. The younger man rose and "begged to be permitted to reply to Professor Owen."[30] Explaining how "the difference between the brain of the gorilla and man" was not nearly "so great as represented by Professor Owen," Huxley reminded the audience that the best available evidence demonstrated that "the difference [of brains] between man and the highest monkey was not so great as between the highest and lowest monkey."[31] Taking one erect step further, Huxley insisted "with regard to the limbs, that there was more difference between the toeless monkeys and the gorilla than between the latter and man." Instead, "the great feature which distinguished man from the monkey was the gift of speech."[32] Darwin's son Francis later described Huxley's discourse as "a direct and unqualified contradiction" to Owen.[33]

If only that were true, but neither Owen nor his comrades were ready to admit defeat, nor would they anytime soon. There was, in fact, a great deal more work—and debate—needed to confirm Huxley's claims and disprove Owen's.

The *Athenaeum*'s account of the Owen-Huxley conflict was nothing short of sensational: "The main interest of the week has unquestionably centered in the Sections where the intellectual activities have sometimes breathed over the courtesies of life like a sou'-wester, creating the waves of conversation with white and brilliant foam." If such embellishment was not enough, the reporter enlarged matters further by breathlessly noting, "The flash, and play, and collisions in these Sections have been as interesting and amusing to the audiences as the Battle at Farnborough or the Volunteer Review to the general British public."[34] Then, as today, few things attract more eyeballs than controversy and hyperbole.

ʃ

IT IS CAPTIVATING to read the daily letters between Huxley and his "dearest wife," Henrietta, or "Nettie," during his meteoric rise to the heights of British science. He signs these missives "Ever yours, Hal" and sends along "hugs and kisses to my babies."[35] Henrietta reciprocates by addressing him as "my darling Hal" and closing with "your true wife, Nettie Huxley."[36] Their correspondence reminds us that despite being a "bulldog" in the public arena, "Hal" was all too human—a devoted family man who missed his beloved wife and three children while away at a business meeting.

On the evening of June 28, Huxley told Henrietta about his "battle royal" with Owen in section D. He defended his aggression by declaring he had "wished to avoid anything like a row . . . however, it was no good. Owen got up and made a very clever speech attacking statements of mine indirectly. So when he had done, I got up, and girding my armor—went at it. I was very well listened to & that was all I wanted. . . . I should like you to see the business of the sections just for one day, but I think you would find that quite enough. I find them duller every time I attend."[37] Such battles were not without huge personal costs. An exhausted Huxley confessed how "I am none the worse for being stirred up a little . . . [but] you could not expect to have a row every day of the week."[38] Soon after reading of her husband's rare moment of vulnerability, Henrietta wrote back to reassure him, "So you have been battling with the Archetype. You were quite right to try to evade the fight [but] he must be fought."[39]

The following morning, June 29, Huxley presented his paper on the development of *Pyrosoma*. By the close of the session, he decided he had had enough of the BAAS meeting for one year.[40] He returned to his rooms at Christ Church College to pack his bags, followed by ducking his head into a few of the "beautiful" college chapels.[41] Satisfied by those explorations, Huxley made his way down George Street to the railway station to catch the 4:05 p.m. train for Reading.[42] Once there, he planned to engage a hansom cab to take him to his brother-in-law's nearby country home, where Henrietta and the children were spending the weekend. As if by divine providence, a thunderstorm interrupted these plans.

Along the soggy and blinding trek to the train terminus, Huxley bumped into Robert Chambers, the popular and no longer anonymous author of *Vestiges of the Natural History of Creation*.[43] Chambers told Huxley how the program committee scrambled to shift the second "Darwinian" paper to an earlier slot—from Monday, July 2, to Saturday at noon on June 30—before some of the members were tempted to return home. To accommodate the expected crowd, the venue was changed from a small lecture room to the spacious library of the new Museum of Natural History. The presentation—the Zoology section's plenary, or main, lecture—was to be delivered by John William Draper, an English-born physiologist, physician, and chemist and one of the founders of the New York University School of Medicine.[44]

In 1891, thirty-one years later, Huxley recalled this moment with such clarity that the reader might think it occurred only a few hours ago: "I had heard of the bishop's [Wilberforce's] intention to utilize the occasion. I knew he had the reputation of being a first-class conversationalist, and I was quite aware that if he played his cards properly, we should have little chance, with such an audience of making an efficient defense." Huxley told

Hooker and Huxley seated in the audience at the BAAS Meeting, 1860.

Chambers, "I did not mean to attend it—did not see the good of giving up peace and quietness to be episcopally [sic] pounded." Whereupon Chambers "broke out into vehement remonstrances and talked about my deserting them. So, I said, 'Oh! If you are going to take it that way, I'll come and have my share of what is going on.'"[45] Huxley made an about-face, returned to Christ Church, dined that evening with Oxford's vice-chancellor, Francis Jeune, and in between carefully calculated sips of fine claret and port wine, steeled himself for the coming battle.[46] Later that evening, Huxley wrote Nettie to explain that his trip to Reading would be delayed: "I am eternally in a bustle here—and in the 'intervals and business,' am as tired as a dog," a state that made it difficult for him to "keep genial." Applying another canine (plus feline) cliché, Huxley reported, "The weather is vile, raining cats and dogs, and it [is] but with difficulty that one can get about to see anything after the business of the sections is over."[47]

ſ

Darwin's other field commander, Joseph Hooker, was contemplating chucking the meeting, too. Hooker described his arrival on Thursday afternoon, June 28, and how he "immediately fell into a lengthened reverie." Hooker longed for the company of both his wife and Darwin. Alone, he "was as dull as ditch water & crept about the once familiar streets feeling like a fish out of water." He promised himself that he "would not go near a Section & did not for two days." Instead, the botanist consumed his time by inspecting "the Colleges buildings & alternate sleeps in the sleepy gardens." Like a naughty schoolboy, Hooker "rejoiced in [his] indolence." That was until he heard how "Huxley and Owen had a furious battle over Darwins [sic] absent body at Section D" before his arrival. As he told Darwin, "You & your book forthwith became the topics of the day, & I d—d the days & double d—d the topics too, & like a craven felt bored out of my life by being woke out of my reveries to become referee on Natural Selection &c., &c, &c.—On Saturday I walked with my old friend of the *Erebus* Capt. Dayman to the Sections . . . & swore as usual I would not go in; but getting equally bored of doing nothing I did."[48]

Hooker's boredom soon ceased to be a problem. As Francis Darwin later dramatized the scientist-gladiators' return to their coliseum: "On Friday [29 June] there was peace, but on Saturday 30th the battle arose with redoubled fury over a paper by Dr. Draper of New York, on the 'Intellectual development of Europe considered with reference to the views of Mr. Darwin.' "[49]

# 11

# Pax Interruptus

The theory of Dr. Darwin, however, on the origin of species by natural selection gave rise to the hottest of all debates.

—The Press *(London), July 7, 1860*[1]

The molecular basis of remembering events, feelings, and sensations constitutes one of the most fascinating topics in neuroscience. Deeply embedded into the temporal lobe of our brain is a complex structure called the hippocampus—the very same anatomical detail that Owen and Huxley bickered over in 1860. The hippocampus acquired its odd name from the Greek word for "seahorse" because of its distinct shape. This is where the brain consolidates information and encodes, updates, maintains, and even alters memories of past events.

This neural complex might be better understood as a room filled with file cabinets. Each page in each file folder represents the memory of an event, sensation, feeling, or idea. Unlike a computer, however, the human brain's memory filing system can easily veer toward the illogical as it sorts out, edits, stores, and recalls these episodes. Some of the file cabinets are in pristine condition, and the memories contained within are sound and robust. Other cabinets are messy, with randomly open drawers and misplaced file folders. A quire of the papers in these files may appear to be in good condition; many more are torn, faded, or smudged; a few are blank, as if recorded in invisible ink or never recorded at all; and some contain idealized or revised stories that seem real yet did not actually happen that way.

Neurophysiological quirks aside, when it comes to the Oxford meeting, only a few contemporaneous newspaper accounts exist—and they contain the biases and most of the more-than-occasional unreliability of mid-nineteenth-century British journalism. None are verbatim courtroom transcripts. Worse, these versions have since been embellished or contradicted by several after-the-fact descriptions in letters and diaries. One intrepid historian took the time to record how, between 1860 and 1880, there were the official BAAS minutes, fourteen newspaper accounts, twelve contemporaneous diaries and letters describing the events of the day, and four after-the-fact diary "hearsay" reports. The same historian dug out eighteen more texts from 1889 to 1921 describing the so-called debate, seventeen long-after-the-fact letters written for publication from 1886 to 1899, and another three unpublished letters written from 1860 to 1890.[2] To complicate matters further, many more pro-Darwin/Huxley accounts have survived compared with relatively few pro-Wilberforce interpretations, imparting an artifactual imbalance when historicizing these events.[3]

The extant primary and secondary sources do not begin to account for the biographies, essays, and monographs that have appeared in the decades since—an enterprise some have sneeringly labeled the Darwin industry.[4] Like a snowball rolling down a snowier hill, the stories that evolved from so many spins and revolutions have become larger than life. In 1900, for example, Leonard Huxley hyperventilated over how "the fierce battle over the *Origin* was not merely the contradiction of one anatomist by another but the open clash between Science and the Church. It was, moreover, not a contest of bare fact or abstraction, but a combat of wit between the individuals, spiced with the personal element which appeals to one of the strongest instincts of every large audience."[5] Alas, the physiological quirks of memory, literary embellishments, and bibliographical idiosyncrasies have left us with a blurred picture of a brief encounter. We are forced—as Eartha Kitt once warbled in what might serve as the siren song for diviners of the past—to "proceed with caution, but, lover, please proceed."[6]

LET'S BEGIN WITH the Christian elephants in the room. With the hindsight of presentism, many have described the devout souls who doubted Darwin as museum relics—an extinct species, a batch of petrified bones without muscle, sinew, blood, and brains. In real time, almost everyone present at the 1860 meeting had long participated in church services and ceremonies beginning in grammar school and extending, at least, through their college education—whether they were believers or not. This was particularly true at Oxford and Cambridge. Every morning, students made their way to chapel, gazed at sculptures and paintings of Jesus Christ, and heard the Word of God. Eight hours later they returned to listen to the dulcet tones of robed choir boys singing the Evensong service; and most every night they knelt at their bedsides to pray.

Each Sunday, across the British Empire, men, women, and children dressed in their best and made their way to church. They read from their Bibles and celebrated the Holy Communion. Who among them had not sung soaring hymns attesting to the mysteries of the Creation? Who had not benefited from the balm of prayer and acts of confession, forgiveness, and repentance? Who would refuse their pastor's promise of salvation and eternal life? Who had not kneeled, stood, and solemnly recited the Nicene Creed: "I believe in one God, the Almighty Father, maker of heaven and earth, and all things visible and invisible."

The central conundrum was that the natural theologians conceived of laws as edicts prescribed by a divine ruler—be they moral edicts, the British system of common and natural law, or those guiding physical science. In this rigid ethos of authority, only the anthropomorphized Creator had the power to originate and determine the outcome of such laws. Natural selection, the theologians claimed, was no divine law at all; it was a randomly spinning wheel of chance, under the vague influence of unpredictable stimuli. Nor did this explanation account for the belief that God created man as perfect, in His image, and without the need of evolution.

Darwin discussed this issue at length in May 1860 with his "American cousin," the Harvard botanist Asa Gray: "With respect to the theological view of the question; this is always painful to me.—I am bewildered.—I had no intention to write atheistically. But I own that I cannot see, as plainly as others do, & as I should wish to do, evidence of design & beneficence on all sides of us. But the more I think the more bewildered I become; as indeed I have probably shown by this letter. . . . Let each man hope & believe what he can." Not content to leave matters there, Darwin explained to Gray his concept of a unifying set of "complex laws" that "may have been expressly designed by an omniscient Creator, who foresaw every future event & consequence."[7]

When reviewing *Origin* for the influential *Atlantic Monthly*, the American botanist echoed his friend's insistence that natural selection was *not* inconsistent with "an intelligent First Cause, the Ordainer of Nature." He appreciated "the philosophical difficulties which the throughgoing implication of design in Nature has to encounter." Gray was, however, careful to add that neither he nor Darwin had the "intention to obviate them." The battle between "the skeptic and the theist is only the old one, long ago argued out—namely, whether organic Nature is a result of design or of chance. Variation and natural selection open no third alternative; they concern only the question how the results, whether fortuitous or designed, may have been brought about."[8]

For most Christians, on both sides of the Atlantic, the binary construction offered by their clergy starkly divided faith from evolution. A more accommodating approach might have been to compare this divide to the example of man's free will and original sin. Nineteenth-century theologians often taught that when the apple-eating Adam and Eve were expelled from Eden, the next set of events occurred precisely as God "knew" they would unfold—even if humans could not understand or predict such processes. Might not the seemingly random, natural world be similarly predicted, adapted, and fine-tuned by a Creator—who introduced environmental, social, and natural changes with the passage of time? After reading Owen's rancorous review of *Origin*, for example, George Rolleston wrote Huxley, "I tell my friends here that if they would only believe that God is Almighty, there would be no difficulty in reconciling Darwin with the established creed. But people will not believe in this, the Second Article of the Apostles' Creed; and they persist in binding down Omnipotence to such a line of operation as they, poor manikins, think they could carry out."[9]

The Church of England as an institution was unprepared to entertain such radical ideas, and many of its clergymen, like Wilberforce, applied rhetorical tricks or threatening sermons to diminish natural selection. For most faithful Christians of this era, moreover, the strictures of the Scriptures were bound too tightly to incorporate Darwin's theory on "descent with modification" into their worldview. To partake of such forbidden fruit, they believed, invited the repercussions of eternal damnation or, at least, the downfall of their church and fellowships.

MID-MORNING Saturday, June 30. The museum porters were setting up additional rows of chairs and benches to accommodate the growing crowd anxious for entry. The venue was the museum's library on the west end of the building. A cluster of sweaty carpenters hammered and nailed together a speaker's platform directly between the two doors of entry. The ambient temperature—86°F—steadily rose after the audience entered the room, inhaling and exhaling hot air.[10] As Charles Lyell wrote the geologist Sir Charles Bunbury a few days later, the museum's library "was crammed to suffocation long before the champions entered the lists. The numbers were estimated at anywhere from 700 to 1000 people. Had it been term-time, or had the general public been admitted, it would have been impossible to have accommodated the rush to hear the oratory of the bold Bishop."[11]

John Stevens Henslow, the president of the section on Botany and Zoology, took the center seat in the first row. Sitting beside him was the American physiologist John Draper, rifling through his notes in preparation for his address. On his "extreme left" was Hooker and John Lubbock, and "nearer the center" was Huxley—all reporting for Darwin duty.[12] In the northwest corner of the long, narrow room huddled a scrum of undergraduates who "had gathered beside Professor Brodie, ready to lift their voices, poor minority though they were, for the opposite party. Close to them stood one of the few men among the audience already in Holy Orders who joined in—and indeed led—the cheers for the Darwinians"[13] That cheering clergyman was William Tuckwell, the "radical parson," Christian socialist, and headmaster of the New College School. In his memoir of a life spent at Oxford, Tuckwell recalled how the battle lines were drawn that afternoon, "with words strong on either side, and arguments long since superannuated; so, all day long the noise of battle roiled. The younger men were on the side of Darwin, the older men against him; Hooker led the devotees, Sir Benjamin Brodie the malcontents."[14] Brodie was sergeant-surgeon to the royal family, beginning with William IV and through much of Victoria's reign. He was widely known for his physiological investigations on bone and joint diseases. Outside the operating theatre, he was a dedicated opponent to Darwin's theory of natural selection and evolution. Two evenings earlier, on June 28, Huxley dined with Brodie's eldest son, also named Benjamin, who was Oxford's professor of chemistry and far more pro-Darwin than his father.[15] Seated near Brodie Sr.—also representing the opposing flank—was Darwin's old captain on the HMS *Beagle*, Rear Admiral Robert FitzRoy, and Professor Lionel Smith Beale, of King's College, London.

Late to the party came the Lord Bishop Wilberforce, and a most dramatic entrance he made. His flowing, floor-length cassock obscured his feet and gave him the appearance of floating above the ground as he walked down the center aisle. After shaking many glad hands, he claimed the seat directly to the right of Henslow. Flamboyantly unfurling the tails of his double-breasted black cutaway coat and exposing his purple vest, the bishop lowered his bulky frame onto a wooden chair. Such pomp and circumstance suggested that Wilberforce was a formal speaker on the program. In fact, Henslow only invited Wilberforce as a courtesy to the sitting bishop in the hosting town. The specifics of the invitation mattered little. Everyone in the room was anxious to learn what old Soapy Sam had up his billowing sleeves.

Over the years, these proceedings have been labeled alternatively as an academic squabble, debate, discussion, brouhaha, or, as Darwin referred to it, "the Battle of Oxford."[16] It is critical to note, however, that this event did *not* constitute a formal debate as practiced by the members of the Oxford or Cambridge Union Societies—with the "Ayes" sitting on one side and the "Noes" on the other. The men and women attending the BAAS meeting weren't armed with fact-filled file cards, nor did they play by the strict rules of engagement practiced by those venerable debating clubs. Although hardly an ironclad settlement of accounts between Christian authority and evidence-based science, the deliberations at Oxford would nonetheless be a familiar form of academic "debate" to anyone attending a scientific meeting today. Professional, and not-so-professional, agendas were unfurled under the cloak of peer review. Once a paper was delivered, there followed moderated questions, clarifications, and a discussion supporting or contesting the quivering speaker at the podium.[17]

The chattering in the room was so loud that it became difficult to hear the session's preliminary announcements—let alone the handful of papers preempted by the Owen-Huxley battle of two days earlier. Other than the meeting's rapporteur, for example, no one seemed to recall Charles Daubeny's request to discuss his latest experiments "proving" the existence of *equivocal generation*—the now antiquated notion of the spontaneous generation of life.[18] Finally, John Stevens Henslow stood up to announce the session's main event—John William Draper's lecture—and the audience snapped to attention. With full professorial affectation, the American physiologist rose from his seat and took command of the lectern. He gazed across the expectant crowd, "turning first to the right and then to the left, of course bringing in a reference to the *Origin of Species*, which set the ball rolling."[19]

John William Draper, 1866.

DRAPER HAD LONG BEEN working on a treatise on the intellectual development of Europe and another hefty volume on the conflict between science and religion—the former not being published until 1863 and the latter in 1874.[20] He was therefore delighted when Henry J. S. Smith, Secretary to the Local Committee of the British Association and the Savilian Professor of Geometry at Oxford, invited him to test run his ideas at the BAAS meeting.[21] Selecting Draper to discuss the philosophy and history of science for the plenary lecture of a Zoology section may seem odd at first glance—but it makes more sense if one considers the subtitle of his lecture, which hung on the peg of Darwin's new book and how Europe's intellectual development mirrored his theory.[22]

The précis of Draper's lecture comprises two full pages of acidic, yellowed paper. When perused today, they threaten to crackle and crumble in one's hands. Representing a simulacrum of these fragile folios, his words are dry, dusty, and brittle. Sadly, the breadth of the physiologist's intellectual quarry was miles long but inches deep. Draper took his colleagues on a whirlwind tour through ancient Greece, Rome, and the Ottoman Empire before delving into a tedious account of the intellectual development of European nation-states. All these processes, Draper insisted, followed Darwin's "immutable law." Even at only seven paragraphs of summary in the *BAAS Report*, the reader's eyes glaze over as if in a drug-induced stupor. The final line summarizing Draper's speech suggests that he went so far as to claim scientific priority over Darwin's book: "From [Draper's] work on physiology, published in 1856, he gave his views in support of the doctrine of the transmutation of species, the transitional forms of the animal and also the human type, the production of new ethnical elements or nations, and the laws of their origin, duration and death."[23] Specifically, in his 1856 textbook, *Human Physiology*, Draper posited that "every living being springs from a germ" and that "we can never say of an organized being that it is in a condition of rest. In truth, it is always in motion. It has a past and a future—coming from one state and going to another."[24]

Draper's address lasted anywhere from an hour to more than ninety minutes, depending on the witness. All agreed he was "hopelessly boring."[25] Merciless fun, too, was made of his American accent, even though Draper was born in Great Britain and did not emigrate to America until he was twenty-one. A graduate of University College, London, he later earned his medical degree at the University of Pennsylvania before landing at the newly created New York University medical school. Isabel Sidgwick—the wife of William Carr Sidgwick, a lecturer in political economy at Merton College, Oxford—panned the poor man: "I can still hear the American accent of Dr. Draper's address, when he asked, '*Air* we a fortuitous concourse of atoms?' and his discourse I seem to remember as somewhat dry."[26] John Green described the lecture as an "hour and a half of Yankee nasalism."[27] The Cambridge ornithologist Alfred Newton called it "a long-winded and dull essay read from a ponderous volume of manuscript resting on a massive desk."[28] Joseph Hooker was even more critical when recounting that the "paper of a Yankee donkey called Draper on 'civilization according to the Darwinian hypothesis' or some such title was being read, & it did not mend my temper; for of all the flatulent stuff and all the self-sufficient stuffers—these were the greatest, it was all a pie of Herbert Spenser [sic] & [Henry Thomas] Buckle without the seasoning of either."[29]

Draper, of course, recalled the afternoon far differently. A few days later, Draper wrote a long letter to his children about his triumphant day. He boasted how "the physiology Section adjourned to hear it 'as a mark of the greatest respect in their power to offer to me.' "[30] From the podium, Draper told his family, he could see that "though very many had to stand, and the paper occupied a full hour in the reading, I was listened to with the profoundest attention." He even crowed that the discussion following his lecture did not adjourn until "late past three and it was impossible for the British Association to have treated me more courteously. They gave me the best day and the best hour and the best room and the best audience."[31]

Draper went on to describe how "the Bishop of Oxford, who is a very fine speaker, rose to make some remarks on 'the very ingenious communication they had just listened to' and on the Darwinian theory." He honestly reported how Soapy Sam "criticized what I had said about the marble resting on the table and also about the impossibility of intellectually elevating men except by changing their physical conditions." Wilberforce should have paid closer attention to Draper's marble line. Indeed, it was a useful analogy for describing the slow changes over time by way of natural selection: "The organic world appears to be in repose because natural influences have reached an equilibrium. A marble may remain motionless forever on a table, but let the table be a little inclined, and the marble will quickly run off, and so it is with organisms on the world."[32] More surprising about Draper's account was that there is no mention of either Thomas Henry Huxley or Joseph Hooker![33] Instead, the New York physiologist concluded by offering an inadvertent understatement of how he brought together the requisite players needed to animate the Darwinian drama that followed: "I cannot now tell you all that has taken place, but I may truly say that I [have] never undertaken anything before which so thoroughly succeeded."[34]

# 12

# Mawnkey! Mawnkey!

After showing how ill-equipped was the Bishop for controversy upon the general question of organic evolution, although it was an open secret that Owen had primed him for the contest, Huxley said: "You say that development drives out the Creator, but you assert that God made you; and yet you know that you yourself were originally a little piece of matter, no bigger than the end of this gold pencil-case."

—*Edward Clodd, 1902*[1]

As president of the Botany and Zoology section, John Stevens Henslow had yet to declare whether he thought natural selection would prove to be correct or cast aside as one of Darwin's dafter notions. A scientist of great renown, many believed that Henslow's presence would temper the mood of the attendees—regardless of which side of the aisle they sat.[2] One admirer was Philip Pearsall Carpenter, a divinity student and invertebrate zoologist who had recently returned to Oxford from America. Carpenter was attending the BAAS meeting to report on a museum that had opened in Washington, DC, a few years earlier, in 1855. Endowed by James Smithson—a wealthy English chemist, mineralogist, and Oxford man (MA, 1786) who had never been to America—it was named the Smithsonian Institution.[3] While waiting for the Saturday session to begin, Carpenter sweetly described Henslow to a friend: "He is now a white-haired old gentleman, with the same beautiful face as ever, giving prizes to village children for wildflowers and snails, beloved by all."[4] Henslow was hardly the neutral observer. He was Darwin's favorite teacher at Cambridge and had recommended him for the post of the naturalist aboard the *Beagle*; he was, thereafter, a close friend and colleague. If these potentially conflicting connections were not enough to raise eyebrows, Henslow's son-in-law was Joseph Dalton Hooker.

After Draper finished his lecture, Henslow politely thanked him and searched the room for anyone willing to initiate a dialogue on what had just been said. No one was eager to go first.[5] The initial pause grew so long and pregnant that it seemed as if deliverance might never come. Finally, Henslow turned to Huxley and inquired if he had something—anything—to say in response. As with Daubeny's paper, two days earlier, Huxley declined, but not before "alleging the undesirability of contesting a scientific subject involving nice shades of [an] idea before a general audience, who could not be supposed to judge upon its merits."[6] This comment, however, fails to wash clean. Thomas Huxley made his name translating complex scientific ideas to all manner of men and women and usually relished such tasks. It would be difficult to find a better place to contest "a scientific subject involving nice shades" than a standing-room-only session of the British Association for the Advancement of Science's annual meeting—where science was routinely popularized and explained to the public. And of course, there is the fact that he aborted his trek out of Oxford the day before when Chambers told him of the Saturday meeting. If Darwin's Bulldog did utter such a statement to this effect, it seems more strategic than literal.

The passing seconds seemed like hours. Eventually, some men of God came to fore, expressing their wishes to be heard. A more cynical observer might suggest that the museum was planted less with botanical specimens than it was with chatty, Christian soldiers ready to attack. The priests scrawled out their names on slips of paper, which attentive pages picked up from their raised hands and rushed to the stage. In the order of receipt, Henslow recognized the clerics—but not before he "wisely announced *in limine* that none who had not valid arguments to bring forward on one side or the other, would be allowed to address the meeting."[7] Apparently, this warning was necessary. As Hooker later told Darwin, on July 2, "I must tell you that . . . 4 persons had been burked [silenced] by the audience & President for mere declamation."[8]

The first to speak—according only to Adam S. Farrar, a preacher at the Chapel Royal, Whitehall, and author of a book of sermons entitled *Science in Theology*—was a "shouting parson from Brompton who gave his name, being one of the Committee of the (newly formed) Economic section of the Association. He, in a voice of thunder, let off his theological venom."[9] If Farrar's retrospective memory is to believed, that "parson" was probably James Booth, a stern Irishman and, for a while, the vicar of St. Anne's Church, where he ran the Wandsworth Trade School for Boys in Garratt Lane, southwest London, until it closed in 1859. A blurred rotogravure of him depicts a heavyset man whose gray face looks like a slab of cold veal, with two slits for eyes, a horizontal slash for a mouth, and a bumpy protrusion serving as his nose. Booth spent his early career applying for and losing fellowships at Trinity College, Dublin. As a clergyman, he tinkered with mathematical formulas and helped bolster the national examination system for English schoolboys. During the 1860 BAAS meeting, he presented several papers to the Mathematics and Physics section and on Friday, June 29, spoke to the Statistical Science section, "On the True Principles of an Income Tax."[10] According to his biographers, Booth's "chief object as theologian was to reconcile the Bible with new scientific knowledge, arguing that science reveals physical truths, the Bible axiomatic moral truths."[11] There is no other record of Booth's comments nor one of him being formally recognized by the chairman. Perhaps it was only recollected by Farrar because the Reverend Booth may have been sitting near Farrar when releasing his "theological venom."

Most of the extant written memories do, however, agree with Farrar's subsequent description as to how "*then* jumped up Greswell, with a thin voice; saying much the same, but speaking as a scholar."[12] Reverend Richard Greswell, the Senior Tutor at Worcester College, Oxford, was a member of both the Royal Society and the BAAS. He was also the local treasurer for the present Oxford meeting. Many found the man to be irritating, but others regarded him as a superb educator, proponent of church schools, a natural theologian to his core, and supporter of Oxford's new natural history museum.[13] In 1889, Reverend John William Burgon included an essay on Greswell in his two-volume valentine to the pious, *The Lives of Twelve Good Men*.[14] Dean Burgon applauded Greswell for showing "no tolerance towards such new theories as those of Darwin: which trespass on the sphere of Revelation, or on the principles of Natural Religion [theology] and spiritual existence."[15]

Less reverential was the Reverend William Tuckwell, who described him as "the Derby dog in the person of old 'Dicky' Greswell of Worcester." That afternoon, Greswell wore a "vast white neckcloth," displayed a "luminous bald head," and hid his "great eyes" beneath thick-lensed spectacles. Completing this disturbing pen portrait, Tuckwell noted how Greswell rose and fell "rhythmically on his toes" as he "opined that all theories as to the ascent of man were vitiated by the fact, undoubted but irrelevant, that in the words of [Alexander] Pope, Great Homer died three thousand years ago."[16] Based on this literary point alone, Greswell "denied that any parallels could be drawn between the intellectual progress of man, and the physical development of lower animals."[17]

The crowd stamped their feet and cheered, "Hear, Hear!"—as if they had suddenly been transported to the floor of the House of Commons. "Old Dicky" heaped on more scorn by asking, "Was it not a fact that [Greece's] masterpieces in literature, the *Iliad* and *Odyssey*, were produced during its national infancy? (Hear, hear.)" Draper's discourse on intellectual development of Europe, he insisted, "was directly opposed to the known facts of the history of man." The Darwinians shouted Greswell down and applauded when he finally resumed his seat.[18]

Next to be recognized was Reverend John Dingle, a bald and heavily bearded vicar of the Lanchester and Burnhope Parish. A day earlier, Dingle offered up an unremarkable paper to the Geology section, "On the Corrugation of Strata in the Vicinity of Mountain Ranges."[19] He was the author of several books exploring the "exercise of God's power in the works of Nature," especially as it pertained to Genesis and "the harmony of Revelation and science."[20]

The Reverend Dingle endeavored to look and sound important by raising "a great fuss about getting a blackboard."[21] Once that iconic tool of teaching was made available for his use, he attempted "to show that Darwin would have done much better if he had taken *him* into consultation." Picking up a piece of chalk and scratching it loudly across the blackboard, he sketched out a few branches and declared, "Let this point A be man. And let that point B be the *mawnkey*."[22]

Looking at the puzzled, blank faces across the vast room, Dingle "found he could not explain himself—because—he had nothing to explain." Stalling for time, Dingle pointed to one of the dots he had drawn on the blackboard and said, "This represents the progress from the first atom."[23] The rowdy undergraduates pounced upon Dingle's weird comments and his weirder pronunciation of the word "monkey." As if striking a match off a leaky gasoline can, the young men's jeers engulfed the library until no one could hear anything but their loud and rhythmic chanting of "*Mawnkey, Mawnkey!*"[24]

Professor Henslow sternly cautioned Dingle, "Confine yourself, please, to Mr. Darwin's theory." Dingle ignored the request, connected the dots he had just drawn, and announced to the room, "That is the line of '*mawnkeys*' ending in man." His schematic chalk stripe inspired the students to stomp their feet, as they continued to chant, "*Mawnkey, Mawnkey!*" Henslow interjected in a louder tone, "We are getting a little beyond the mark,"—which only fueled the "incessant amiable clapping." In his memoirs, Philip Carpenter quipped, "The poor parson did not know what to make of it. We all told him he had better sit down. He looked as much as to say, 'When I ope' my mouth at Twaddletown, no dog barks.' We looked—'Oxford and British Association are not Twaddletown!' At last, he accepted our polite offer of a chair, and fumed to himself."[25]

When the "Mawnkey" chanting finally diminished, Henslow reiterated that only scientific points were to be entered into the discussion.

Benjamin Brodie, circa 1860.

Sir Benjamin Brodie was pleased to oblige, even though he required the support of his walking stick to painfully rise. The seventy-seven-year-old surgeon declared he had no intention of commenting on Draper's paper and instead rebuked Darwin's book as blasphemous rot. Looking out through cataract-blurred eyes, he addressed the audience: "Where was the demonstration that his primordial germ had existed? Man had a power of self-consciousness—a principle differing from anything found in the material world, and [I can] not see how this could originate in lower organisms." Figuratively pushing aside Darwin with a gesture of his wrinkled but skilled hands, he insisted, "This power of man, being identical with the Divine Intelligence, to suppose that it could originate with matter involved the absurdity of supposing the source of Divine power was dependent on the arrangement of matter." The *Oxford Chronicle* reported that Brodie's followers responded with "loud cries of 'Hear, Hear,' and much applause."[26]

As the surgeon stepped down from the stage, the audience's attention shifted to the large cleric seated next to Professor Henslow. Thousands of

Benjamin Disraeli, by Carlo Pellegrini ("Ape").

eyes watched as the Bishop of Oxford shifted his bulky haunches from side to side on a hard chair—perhaps so that he might majestically turn the other cheek. Amid the palpable anticipation, Samuel Wilberforce motioned to Henslow that he was ready to speak. To "immense applause," Russell Carpenter jotted down in his diary, "the parsonic element had gathered strong for their Goliath."[27]

From 1869 to 1889, a Neapolitan-born artist named Carlo Pellegrini drew caricatures of eminent Victorians for *Vanity Fair*, the popular weekly London-based magazine. Pellegrini claimed to have descended from the Medici family and signed his work with the French word "Singe," which means "monkey." This pseudonym subsequently evolved into "Ape." It would be too good to be true to suggest that this *nom de crayon* was a nod to the definitive portraits he drew of Bishop Wilberforce and Thomas

Huxley—and it is. The name referred to the mischievous style of his pictures rather than the descent of man. Initially, "Ape" parodied politicians, beginning with a caricature of the powerful prime minister, Benjamin Disraeli, which appeared in the January 30, 1869, issue of *Vanity Fair.* So successful was this colorful cartoon—replete with the requisite stereotypical exaggeration of a Jewish nose—that Pellegrini was commissioned to draw many more "Men of the Day" for the magazine. Over the next several decades, *Vanity Fair* published more than two thousand caricatures by Pellegrini and a cohort of artists and cartoonists. Several other London-based magazines of the era followed suit in publishing caricatures of great and recognizable men. Indeed, the lampooning of a public figure's face, features, and personality became the mark of arrival in London's high society.[28]

Pellegrini's 1869 *Vanity Fair* portrait of Samuel Wilberforce may well be his masterpiece. Above the stinging caption, "Not a Brawler," the artist portrayed the bishop in a black and gray cassock draped over a white cotton shirt with ruffle-cuffed lawn sleeves—each puffy enough to become a parachute for a small monkey. Gazing directly at the viewer with a judgmental smirk on his well-fed face, the bishop clasps his hands together at the level of his heart, as if washing them—à la Soapy Sam.[29]

"Argumentative, rhetorical, [and] amusing." These were the three words Reverend William Tuckwell used to describe the Lord Bishop's mood as he rose from his chair. Soapy Sam did so with the gravitas he had perfected on so many thousands of Sundays past.[30] Once fully erect, Wilberforce towered over the men seated near him. He grinned as he was "loudly cheered" by his flock of faithful men and women. After dismissing Draper's metaphor of marbles, tables, and equilibrium, the bishop solemnly announced to the audience that he had "given the theory advanced by Mr. Darwin his most careful and anxious consideration." What he failed to confess, however, was his scathing review of *Origin* that would soon appear in the summer issue of the *Quarterly Review*.[31] Reading the room like the expert orator he was, Wilberforce sensed that his audience was in no mood for a sermon. In the company of so many men of science, he trotted out his First Honors degree in mathematics from Oxford and previous successes at past BAAS meetings to suggest an intellectual prowess that was not his to declare.

The bishop explained how he had applied Baconian "inductive science," philosophy, and logic and—lo and behold—Darwin's theory "entirely broke down" and "fell to pieces."[32] Wilberforce blithely dismissed the thousands of facts Darwin had gathered and interpreted for his book—all of which, Wilberforce claimed, "utterly failed to prove his theory." This was because, he asserted, "the permanence of specific forms was a fact confirmed by all observation, the few exceptions that existed being confined to a few cases in certain species of plants. Take, for instance, the remains of animals, plants, and man, found in those earlier records of the human race—the Egyptian catacombs," Wilberforce rambled. "Now, anatomists tell us that even in mummies, 4,000 years old, there is not the slightest physiological difference as compared with the race now—('Hear, hear!' yelled the crowd)—and so it was with animals and plants. All spoke of their identity with existing forms and of the irresistible tendency of organized beings to an unalterable character" (applause).[33] Such facile statements, of course, were a gross manipulation of what Darwin wrote, and Wilberforce knew it. To begin, there was Charles Lyell's widely accepted geological premise that the world was far older than described in the Bible. More specific was Darwin's careful explanation in *Origin* how it required eons to affect the slow origin or transmutation of species—as opposed to the evolutionary eyeblink of a few thousand years when the Pharaohs and their ilk were mummified.[34]

The bishop insisted, "in a light, scoffing tone, florid and fluent," that *Origin* was a dangerous flight of fancy.[35] "Even in the great case of the pigeons, quoted by Mr. Darwin," Wilberforce selectively asserted, "he admitted that no sooner were these animals set free, than they returned to their primitive type."[36] After the bishop's flock parroted back the obligatory "Hear, hear," Wilberforce appropriated Huxley's criticism as to how "everywhere sterility attended hybridism, as was seen in the closely allied forms of the horse and the ass."

What really irked the bishop, however, was Darwin's "most degrading assumption—('Hear, hear!' cheered the crowd)—that Man who, in many respects, partook of the highest attributes of God—('Hear, hear!' again)—was a mere development of the lowest forms of creation."[37] Wilberforce faded and feigned that "he could scarcely trust himself to speak upon the subject, so indignant did he feel at the idea." Not surprisingly, he was not nearly so fragile when discussing this very point in the *Quarterly Review*. Inconsistent with God's scheme of creation, he argued on those pages, was "Mr. Darwin's daring notion of man's further development into some unknown extent of powers, and shape, and size, through natural selection acting through that long vista of ages, which he casts mistily over the earth upon the most favored individuals of his species."[38]

Joseph Hooker later told Darwin how "all the world was there to hear Sam Oxon—Well Sam Oxon got up & spouted for half an hour with inimitable spirit ugliness & emptyness [sic], & unfairness, I saw he was coached up by [Richard] Owen & knew nothing & he said not a syllable but what was in the Reviews."[39] In real time, Hooker simmered and spluttered while Wilberforce "ridiculed Darwin badly, and Huxley savagely but all in such dulcet tones, so persuasive a manner, and in such well-turned periods, that I who had been inclined to blame the President [Henslow] for allowing a discussion that could serve no scientific purpose, now forgave him from the bottom of my heart."[40] Sitting nearby, Reverend William Tuckwell accused the bishop of plagiarizing his comments from a "mountebank sermon" delivered by the anti-Darwinist John William Burgon, a collection of words he regarded as "a piece of clever, diverting, unworthy clap-trap."[41] Hooker and Tuckwell may have voiced minority opinions. Years later, Leonard Huxley set the scene as if it played on London's West End, instead of in a staid museum library: "The very windows by which the room was lighted down the length of its west side were packed with ladies, whose white handkerchiefs, waving and fluttering in the air at the end of the Bishop's speech, were an unforgettable factor in the acclamation of the crowd."[42]

Taking the verbal equivalent of a victory lap, Wilberforce chummed the audience by declaring he had no desire for "timidity in scientific investigation ('Hear, hear!'). Religion had nothing to fear ('Hear, hear!')." His chief concern centered on "the hasty adoption of unsound hypotheses and unproved assertions for the weighty realities of scientific truth" (applause). It was not that "science and revelation were inimical to the other," Sam soaped on, "but that what appeared irreconcilable in the present state of scientific knowledge would in the fullness of time be made manifest and redound to the triumph of both" (prolonged cheering).[43] In a mellifluous voice, he gloated that "Mr. Darwin's conclusions were an [sic] hypothesis, raised most unphilosophically to the dignity of a causal theory." Fortunately, he "was glad to know that the greatest names in science were opposed to this theory, which he believed to be opposed to the interests of science and humanity."[44]

There exist several conflicting accounts of how the Bishop of Oxford phrased the next, and most momentous, query of his storied, public speaking career.[45] Hooker characterized it as the climacteric that changed everything: "Unfortunately, the Bishop hurried along on the current of his own eloquence, so far forgot himself as to push his attempted advantage to the verge of personality in a telling passage in which he turned round and addressed Huxley."[46] Wilberforce was described as staring down at Huxley—who was seated far too close to his tormenter—and mocking him as "Professor Huxley who is about to demolish me." Taking a few beats to great dramatic effect, the bishop pounced—according to the *Oxford Chronicle*—and asked "if [Huxley] had any particular predilection for a monkey ancestry, and, if so, on which side—whether he would prefer an ape for his grandfather and a woman for his grandmother or a man for his grandfather, and an ape for his grandmother?" The laughter in the room nearly shook the library's newly installed roof beams. And yet, there were many in the room—for and against Darwin—who rejected the bishop's question as vulgar, deceitful, and petty. Wilberforce displayed bad behavior and poor character in front of men who fetishized the veneer of good manners, common decency, and a gentlemanly demeanor.[47]

The undergraduate and budding historian John Green recorded a detailed memory of this incredible instant: "Up rose Samivel—" contemptuously using the "Old London" accent popularized by Dickens's *The Pickwick Papers*.[48] The bishop "proceeded to act the smasher; the white chokers, who were abundant, cheered lustily, a sort of 'Pitch it into him' cheer, and the smasher got so uproarious as to pitch into Darwin's friends—Darwin being smashed—and especially Professor Huxley. Still the white chokers cheered, and the smasher rattled on. . . . Let me say that such rot never fell from episcopal lips before." The Darwinians in the room demanded, "Let the learned Professor speak for himself . . . [when] arose Huxley, young, cool, quiet, sarcastic, scientific in fact and in treatment, and gave his lordship such a smashing as he may meditate on with profit over his port at Cuddesdon [the Bishop of Oxford's palace]."[49] Reverend Tuckwell, on the other hand, simply recalled Huxley as "white with anger."[50]

Thomas Huxley later wrote a stirring account of what segued into one of the most celebrated instances of scientific discourse: "So I came and chanced to sit near old Sir Benjamin Brodie. The bishop began his speech, and to my astonishment very soon showed that he was so ignorant that he did not know how to manage his own case. My spirits rose proportionately and when he turned to me with his insolent question, I said to Sir Benjamin in an undertone, 'The Lord Hath delivered him into mine hands.'" After stage-whispering this well-chosen passage from the Old Testament's *Book of Samuel*, "that sagacious old man stared at me as if I lost my senses. But in fact, the Bishop had justified the severest retort I could devise, and I made up my mind to let him have it. I was careful, however, not to rise to reply, until the meeting called for me—then I let myself go."[51]

With the exception of the word, "equivocal," Huxley later endorsed John Green's transcription of his response to Wilberforce:

> I asserted, and I repeat—that a man has no reason to be ashamed of having an ape for his grandfather. If there were an ancestor whom I should feel shame in recalling, it would rather be a man, a man of restless and versatile intellect, who, not content with an *equivocal* success in his own sphere of activity plunges into scientific questions with which he has no real acquaintance, only to obscure them by an aimless rhetoric, and distract the attention of his hearers from the real point at issue by eloquent digressions and skilled appeals to religious prejudice.[52]

Another eyewitness, A. G. Vernon-Harcourt, a Reader in Chemistry at Oxford—who studied chemical kinetics and invented a device to safely administer chloroform anesthesia—recollected Huxley's ape line differently in a letter to Huxley's son Leonard: "Your father first explained that the suggestion was of descent through thousands of generations from a common ancestor, and then went on to this effect—'But if this question is treated, not as a matter for the calm investigation of science, but as a matter of sentiment, and if I am asked whether I would choose to be descended from the poor animal of low intelligence and stooping gait, who grins and chatters as we pass, or from a man, endowed with great ability and a splendid position, who should use those gifts to discredit and crush humble seekers after truth, I hesitate what answer to make.' " When writing to Huxley's son Leonard decades later, Vernon-Harcourt insisted on adding the caveat, "No doubt your father's words were better than these, and they gained effect from his clear and deliberate utterance but in outline and in *scale*, this represents truly what was said." According to Vernon-Harcourt, the crowd erupted with such loud cheering and clapping that they nearly drowned out Huxley's witty response.[53]

In 1895, George Johnstone Stoney—the Irish physicist who introduced the term *electron* to denote the basic unit of electricity—recalled how "the audience was unpleasantly partisan, a majority on the Bishop's side, the minority on Huxley's, and each moiety applauding every hit made by its champion. There seemed to be very little impartiality." Wilberforce, Stoney wrote, "roused up the feelings of his larger section of the audience—exercising an art which [he] possessed in extraordinary perfection to regard it as ridiculous and degrading to suppose that man is descended from an ape! When he sat down amid vociferous and excited applause, Huxley rose very slowly; and the first words of his rejoinder were, 'I had rather be the offspring of two apes than be a man and afraid to face the truth.'—an announcement which was followed by the counter-cheers of his minority."[54]

Three years later, in 1898, Isabel Sidgwick offered her account to the readers of *Macmillan's Magazine*: "On this Mr. Huxley slowly and deliberately arose. A slight tall figure stern and pale, very quiet and very grave, he stood before us, and spoke those tremendous words,—words which no one seems sure of now, nor I think, could remember just after they were spoken, for their meaning took away our breath, though it left us in no doubt as to what it was."[55]

Perhaps the most pertinent document, however, was the letter Huxley wrote, on September 9, 1860, to his friend Frederick Dyster—a surgeon and naturalist who attended the July 1858 meeting of the Linnean Society, when Darwin's and Wallace's papers on natural selection were first read. This account carries an imperfect historical precedence of being written by Huxley ten weeks after the event:

> I had listened with great attention to the Lord Bishop's speech but had been unable to discover either a new fact or a new argument in it—except indeed the question raised as to my personal predilections in the matter of ancestry.—That it would not have occurred to me to bring forward such a topic as that for discussion myself, but that I was quite ready to meet the Right Rev. prelate even on that ground. If then, said I, the question is put to me whether I would rather have a miserable ape for a grandfather or a man highly endowed by nature and possessed of great means of influence and yet who employs these faculties and that influence for the mere purpose of introducing ridi-

> cule into a grave scientific discussion, I unhesitatingly affirm my preference for the ape. Whereupon there was unextinguishable laughter among the people, and they listened to the rest of my argument with the greatest attention. Lubbock and Hooker spoke after me with great force and among us we shut up the bishop and his laity. I happened to be in very good condition and said my say with perfect good temper and politeness.—I assure you of this because all sorts of reports [have] been spread about, e.g., that I had said I would rather be an ape than a bishop, etc.[56]

The final point of his letter was of particular importance. Huxley was too savvy to fall into such a rhetorical trap. Privately, however, he could not help but boast to Dyster of the pleasure he took in humiliating the bishop, "so that . . . Samuel will think twice before he tries a fall with men of science again. If he had dealt with the subject fairly and moderately, I would not have treated him in this way—But the round-mouth, oily, special pleading of a man who is ignorant of the subject, presumed on his position and his lawyer faculty gave me a most unmitigated contempt for him."[57] No matter the precise wording of their encounter, Darwin's Bulldog took an enormous bite out of Soapy Sam's hide.

According to Reverend Tuckwell, there was "a gasp and shudder through the room, the scientists uneasy, the orthodox furious, the bishop wearing that fat, provoking smile which once . . . impelled Lord Derby in the House of Lords to offer up an unparliamentary quotation from *Hamlet*."[58] That dramatic line, incidentally, was the stuff of Shakespearian perfection: "A man may smile, and smile, and be a villain."[59] Philip Carpenter later recalled, the "time was, when a man might have been burnt at Oxford for such impertinence" and how grateful he was that "the English people have left off burning." He could not help slyly adding, however, "There were three native-born Americans burnt, to my certain knowledge, the short time I was in the [United] States."[60]

Isabel Sidgwick recalled how "one lady fainted and had to be carried out" of the room. That woman was Lady Jane Kirk Purnell Brewster, the wife of David Brewster, then principal of St. Andrews and a well-regarded astronomer, physicist, and Christian, who was deeply opposed to Darwin's book.[61] During the delay of removing the seemingly unconscious Lady Jane from the library, the audience perspired and shook their heads in wonder, some overjoyed by Huxley's rebuttal, others mortified.

Henry Fawcett, a professor of political economy at Cambridge and, from 1880 until his death in 1884, Britain's Postmaster General, adjudicated that "the retort was so justly deserved, and so inimitable in its matter, that no one who was present can ever forget the impression it made."[62]

Charles Lyell agreed. On July 4, he wrote that "many blamed Huxley for his irreverent freedom; but still more . . . declared that the bishop got no more than he deserved."[63] George Rolleston later wrote Huxley he was so offended by Wilberforce's snarky behavior that he "intend[ed] to pass a slap at the Base Bishop before he recovers the cudgeling you have given him."[64]

Only Hooker remained unimpressed by the histrionics. Explaining the exchange to Darwin, he wrote, "Huxley answered admirably & turned the tables, but he could not throw his voice over so large an assembly, nor command the audience; & he did not allude to *Sam's* weak points nor put the matter in a form or way that carried the audience." Hooker did, however, admit that "the battle waxed hot."[65]

HUXLEY DID NOT conclude his exordium with an easily misquotable quip. As George Johnstone Stoney recollected, in contrast to Hooker's reportage to Darwin: "He then proceeded, in a quietly scarifying manner, in contrast to the Bishop's passionate appeal, to correct the Bishop of Oxford's statements where erroneous and to reply to the parts of his arguments which remained."[66] According to the July 21 issue of the *Oxford Chronicle and Berks and Bucks Gazette*, Huxley took Wilberforce to task for "obscuring the light of scientific truth." Darwin's book was no "mere hypothesis," he defended, it was "an explanation of phenomena in natural history holding the same relation as the undulating theory to the phenomena of light. Did anyone object to that theory because an undulation of light had never been arrested and measured?" The scientist appropriately demanded that Wilberforce and his followers "bring forward any important fact against his theory," which was "the result of laborious research, and abounded in new facts bearing upon it. Without asserting that every part of the theory had been confirmed," Huxley added, "[Darwin had produced] the best explanation of the origin of species which had yet been offered." To which his supporters again chattered "Hear, hear."[67]

The zoologist plunged his rhetorical shiv an inch deeper as he repeated his objection "against this subject being dealt with by amateurs in science, and made the occasion of appeals to passion and feeling."[68] Huxley scolded Wilberforce for his flawed "psychological distinction between man and animals, it must be remembered man himself was a once a monad—a mere atom of matter—and who could say at what moment of his development he became consciously intelligent." Again, the crowd exclaimed "Hear, hear!"

"The question," Huxley explained, "was not so much one of a transmutation or transition of species as of the production of forms which become permanent." This was a process that breeders of livestock had long known and practiced. Instancing "the short-legged sheep of America," Huxley pointed out how these animals "were not produced gradually but originated in the birth of an original parent of the whole stock, which had been kept up by a rigid system of artificial selection."[69] With that, Huxley descended from the podium. Michael Foster, the Cambridge physiologist, later recalled how Huxley was "received coldly" by much of the audience when he first stood up to speak, "but as he made his points the applause grew and widened, until, when he sat down, the cheering was not very much less than that given to the bishop. To that extent he carried an unwilling audience by the force of his speech."[70]

ʃ

GIVEN THAT the BAAS was a formal British conclave, there followed a round of amendes honorables for the words spoken more out of passion than elucidation. The bishop "regretted that Professor Huxley had taken umbrage in what he had said" and disingenuously claimed he "did not know that he had said anything which could possibly give offense to Mr. Darwin's greatest friends." Regarding his query on parentage and apes, Soapy Sam claimed to have surrendered to the "merriment of the audience and it was merely a passing allusion." Unable to leave well enough alone, Wilberforce "ridiculed Professor Huxley's appeal to authority in connection with his remarks on amateurs in science," particularly when there were many eminent men of science, such as Brodie, Owen, John Herschel, and William Whewell, who disagreed with Darwin's theory.[71] With that, the bishop turned the table by asking how "the Professor could talk [of Christian doctrine] as he had done about authority he did not know." His side of the room responded with "laughter and cheers." The bishop "then noticed the Professor's concluding remarks, denying the cogency of the illustrations, and after experiencing some interruptions in his scientific dicta, sat down amid loud cheers."[72]

Huxley, too, apologized for his harsher remarks and assured the audience that the bishop "had no desire to mislead, but he thought he misapprehended his remarks upon authority." What Huxley intended to say was that the bishop's expertise "derived from a reputation in another sphere." To which his side of the room predictably responded with a "Hear, hear" and laughter.[73]

# 13

# The Rebuttals

For myself, also, I rejoice profoundly; for thinking of the many cases of men pursuing an illusion for years, often & often a cold shudder has run through me & I have asked myself whether I may not have devoted my life to a phantasy. Now I look at it as morally impossible that investigators of truth like you & Hooker can be wholly wrong; & therefore I feel that I may rest in peace.

—*Charles Darwin to Charles Lyell, November 23, 1859*[1]

A herd of attendees lined up to comment on the spectacle they had just witnessed. Much like overeager pupils who constantly raise their hands in class, the purpose of participation was inspired less by a need for clarification than to ensure that their presence was known. All these men wanted the historical record to reflect that they, too, played a role in the Darwinian deliberations of that afternoon.

The first fellow to be recognized was James Bird, who a day earlier, June 29, presented a paper to the Chemistry section on "the deodorization of sewage."[2] Bird served as a surgeon in the Bombay Army, where he fought in Bengal during the 1819 Kaira campaign and at the capture of Kittur in 1827.[3] In his spare time, he was an active member of the Royal Asiatic Society, authored books on cave sculptures in western India, and translated Persian texts into English.[4] An avid collector of scarce Roman, Greek, Byzantine, English, and Asian coins, his trove was later purchased at auction by the British Museum.[5] Bird could have easily served as a model for such loose change; his profile—framed by wavy, gray hair and a fine, aquiline

James Bird, MD.

nose—was both Byronic and beautiful to behold. Upon retirement from military duty in 1847, the doctor returned to London to lecture on tropical infections at St. Mary's Hospital, where he pontificated with great confidence on antiquated treatments for cholera, military medicine, and "the diseases of Europeans in hot climates."[6]

Despite his sterling credentials, Bird had a habit of betting on the wrong scientific horses. He was present for the October 1849 meetings of the Westminster Medical Society when John Snow—the founder of modern epidemiology—famously announced his theory that cholera was transmitted not by miasma or foul air, as thought by most doctors of the day, but instead by ingesting water contaminated with some type of "morbid poison." Snow's lecture on the "pathology and mode of communication of cholera" predated the germ theory of disease by decades. After Snow completed his talk, Bird stood up to declare that while he was "not prepared to deny altogether the truth of Snow's views that it could be multiplied through the medium of water, impregnated with the poisonous dejections of cholera patients, he could not believe that such medium of communication had more than a partial effect." Based on his thirty years of experience with cholera in India, Bird sounded more like an ancient Hippocrates than the modern-day epidemiologist: "While endemic influences of low, damp situations, vegetable and animal effluvia, bad water, imperfect ventilation, and deficient food acted as predisposing causes in giving rise to this intractable malady among the people, an epidemic atmospheric constitution was necessary for its general diffusion. The atmosphere is the principal medium by which cholera is disseminated though the human recipient of the morbific *miasm* occasionally acts, as in yellow fever and influenza, a secondary agent in propagating it."[7] Five years later, in 1854, Snow mapped a cholera outbreak in the Soho district of London, which definitively proved that the source of "morbid" contamination was sewage-tainted water drawn from the Broad Street pump and, ultimately, the sewage-tainted Thames River.[8]

Now, in 1860, after a nod from Henslow, Dr. Bird took the floor of a different epoch-changing debate, with equally wrong results—making him the poster boy of bad judgment for *both* germ theory and evolutionary biology. At Oxford, Bird insisted, without evidence, that Mr. Darwin's theory was statistically unsound. Worse, the research Darwin presented in his book allowed one to prove almost anything and everything—a criticism that incited a swarm of buzzing. One audience member loudly demanded, "Question!" It soon became clear that Bird had nothing more to add to the conversation. Within a few seconds, "the learned doctor's remarks were cut short by the impatience of the audience."[9]

The next inquisitor was a "gray-haired, Roman-nosed, elderly gentleman" in dress-blue, gold-trimmed, naval regalia—Rear Admiral Robert FitzRoy, Darwin's skipper on the HMS *Beagle*.[10] He was slated to present a paper on "British Storms, Illustrated with Diagrams and Charts" for the Meteorology session on Tuesday, July 3.[11] Before he attended to his stated business, FitzRoy boomed, he wanted to make a public apology for having facilitated Darwin's views by taking him under his command. During their voyage, he "had often expostulated with his old comrade for even entertaining views that were contradictory to the First Chapter of *Genesis*." Apparently, his protests failed.[12] George Johnstone Stoney later recalled what followed in excruciating detail. The admiral "stood up near the center of the crowded Sheldonian Theatre [sic], and lifting an immense Bible first with both and afterwards with one hand over his head." With great solemnity, FitzRoy "implored the audience to believe God rather than man—God, who vouchsafes Himself to speak to every man in that book. He urged them to reject with abhorrence the attempt to substitute human conjectures and human inventions for the explicit revelation which the Almighty has Himself made in that book of the great events which took place when it pleased Him to create the world and all that it contained."[13]

After FitzRoy's tempest, the room fell completely quiet. "Even his own side of the audience, who were the majority," wrote Stoney, "regretted the incident and felt it was out of place." Seated next to Humphrey Lloyd—the physicist and provost of Trinity College, Dublin—Stoney recalled, "Our perception that the issue sought to be raised was an *ignoratio elenchi* [an extraneous or irrelevant point] was almost smothered by our sense of the deplorable character of the whole incident."[14] The oppressive silence was broken by a few impulsive young men shouting, "Question!" Befuddled by their cries, the admiral feebly attempted to deny "Professor Huxley's statement that Mr. Darwin's work was a logical arrangement of facts." But, William Tuckwell recalled, FitzRoy "had nothing more to say than Darwin's book had given him acutest pain."[15] The *Oxford Chronicle* reported that "the interruptions became so boisterous that the chairman requested him to sit down."[16]

It is impossible to discern what was on FitzRoy's mind at the time of his astonishing public display. A few weeks later, on July 16, Darwin wrote to Henslow about "poor FitzRoy with the Bible incident . . . I think his mind is often on verge of insanity."[17] Darwin's diagnosis was not far off the mark. After a series of financial losses and spats with government officials—who alleged he had exceeded his remit at the Met Office—FitzRoy fell into a deep depression. On the morning of April 30, 1865, he locked the door to his dressing room and slit his throat with a shaving razor. His aim was true, and he bled to death within minutes. In acknowledgment of his work for the British Empire, Queen Victoria allowed his widow and daughter to live in a "grace and favour" apartment at the Hampton Court Palace. FitzRoy's friends started a subsistence testimonial fund for the benefit of his heirs and raised £6,100. The largest individual donor was Charles Darwin. He pledged a gift of £100.[18]

Once the naval officer acquiesced to the crowd's demands to sit down and shut up, Henslow recognized Lionel Smith Beale, of King's College, London. Beale and Huxley had some history of their own. In 1852, Beale was appointed professor of physiology at King's much to Huxley's disappointment, because he was the other finalist for the position.[19] Eight years later, at the BAAS meeting, Beale claimed to be "quite unable to decide the question on one side or the other."[20] This was especially odd, since Beale's "philosophical and biological views were strongly opposed to the wave of materialism that followed in the wake of Darwin and Huxley."[21] The doctrine of materialism held that all physical matter—organic and inorganic—could be explained in terms of chemical and physical phenomena. In keeping with his opposition to this foundational point of modern science, Beale pointed out "for Professor Huxley's consideration, some of the difficulties which Mr. Darwin's theory had to deal with . . . [such as] those *vital* tendencies of allied species which seemed independent of all external agents."[22] Here, he was referring to the antiquated doctrine of vitalism, whereby living beings—in particular, humans—were thought to contain unique, vital "forces, properties, powers or principles which are neither physical nor chemical" and were different from all other properties of inorganic matter. In other words, they had a soul.[23] Predictably, the anti-Darwinist contingent cheered and pounded their hands together.[24]

Henslow again admonished the assembly to come to order.[25] Once the room quieted, he recognized John Lubbock, Darwin's twenty-six-year-old neighbor—a banker, amateur biologist with an expertise in bees, wasps, and ants, and, from 1870 to 1880, a Liberal Party member of Parliament. Best remembered for introducing the 1871 law creating the British tradition of "bank holidays," Lubbock was long of leg and slight of build, with a face notable for bushy, brown side whiskers and a rapidly receding hairline.[26] In 1878, the *Vanity Fair* caricaturist Leslie "Spy" Ward drew Lubbock's head to look like Thomas Edison's yet-to-be developed light bulb—big, bright, and tapering at the neck, just like the metal collar that screws into

Sir John Lubbock, by Leslie Ward ("Spy").

a socket.[27] Elegantly dressed, the banker-biologist stood, with his tall, silk hat in hand, and repeated Huxley's comment that "Mr. Darwin's book was the most logical and powerful arrangement of facts that had ever been given upon the subject." Considering that no better theory had yet been proposed, he "expressed his willingness to accept his hypothesis in the absence of any better." With a whiff of condescension, Lubbock voiced "surprise at the Bishop of Oxford's reference to the Egyptian mummies, since to the naturalist as to the geologist, time was not an essential element in these changes. Time alone produced no change." The Darwinists responded with cheers until the "Derby dog," Reverend Richard Greswell, popped up like an unnecessary comma. He "indignantly" denied Huxley's assertion "that man was originally an atom of matter." Without doubt, Greswell shouted, he was never atomic. The younger members of the crowd whooped, "Oh, Oh," followed by "loud laughter amid which the reverend gentleman sat down."[28]

ſ

A WEARY HENSLOW looked to his son-in-law, Joseph Hooker, for help in quieting the decidedly unscientific disruptions. Hooker knew he had a tough act to follow in Huxley but must have felt a strong desire to demonstrate that *he* was Darwin's most thoughtful colleague. Few historians quote the botanist's subsequent comments as freely as they paraphrase Huxley's glib retort to Wilberforce's question. This is unfortunate because Hooker's argument was a superb example of how scientists use evidence rather than rhetoric to make their positions understood. Hooker's words were intricate yet evocative, gaining speed and power with each successive phrase and measure.

The botanist refuted Wilberforce's assertion that "all men of science were hostile to Mr. Darwin's hypothesis." He, for one, was in favor of it. Hooker modestly added that he "could not presume to address the audience as a scientific authority," but since his opinion had been solicited, he "would briefly give it." Unfortunately, the bishop's "eloquent address," he said, "completely misunderstood Mr. Darwin's hypothesis." This was most obvious when "his Lordship intimated that [Darwin] maintained a doctrine of the transmutation of existing species one into another and had confounded this with that of the successive development of species by variation and natural selection." The bishop's critique "was so wholly opposed to the facts, reasonings, and results of Mr. Darwin's work, that [Hooker] could not conceive how anyone who had read it could make such a mistake,—the whole book, indeed, being a protest against that doctrine."[29]

Hooker then thrashed Wilberforce's interpretation of the "general phenomena of species," in which Wilberforce claimed "that these did not present characters that should lead careful and philosophical naturalists to favor Mr. Darwin's views." Hooker explained how *his* years-long study of the "Vegetable Kingdom was diametrically opposed" to the bishop's contention. At a minimum, half of the known plants could be categorized into groups whereby each species was connected to the others in the same group "so much so that, if each group be likened to a cobweb, and one species be supposed to stand in the center of that web, its varying characters might be compared to the radiating and concentric threads, when the other species would be represented by the points of union of these." In other words, Hooker continued, "the general characteristics of orders, genera, and species amongst plants differed in degrees only from those of varieties and afforded the strongest countenance to Mr. Darwin's hypothesis."[30]

Most important was how Hooker made it clear that by accepting "Mr. Darwin's views," he was *not* adopting them as a "creed," despite the bishop's false charges to the contrary, because "he knew no creeds in scientific matters." As a young botanist, he began his scientific work under the premise "that species were original creations; and it should be steadily kept in view that this was merely another hypothesis." The Biblical explanation, when taken "in the abstract, was neither more nor less entitled to acceptance than Mr. Darwin's: neither was, in the present state of science, capable of demonstration, and each must be tested by its power of explaining the mutual dependence of the phenomena of life."[31] Hooker admitted he had initially adhered to the natural theology-based hypothesis that "species were original creations," even though most modern studies of botany "had developed no new facts that favored it, but a host of most suggestive objections to it."

For the past fifteen years, Hooker patiently explained, he had applied Mr. Darwin's ideas "to botanical investigations of all kinds in the most distant parts of the globe, as well as for study of some of the largest and most different Floras at home." Now, after closely reading *Origin*, the botanist had "no hesitation in publicly adopting his hypothesis" because it offered "by far the most probable explanation of all the phenomena presented by the classification, distribution, structure, and development of plants in a state of nature and under cultivation." As a result, Hooker pledged to "continue to use [Darwin's] hypothesis as the best weapon for future research, holding himself ready to lay it down should a better be forthcoming, or should the now abandoned doctrine of original creations regain all it had lost in his experience."

Neither shocking nor loutish, Hooker's commentary commanded respect. Indeed, it was the self-effacing Hooker, rather than bombastic Huxley, who best tipped the scale that day.[32] As Lyell wrote to a friend a few days after the event, "The bishop had been much applauded in the section, but before it was over the crowded section (numbers could not get in) were quite turned the other way, especially by Hooker."[33] Even Huxley—who up until that day had never complimented Hooker to his face—praised the botanist for his "splendid" eloquence and cracked that "he did not know before what stuff [Hooker] was made of."

Hooker's July 2 letter to Darwin described every succulent second of his dismissal of the bishop:

> The excitement increased as others spoke—my blood boiled, I felt myself a dastard; now I saw my advantage—I swore to myself I would smite that Amalekite Sam hip & thigh if my heart jumped out of my mouth & I handed my name up to the President (Henslow) as ready to throw down the gauntlet. . . . It moreover became necessary for each speaker to mount the platform and so there I was cocked up with Sam at my right elbow, & there & then I smashed him amid rounds of applause—I hit him in the wind at the first shot in 10 words taken from his own ugly mouth—& then proceeded to demonstrate in as few more, 1) that he could never have read your book & 2) that he was absolutely ignorant of the rudiments of Bot[antical] Science—I said a few more on the subject of my own experience, & conversion & wound up with a very few observations on the relative position of the old & new hypotheses, & with some words of caution to the audience—Sam was shut up—had not one word to say in reply & the meeting *was dissolved forthwith* leaving you master of the field after 4 hours battle.[34]

ſ

THE AFTERNOON DAYLIGHT, too, was fast dissolving. As "the sacred dinner hour drew near," Professor Henslow "dismissed [the audience] with an impartial benediction."[35] Before the crowd dispersed into the evening, Charles Daubeny announced a gathering at his residence in the botanical garden for those interested in continuing the discussion.[36] Decades later, Isabel Sidgwick told Huxley's son Leonard that the pro-Darwinists were in the distinct minority that afternoon: "I never saw such a display of fierce party spirit, the looks of bitter hatred, which the audience bestowed—(I mean the majority) on us who were on your father's side—as we passed through the crowd we felt that we were expected to say 'how abominably the Bishop was treated'—or to be considered outcasts and detestable."[37] At Daubeny's soiree, Sidgwick recalled, "those who attended were decidedly in the Huxley-Hooker camp, where everyone was eager to congratulate the hero of the day. I remember that some naive person wished 'it could come over again;' and Mr. Huxley, with the look on his face of the victor who feels the cost of victory, put us aside saying, 'Once in a life-time is enough, if not too much.'"[38] That said (or not said), Huxley did gush to Frederick Dyster, ten weeks after the BAAS meeting, over how much he enjoyed his encounter with the bishop: "I believe I was the most popular man in Oxford for [a] full hour and twenty hours afterwards."[39]

Hooker would later finesse his July 2, 1860, testimony for Francis Darwin's 1887 biography of his father, under the guise of anonymity: "There was a crowded *conversazione* in the evening at the rooms of the hospitable and genial Dr. Daubeny, where the almost sole topic was the battle of the *Origin*, and I was much struck with the fair and unprejudiced way in which the black coats and white cravats of Oxford discussed the question, and the frankness with which they offered their congratulations to the winners in the combat."[40] It must be noted, however, that Hooker's ex post facto account came with a caveat: "It is impossible to be sure of what one heard, or of impressions formed, after nearly 30 years of active life," he said in a letter to Francis Darwin. "I do not like it altogether. I should like

Huxley to see it if you put it in print. Pray Anglicize it where necessary. . . . I have been driven wild formulating it from memory."[41]

For Huxley, as his son Leonard later regaled, the event became a permanent part of his origin story: "It was now that he first made himself known in popular estimation as a dangerous adversary in debate—a personal force in the world of science that could not be neglected. From this moment, he entered the front fighting line in the most exposed quarter in the field."[42] Thanks to Wilberforce's conduct unbecoming an English gentleman, Huxley shed years of hardship and obscurity as he donned the mantle of "Darwin's Bulldog." For the remainder of his career, Thomas Huxley was the public face of evolution—the perfect position for a scientist with a huge chip on his shoulder and the gift of gab, who relished new intellectual vistas and a good, ripping fight.

# 14

# The Dogs Bark but the Caravan Moves On

*We are all taught, and being taught, believe*
*That man, sprung from an Ape, is Ape at heart . . .*
*He bought white ties, and he bought dress suits,*
*He crammed his feet into bright tight boots—*
*And to start in life on a brand-new plan,*
*He christened himself Darwinian Man!*
*But it would not do,*
*The scheme fell through . . .*
*While a Man, however well behaved,*
*At best is only a monkey shaved!*

—*William S. Gilbert,* Princess Ida, *1884*[1]

The preceding narrative begs the question, Who was the victor at Oxford in 1860? That answer depends almost entirely on which experience, text, or source one chooses to believe. With the passage of time, however, opinion began to sway in Darwin's direction. A mere six months after the event, in December 1860, *Macmillan's Magazine* reported that Darwin's book excited more public curiosity than any other scientific treatise published during the nineteenth century: "It has for a time divided the scientific world into two great contending sections. A Darwinite and an anti-Darwinite are now the badges of opposed scientific parties. Each side is ably represented."[2] The magazine reiterated this opinion in 1888: "On the whole it seemed to be a drawn battle for both sides stuck to their

Gilbert and Sullivan's *Princess Ida* at the Savoy Theatre, January 1881.

guns."[3] A few years later, in 1892, William H. Freemantle, an Anglican priest and botanist, claimed that "those most capable of estimating the arguments of Darwin in detail saw their way to accept his conclusions."[4] Leonard Huxley, who quotes Freemantle in his 1900 biography of Thomas Huxley, tried to bridge the gap: "Instead of being crushed under ridicule, the new theories secured a hearing, all the wider, indeed, for the startling nature of their defense."[5] For Wilberforce's onward marching Christian soldiers, however, such declarations were premature and sacrilegious.

HUXLEY SERVED a dish of cold revenge to Owen at the 1862 BAAS meeting held at Cambridge University. Their dispute remained focused on the hippocampi and big toes of apes and men. Several Darwinians contradicted Owen's claims that afternoon. Perhaps most spectacular was the anatomist

William H. Flower, who announced, "I happen to have in my pocket a monkey's brain."[6] Responding to this astonishing claim, Huxley—now the president of the Zoology section—slammed down "the futility of discussions like the present." Instead, he insisted, "the Darwinian hypothesis must be worked out by patient inquiry and be either confirmed or confuted by investigations and facts, which could hardly at present be gone into. All the necessary facts had not yet been discovered, and, if discovered their significance could hardly be put clearly before a general audience."[7]

A week later, on October 9, however, Huxley wrote Darwin a far different account of the session: "All the people present who could judge saw that Owen was lying & shuffling: the other half saw he was getting the worst of it but regarded him I think, rather as an innocent old sheep, being worried by three particularly active young wolves.—He rolled his eyes about & smiled so sweetly every time the teeth set sharp into his weasand."[8] *Macmillan's Magazine* subsequently proclaimed that the 1862 meeting "was very different two years after when the hostile forces were again arrayed at Cambridge. Then the Anti-Darwinians were smitten along the whole line, and their rout was evident to all."[9]

Less discussed was how offensive the Huxley-Wilberforce confrontation proved for both sides of the aisle. On July 1, 1860, the poet and lawyer Arthur Munby wrung his hands over how "even the proprieties of the Association have been outraged." Munby walked to the Oxford University Church of St. Mary's that Sunday morning to hear the Reverend Frederick Temple preach on the harmony of science and revelation. Temple insisted that the two were not foes and that harmony was to be found "in a different direction; not in petty details of fact . . . but in the deep identity of tone, character, and spirit which pervade both the books."[10] Just as when Bishop Wilberforce gave a very different homily to the BAAS membership at the same church in 1847, the pews were packed with "dons and savans, for whom the sermon was intended." Munby found Temple's 1860 address to be "eloquent, learned, forcible—yielding much, but not preserving the remainder: confessing the horrid hiatus, and yet flinging across it his airy bridge—which fell, alas how short!"[11]

After church, Munby made his way to All Souls College to take tea in the rooms of England's future permanent Under-Secretary of State, Godfrey Lushington, along with physicist James Clerk Maxwell, the lawyer Richard B. Litchfield, and Godfrey's brother Vernon. At table, the bewildered Litchfield and Godfrey Lushington moaned a duet of woe: "And this is all! We go to hear a great divine, proving before these savans that science and revelation are *not* at war; and we come away misdoubting worse than ever . . . will all this do away with the facts of astronomy, geology, and Darwin?" And when it came to Huxley's preference of an ape for a father, Litchfield and Lushington cried out, in unison: "To such straits have we come! It's no use—defense is no longer possible—the controversy has been pushed to the last point; and that will soon be given up."

Munby bravely declared: "Let any number of insensate laws and necessary God-excluding developments be proved for nature." He believed that the separation of God's love from powerful laws of nature would only strengthen the "imagination" of a "Father in heaven, and a Christ, too." Nonetheless, he was shaken by the whole experience: "How saddening this sermon was and all our talk about it, and its subjects which recurred again & again all day! What grim laughter it provokes to see, here and at Cambridge, one's friends, fellows of colleges and the like, living in the midst of a system which they fret at & despise and think rotten to the core! Dear lovable men—wise and thoughtful, full of love and good works, and yet hopeless—in appearance, but not surely in reality, without God in the world."[12]

Alfred Newton was Cambridge University's leading ornithologist, the author of the highly regarded *Dictionary of Birds*, and a longtime proponent of Darwinian thought.[13] In 1865, Newton asked both Huxley and Darwin to support his promotion to the Chair of Comparative Zoology position that opened up at Cambridge. They both declined to support him

Alfred Newton.

on the grounds that they felt his work too narrow, too limited to birds alone, and yet they did so in such an elegant manner that Newton—who received the appointment despite the snub—never held it against them.[14] Newton saved his resentment for Richard Owen and the ways he lorded his power over others. The same year, Newton and Owen got into a feud over the recent discovery of fossilized remains of the dodo bird in Mauritius, where Alfred's brother Edward was the colonial administrator. Owen and Newton understood the importance of describing this rare, extinct creature. Owen predictably and "unscrupulously exploited his position as one of Alfred Newton's referees for the Cambridge professorship, intercepting the specimen intended for Newton and producing the first written description of the dodo."[15]

In 1888, Newton offered a critical postscript to the June 30 debate, which often goes ignored: "The principal discussion which took place on Saturday, June 30th, 1860, was adjourned until the following Monday. In the time which intervened, some arrangement was, I suppose, made by the leading men of the Association to let drop the matter, which had excited such strong feelings. At all events the discussion was not renewed: a wise termination, no doubt, but disappointing to a good many besides myself."[16] Newton was no fan of Wilberforce and was deeply offended by the bishop's "fatuous" and "false" comments on Darwin's work, especially when such "a distinguished man . . . could not even make sense of the 'brief' with which he has been furnished by a learned authority [Owen] who ought to have known better." Recalling an anonymous anti-Darwin essay in the summer 1860 issue of the *Quarterly Review*, he was flabbergasted to discover how much of its "phraseology" was identical to what Wilberforce had uttered a few days earlier—along with his "taunting but nonsensical" questions—during "the ever-memorable discussion of the meeting of the British Association at Oxford." Newton must have bristled as he wrote, "It is fortunate, for the reputation of some of the speakers, that no accurate report of that discussion seems to exist."[17]

Adam Storey Farrar best explained the bishop's breaches of public conduct, albeit thirty-five years after the event, in a letter to Huxley's son Leonard:

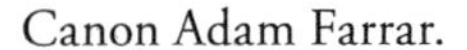

Canon Adam Farrar.

> Your father's reply showed there was vulgarity as well as folly in the bishop's words; and the impression distinctly was, that the Bishop's party, as they left the room, felt abashed, and recognized that the Bishop had forgotten to behave like a proper gentleman. *The victory of your father was not the ironical dexterity shown by him, but the fact that he got a victory in respect of manners and good breeding.* You must remember that the whole audience was made up of gentlemen, who were not prepared to endorse anything vulgar. The speech which really left its mark scientifically on the meeting was the short one of Hooker.[18]

ASIDE FROM THE lengthy accounts of the meeting in the local Oxford newspapers, most of the published reports in more distant publications were only a line or two embedded in longer accounts of the BAAS proceedings and without contextualizing the event as a debate between faith and science.[19] Several editorials that appeared in a handful of national papers represent a palette of mixed reviews on the meeting's impact. For example, the weekly literary review *Athenaeum*—which covered the event extensively—concluded that each side "found foemen worthy of their steel and made their charges and countercharges very much to their own satisfaction and the delight of their respective friends."[20] The radical London *Morning Star* chimed that "the new Darwinian theory, whatever may be its real merits in a scientific view, has no small number of supporters amongst the members of the Association."[21]

Taking a different tack, the London *Press*—which appealed to conservative, clerical, and upper-class readers—opined, "The men of science were the aggressors, or, if you will, the reformers, while the divines defend the bulwarks . . . in the debate the bishop brought all the well-known powers of his eloquence to substantiate the permanence of species. On the other hand, men eminent in science, falling under the direct influence of facts, probably, moreover, in some measure overwhelmed and confounded by the indefinite and all but infinite multiplication of species, gave at least a provisional and partial adhesion to the hypothesis of Dr. Darwin, which may possibly relieve them from an ever-increasing perplexity." The editorial closed by complimenting Oxford University for its hospitality: "to see, as it were, the greeting of ages medieval with times present; to mark how well it is possible for the Christian, the classic, and the scientific to co-operate in the one grand end—the advancement of man and the glory of GOD."[22]

*John Bull*—the weekly defender of British conservatism and the high Anglican church—defiantly spun the story to conclude: "The Darwinian theory of the origin of species was fully and ably discussed in the section of Zoology. Here Professors Huxley (and to some extent) Hooker were opposed by a powerful phalanx, led by the Bishop of Oxford, and composed of Sir B. Brodie, Professors Owen and Beale, Mr. Greswell, and Admiral FitzRoy. The impression left on the minds of those most competent to judge was that this celebrated theory had been built on very slight foundations, and that a series of plausible hypotheses had been skillfully manipulated into solid facts while a vast array of real facts on the opposite side had been ignored."[23]

The *Christian Remembrancer*, a quarterly review whose title encapsulates its editorial stance, regretted that "men who enjoy an audience cannot easily refrain from yielding to the sympathy which its presence begets in them." In phrases one imagines were repeated in sermons across the land, the columnist roared: "We only seek to sound a trumpet to all who are willing to man the walls, which the insidious plausibilities of modern science are threatening; for the surest ground of all defense of truth, whatever its place in the scale of faith, is to recall men's mind to the great ultimate verities of their moral responsibility, of God, the author of our Faith, the creator and pillar of all things."[24]

A socially progressive "weekly journal of politics, literature, music, and the fine arts, ecclesiastical, home, foreign and colonial news" called the *Guardian*—not to be confused with the daily newspaper *Manchester*

*Guardian*—was so offended by Huxley that it refused to print his name! Instead, the unnamed editorialist bellowed and misquoted, "When professors lose their tempers and solemnly avow they would rather be descended from apes than Bishops; and when pretentious sciolists seriously enunciate follies and platitudes of the most wonderful absurdity and draw upon their heads crushing refutations from the truly learned, there is mingled with our more serious feelings a sensation of amusement, which, despite of the proprieties, is apt to break out into a very decided cachinnation, if not into an absolute guffaw."[25]

ACROSS TOWN, on June 29, within the paneled recesses of the Cuddesdon Palace—the evening before his now famous battle—Soapy Sam dined in comfort and confidence with his guest, Richard Owen.[26] In their defense, neither men were fools even if they often proved to be not very nice. Time—as it eventually does to us all—passed them by, blinding their response to new knowledge. The inability to adapt and embrace the new science reduced these once prominent men into fossils, like the lumbering *Megalosaurus* in Dickens's *Bleak House*.

Wilberforce's unawareness of how badly he represented himself is recorded in his diary entry for June 30: "At Zoological [section, I was] called up by Henslow on Darwinian theory and spoke at some length in controversy with Huxley." So confident of his victory, he proceeded to record a lengthy description of riding his horse later that day.[27] On July 3, Wilberforce wrote Sir Charles Anderson with a flourish, "On Saturday Professor Henslow . . . called on me by name to address the Section on Darwin's theory. So, I could not escape and had quite a long fight with Huxley. I think I thoroughly beat him."[28] The following day, July 4, Balfour Stewart, the director of the Kew Observatory, distinguished meteorologist, devoted churchman, and follower of psychic phenomena, agreed with Wilberforce. Stewart wrote his former teacher, the physicist and natural philosopher James David Forbes, "The Bishop had the best of it."[29]

Fifty-five miles away, still water-curing in Surrey, Charles Darwin's "continuous bad headach[e] [sic] for 48 hours" finally abated after reading *his* friends' account of their victory. To Hooker, on July 2, he gushed, "I was low enough & thinking what a useless burthen I was to myself & all others, when your letter came & it has so cheered me. Your kindness & affection brought tears into my eyes. . . . How I should have liked to have wandered about Oxford with you, if I had been well enough; & how still more I should have liked to have heard you triumphing over the Bishop.— I am astounded at your success & audacity. It is something unintelligible to me how anyone can argue in public like orators do. I had no idea you had this power."[30]

The following day, on July 3, Darwin wrote Huxley how Hooker had informed him of "the awful battles which have raged about 'species' at Oxford. He tells me you fought nobly with Owen, (but I have heard no particulars) & that you answered the B. of O. [Battle of Oxford] capitally." Darwin's joy leaps off the page as he quipped, "I often think that my friends (& you far beyond others) have good cause to hate me, for having stirred up so much mud, & led them into so much odious trouble.—If I had been a friend of myself, I should have hated me. (how to make that sentence good English I know not.) But remember if I had not stirred up the mud someone else certainly soon would." Praising his friend's pluck, Darwin confessed he "would as soon have died as tried to answer the Bishop in such an assembly." Despite the Bishop's "ridicule" and "savage" behavior, Darwin predicted that Wilberforce would "soon get weary of subject & let us have some peace . . . though, on other hand, I do believe this row is best thing for subject."[31] Two days later, July 5, Darwin could not help but tease Huxley some more. "But how durst you attack a live Bishop in that fashion? I am quite ashamed of you! Have you no reverence for fine lawn sleeves?"[32]

As if he needed any more evidence of the trouncing, the Scottish biologist Hugh Falconer wrote Darwin on July 9 to describe how "the Saponaceous Bishop got basted and larded by Huxley. Owen also came in for such a set down by Huxley, as I have never witnessed within my experience of Scientific discussion. Your interests I assure you were most tenderly watched over by your devoted *Elèves*."[33]

On July 20, Darwin wrote Huxley again, definitively stating: "From all that I hear from several quarters, it seems that Oxford did the subject great good.—It is of enormous importance the showing the world that a few first-rate men are not afraid of expressing their opinion. I see daily more & more plainly that my unaided book would have done *absolutely* nothing."[34]

Wilberforce's odious performance led to many more jokes at his expense. At July's end, for example, Darwin's eldest son, William, told his father a story circulating around Cambridge—one that Darwin promptly passed on to Huxley. The anecdote centered on the economist Henry Fawcett, who, in 1858, lost his sight in a shooting accident. Upon exiting the Museum of Natural History, Fawcett and a colleague found themselves directly behind Bishop Wilberforce. The unnamed colleague asked Fawcett "whether he thought the Bishop had ever read the *Origin*." To which Fawcett replied, perhaps too loudly, "Oh no, I would swear he has never read a word of it." At this point, William wrote, "the Bishop bounced round with an awful scowl and was just going to pitch into him, when he saw that he was blind, and said nothing."[35]

Wilberforce was obliquely ridiculed in the House of Commons. The Scottish politician Mountstuart E. Grant Duff described how his colleague, Richard Monckton Milnes—the well-known champion of women's rights and unrequited suitor of Florence Nightingale—poetically mused from the floor: "Huxley asserted 'that the blood of guinea pigs crystallises in rhombohedrons.' Thereupon the Bishop sprang to his feet and declared that 'such notions led directly to Atheism!' "[36]

Thomas Huxley never could contain his animus for Samuel Wilberforce. In autumn 1860, he wrote to Charles Kingsley—the cleric, Christian socialist, Oxford professor, and author of the popular, pro-evolution, children's novel *Water-Babies: A Fairy Tale for a Land Baby*: "If that great and powerful instrument for good or evil, the Church of England, is to be saved from being shivered into fragments by the advancing tide of science—an event I should be very sorry to witness, but which will infallibly occur if men like Samuel of Oxford are to have the guidance of her destinies—it must be by the efforts of men who, like yourself, see your way to the combination of the practice of the Church with the spirit of science."[37]

In January 1861—only four months after losing his son Noel to scarlet fever—Huxley sent Wilberforce a copy of his recently published paper, "On the Zoological Relations of Man with the Lower Animals." Huxley's specific aim was to clarify the anatomical disagreements he and Richard Owen had gotten into "during the late session of the British Association at Oxford." The attached note—embossed with the crest of their gentleman's club, The Athenaeum, and bordered in black to reflect the mourning period for his son—announced: "Professor Huxley presents his compliments to the Lord Bishop of Oxford—Believing that his Lordship has as great an interest in the ascertainment of the truth as himself, Professor Huxley ventures to draw the attention to the Bishop to a paper in the accompanying number of the *Natural History Review*."[38] A few weeks later, Wilberforce tersely thanked the scientist for the article, adding in a courtly but doubtful manner that he "shall have great pleasure in studying it."[39] The bishop's chilly response was a far cry from the reaction of Huxley's students to the paper. After lecturing on the topic several weeks later, the zoologist boasted to his wife, "My working men stick by me wonderfully, the house being fuller than ever last night. By next Friday evening they will all be convinced that they are monkeys."[40]

Some have erroneously claimed that because Wilberforce and Huxley later served on some of the same public committees they shook hands and made things up like the good fellows they aspired to be. For both men, however, the rage of their mutual moment was never really extinguished. For example, on January 20, 1862, Huxley recounted to Darwin how Charles-Édouard Brown-Séquard, the famed neurophysiologist, "told me it was worthwhile to come all the way to Oxford [from the United States] to hear the bishop pummeled."[41] The following spring, Wilberforce crusaded to ban dangerous science books—including Huxley's 1863 book, *Man's Place in Nature*—that he felt threatened Christianity and the Church of England.[42] An April 10 editorial published in the conservative *Daily Telegraph* decried Wilberforce as "parochial" and a "humbug," who sought to preside over "a time when the works like those of Lyell, Darwin, and Colenso are torn from the hands of Mudie's shopmen, as if they were novels."[43] Decades later, Huxley found vindication from such doltish censorship by observing: "All the propositions laid down in the wicked book[s], which [were] so well anathematized a quarter of a century ago, are now taught in the text-books. What a droll world it is!"[44]

Huxley's cruelest swipe against Soapy Sam occurred thirteen years after verbally burying him at Oxford. On July 18, 1873, Wilberforce was killed in a freak accident. The bishop was "in the midst of the keen but tranquil enjoyment of a summer evening ride . . . rejoicing in the fine weather [and] keenly noticing the beauty of the scenery at every point of the way." His "characteristic love of trees" prompted him to slow down so that he might identify "the different kinds and the soils which suited them." His horse stumbled on "a slight dip in the smooth turf," threw Wilberforce onto his head, and rolled over him. The stallion survived but the bishop did not. "In that short interval of time," the *Manchester Guardian* lamented, "all has vanished, all things earthly, from that quick eye and that sensitive and sympathetic mind."[45]

Huxley first heard the news of Wilberforce's demise while traveling through the Auvergne-Rhône-Alpes region. Cranky from "a day or two's diarrhoea"—caused by a neighbor's drainpipes being "adrift close to the hotel"—he left France with his own pipes intact for the "fresh air" of Baden-Baden, Germany. After a long train ride and checking into his hotel suite, on July 30, Huxley wrote to John Tyndall, the physicist and professor of natural philosophy at the Royal Institution. The letter included a perfectly nasty epitaph for someone who helped make *his* name so famous:

*"Poor dear Sammy! His end has been all too tragic for his life. For once, reality and his brain came into contact and the result was fatal."*[46]

ſ

WILBERFORCE'S SON REGINALD, a barrister and squire of a 1,000-acre estate in Sussex, continued the poisonous feud by attempting to paper over the Oxford event. In 1879, Reginald took over the completion of his father's biography after the untimely death of its first volume's author, Arthur Ashwell. Reginald's "volume II" was published by John Murray in 1881. Therein, he blithely described the 1860 meeting, "where a discussion took place on the soundness or unsoundness of the Darwinian theory. The bishop, who . . . had just reviewed Mr. Darwin's work *On the Origin of Species by Means of Natural Selection*, made a long and eloquent speech condemning Mr. Darwin's theory as unphilosophical and as founded on fancy, and he denied that any one instance had been produced by Mr. Darwin which showed the alleged change from one species to another had ever taken place." Reginald then described how his father "made a great impression [and said] that whatever certain people might believe, he would not look at the monkeys in the Zoological as connected with his ancestors." In the next sentence, he referred to Huxley only as a "certain learned professor" who responded, "I would rather be descended from an ape than a bishop." Once told of what Reginald Wilberforce had written, Huxley—who always denied publicly uttering this ungentlemanly and blasphemous insult—demanded a correction. A year later, when volume III of the biography appeared, Reginald begrudgingly revised the misquote but buried it, again without including Huxley's name. The adjusted version appears in an eye-straining, agate font, under the heading, "Errata," among the pages of the book's front matter. It is safe to assume that it was ignored by a majority of those readers interested enough to plow through 1,464 pages of all things Samuel Wilberforce.[47]

In 1887, Reginald—again in the name of the father—sparred with Huxley. This round took place in one of the most public spaces in all Great

Britain—the "Letters to the Editor" section of the *Times*. The occasion was the publication of Francis Darwin's popular biography of his father, for which Huxley wrote a chapter on "the reception of the *Origin of Species*"—an essay that forever linked Charles Darwin's scientific stature to that "of Isaac Newton and Michael Faraday."[48] Specifically, Reginald contested Huxley's derision of Bishop Wilberforce's published review of *Origin*. Although Huxley referred to the author of the *Quarterly Review* essay as "foolish and unmannerly," he did *not* identify the bishop by name. Huxley did, however, fulminate that "the world has seen no such specimen of the insolence of a shallow pretender to a Master in Science as this remarkable production, in which one of the most exact of observers, most cautious of reasoners, and most candid of expositors of this or any other age is held up to scorn as a 'flighty' person . . . and whose 'mode of dealing with nature' is reprobated as 'utterly dishonorable to Natural Science.' "[49]

Inadvertently revealing his father as author of the *Quarterly Review* piece, Reginald plaintively asked "the reason why these dead ashes are fanned into flame? Why is Professor Huxley now to characterize Bishop Wilberforce as a 'shallow pretender' and a man with a 'want of intelligence'? . . . Is it the memory of the debate in the Sheldonian Theatre [sic] in 1860? Did the lash of Bishop Wilberforce's eloquence sting so sharply that, though 27 years have passed, the recollection of the castigation then received is as fresh as ever?"[50]

Three days later, on December 1, 1887, Huxley responded, con brio, "When Mr. Wilberforce succeeds in convincing reasonable men that a person who indulges in 'flighty anticipations' is not a 'flighty' person, I shall be happy to admit myself guilty of misquotation." He disingenuously pled ignorance over the true author of the *Quarterly Review* article because up until Reginald Wilberforce's letter to the editor, "there was no proof of the fact" and he had no intention of making a report that could well be a "baseless slander." All Huxley "knew" of the review's provenance was that it was written by "a bitter scientific antagonist of Mr. Darwin." Having put that issue to rest, he pointed out some more errors in Soapy Sam's 1860 review before going in for the kill. Decades after humiliating Wilberforce *Pere*, Huxley destroyed the son with one of his signature jabs: "Those who were present at the famous meeting in Oxford, to which Mr. Wilberforce refers, will doubtless agree with him that an effectual castigation was received by somebody. But I have too much respect for filial piety, however indiscreet its manifestations, to trouble you with evidence as to who was the agent and who the patient in that operation."[51]

EPILOGUE

# After Myth

Great is the power of steady misrepresentation; but the history of science shows that fortunately this power does not long endure.

—*Charles Darwin, 1872*[1]

In order to become a scientific fact, valid discoveries and ideas must be supported by the scrupulous development and testing of hypotheses, steady accretion of evidence, reproducibility of experimental results, and new forms of proof facilitated by better technology. It might be useful to imagine this process as a corps of scientists building a house of cards—each card representing a different laboratory, perspective, approach, and fact. In toto, they form a complicated but fragile structure that needs constant attention, revision, and, often, abandonment to build a plausible structure. Conversely, those seeking to smash such theories search for the one card they can pull out of the rickety edifice and, after the whole thing topples down, happily exclaim, "See? I told you it was all wrong!" These sophomoric attacks on the production of knowledge are aided by the impossibility of proving a negative and—with all due respect for Sir Isaac Newton's dictum of finding true causes for natural phenomena—how few people comprehend that the absence of evidence does not necessarily translate into evidence of absence. This process may seem obvious to many modern-day readers, but in Darwin's day, such evidentiary matters—indeed, the very substance of scientific proof—were just being proposed and developed. Indeed, one of Huxley's greatest scientific contri-

Charles Darwin, formal portrait.

butions was his eloquent insistence that Darwin's hypothesis be confirmed by many more "investigations and facts."[2]

Public debates over scientific theories—and the myths such occasions often yield—tend to be more captivating than the painstakingly slow work of the laboratory. "As a general rule," the *Christian Remembrancer* sermonized in 1860, these rhetorical battles "promise more than they perform. They draw out the antagonism of the question in a personal form by virtue of the presence and the living words of thinkers and speakers; unfortunately, it does not often happen that the thinkers speak, or the speakers think."[3] This observation helps explain why the most lasting aspects of the Oxford meeting were Wilberforce's sneering question and Huxley's devastating retort—even if those comments tend to be paraphrased rather than quoted literatim.

ʃ

THE VICTORIOUS, pro-Darwinian narratives of the 1860 BAAS meeting did not really begin to take a firm hold of the public's imagination until after the late 1880s. This was due to two important literary events. In 1887, Francis Darwin published an authorized, authoritative biography of his father—a three-volume, blockbuster of a book that went through several editions and printings. Thirteen years later, in 1900, the evolution of evolution's history was expanded by Leonard Huxley's hagiographic and best-selling two-volume biography of his pater.[4] These books, as important and valuable as they are, had an obvious mission. They set their fathers' admirable careers—including their roles in "the Battle of Oxford"—firmly within the frames of awareness and admiration.[5]

It must be difficult being the child of a famous personage. Entitled and expecting to assume their parents' crowns, the offspring receive all types of favors and privileges—simply because of their surnames. Like the moon, light is reflected *on* them rather than *from* them. Even exemplary ones—such as the distinguished Darwin and Huxley children and grandchildren—grew up in the shadow of their parents and are constantly measured by the yardstick of their parents' achievements. Genius and intelligence shine at far different wavelengths, and the former is rarely acquired or inherited. Too often, the children of the famous are spoiled second-raters who must protect the family brand and maintain their waning pseudo-royalty.

During the Victorian and Edwardian eras, the best way to perpetuate and burnish fame across generations was to hoard a successful parent's papers, delay making them accessible to historians, and produce reverential volumes of their "Life and Letters." This is exactly how things played out with the industrious Darwin and Huxley clans. Less popularly selling but nevertheless family literary affairs included a biography of Emma Darwin by her daughter, Henrietta Darwin Litchfield; Reginald Wilberforce's biography of his father, Samuel; and a book on the life and work of Joseph Hooker compiled by Hooker's second wife, Lady Hyacinth Jardine Symonds Hooker, and written by Leonard Huxley. Charles Lyell had no children, but his sister-in-law, Katherine Murray Lyell—the eldest daughter of Scottish social reformer Leonard Horner—produced a biography of the geologist. Richard Owen's grandson (also named Richard) scribed a life of Darwin's archenemy, which featured a tame and respectful summation on Owen's "position in the history of anatomical science," written by none other than the Right Honorable Thomas H. Huxley![6] Except for the Thomas Huxley biography by his son Leonard, these books were all published by John Murray, who appears to have made money from almost every side of the evolution debate.[7] Adding to these slanted histories, a brigade of scientists, historians, novelists, playwrights, screenwriters, and journalists have produced carloads of books and articles, with no end in sight.

ʃ

WITH EACH SUCCESSIVE AMPLIFICATION, the "Battle of Oxford" has informed and misinformed. The nuances, details, and far from perfect memories have been repeated and burnished by cadres of scientific victors. Legends flourish because they feature well-defined heroes and antiheroes who follow narrative arcs that are easy to understand and repeat—especially when compared with complex cultural trends and obtuse scientific theories.

Contained within the simplicity of such tales—be they drawn from the parables of Scriptures, the lecture room, or those created by modern-day fabulists—lies a dangerous morass. Merely reducing the Oxford saga into a demi-glace of apes, ancestry, and snappy quips creates a savory version of history that we follow at our own peril. That should not mean we cast a blind eye on these tales, either. Like it or not, such myths do exist and flourish! The widespread mythology of the events we have discussed here is important precisely because it has been consumed and repeated by so many millions of scientists and biology watchers.

Upon dissecting, analyzing, and comparing the extant texts of the day, we discover that the protagonists and antagonists were men with unquenchable thirsts for knowledge, ambitious professional agendas, competitive instincts, and distinctive personalities. Some revealed unremitting enmity or distrust for those trespassing on their expertise (Owen, Wilberforce, Huxley, and, to some extent, Darwin and Hooker); poor health, writer's block, and literary procrastination (Darwin); the devastation of losing a young child or spouse (Darwin, Huxley, and Wilberforce); and an allegiance to theological doctrine (Wilberforce and Owen). They all desired their ideas be heard in formal adjudicative proceedings and illustrated the English adage "Great men's faults are never small."[8]

Scientists, physicians, and professors of the twentieth and twenty-first centuries have willingly (and often) employed the Darwin mythology to advance their research agendas and budgets—long before their work produced anything of social or scientific value. The triumphant accounts did succeed in helping scientists to oversell the value, expense, and speed of modern biological discovery—just as medical doctors of the same period often overhyped their ability to prevent, heal, and cure disease to raise money for research, medical schools, and teaching hospitals—long before such miracles were possible.[9]

Dramatists and novelists, too, have elevated scientists as forward-thinking heroes of integrity, often at the expense and contrast of pious fops stuck in the muck and mire of the past. Few evolutionary fictions are more famous than Jerome Lawrence and Robert E. Lee's 1955 Broadway play and 1960 Hollywood film *Inherit the Wind*, which depicts the Scopes Monkey Trial of 1925.[10] This courtroom drama features an eloquent, combative, and crusading atheistic attorney—based on Clarence Darrow, the famed "attorney for the damned"—who defends a schoolteacher on trial for teaching evolution in a Tennessee high school, which was then against state law. That teacher—based on John Scopes—could not speak on his own behalf because of court procedure rather than lactose intolerance, but, like Darwin, he needed a bulldog to defend him. The play's antagonist was a gluttonous, Christian prosecuting attorney—based on the perennial presidential candidate, populist, and former US congressman, William Jennings Bryan.

The resonances of these characters to Huxley and Wilberforce are loud and clear. The playwrights were well versed on the lives and writings of Darwin and Huxley. Darrow and Bryan, too, referred to these texts in their courtroom presentations.[11] In real life, and on stage, Darrow loses his case—essentially because Scopes broke a stupid law—but the play's ending made it clear that science and evolution would supersede faith in the long game of history. Even the play's title is, ironically, taken from Proverbs 11:29: "He that troubleth his own house shall inherit the wind: and the fool shall be servant to the wise in heart." Instead of the Bishop of Oxford ceding the high ground with a vulgar comment about Huxley's ancestry, Darrow riles "the Great Commoner" into a peak of confusion by grilling him on biblical history. On the witness stand, Bryan loses both his composure and dignity. Shortly after testifying, he melodramatically dies of a "busted belly," elevating him to a sort of of martyrdom. Rife with historical inaccuracies and fictional flourishes, *Inherit the Wind* remains for many a popular and dramatic introduction to the ongoing battle between evolution and religion.

In the century since the Scopes trial, edition after edition of biology textbooks have recounted both it and, far more frequently, the triumphant tale of Huxley at Oxford as vehicles for teaching evolution to impressionable readers.[12] Prominent museum exhibits, too—including those mounted at the Oxford Museum of Natural History and the American Museum of Natural History—have helped keep the Oxford debate alive.[13] The Huxley-Wilberforce exchange even caught the attention of the once ubiquitous National Public Radio personality Garrison Keillor, who, with the benefit of 20/20 hindsight, breezily told his millions of listeners: "In the end, though each side claimed victory, most accounts chalk it up as a win for the Darwinians."[14] As a character in John Ford's 1962 film *The Man Who Shot Liberty Valance* famously articulated, the legend has long since been printed, and we must contend with that tangled history as well as the historical events that may or may not have actually happened.[15]

At the risk of being accused of advancing a Whig form of history, one cannot ignore the sweeping progress modern-day scientists have made in demonstrating Darwin's theories on natural selection and the descent of man.[16] A recent example of this success was the 2022 Nobel Prize in Physiology or Medicine, which was awarded to Svante Pääbo for his paleogenomic "discoveries concerning the genomes of extinct hominins and human evolution." Pääbo's work allows for the comparative genetic analysis of humans to their closest known evolutionary relatives, the Neanderthals—

who lived more than eight hundred thousand years ago![17] Such scientific progress makes it all too easy for casual observers to claim victory for Darwin, Huxley, and Hooker on their big day in Oxford. It is easier still to gloss over the massive amount of time, labor, fieldwork, discovery, and elucidation of the fossil record, not to mention new laboratory technologies, it took to corroborate the ideas first elaborated in the extraordinary mind of a solitary, Victorian gentleman.

ʃ

JUST BEFORE the seventy-three-year-old Darwin took his final breath on April 19, 1882, he told Emma, "I am not the least afraid of death—Remember what a good wife you have been to me—Tell all my children to remember how good they have been to me."[18] His loving last words were a far cry from a vicious rumor spread by a Toronto clergyman claiming that on his deathbed Darwin "whined for a minister and renouncing Evolution, sought safety in the blood of the Savior."[19]

During her long reign, Queen Victoria knighted many men, but Darwin never made her honors list. Darwin's funeral on April 26 was attended by thousands of influential and important admirers. Yet even his final resting

Darwin's funeral.

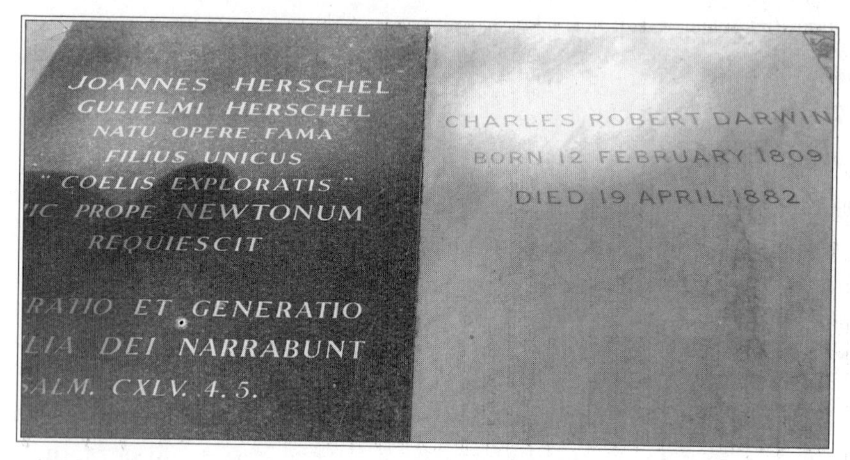


The Westminster Abbey graves of Darwin and Herschel.

place, in the north aisle of the nave of Westminster Abbey—a few stones from Sir Isaac Newton and right next to Sir John Herschel—required a major lobbying effort by his scientific peers.[20] Despite their heated disagreement over Darwin's theories (Herschel derided natural selection as "the law of higgledy-pigglety"), there is historical justice in having Herschel—who coined the phrase "that mystery of mysteries"—spending eternity so physically close to the man who solved it.[21]

In the decades that immediately followed, a generation of biologists and theorists nearly drove their precious cargo off the rails. As the nineteenth century ended, many people of privilege worried about the future of their gene pool, the fall of brutal imperialism, massive human migration, increasingly diverse populations slow to assimilate, and expanding urban centers ill-equipped to master these problems. Under the twin spells of arrogance and certitude, some tried to apply Darwin's ideas to restoring the fractured social sphere in their own image.

It was Herbert Spencer who coined the axiom "survival of the fittest" as shorthand for a notion he called "social Darwinism." His speculative but wildly popular doctrine posited that human society was subject to the same laws of natural selection as the plants and animals, with "the fittest" referring to men and women who looked just like him.[22] Repurposing Dar-

Herbert Spencer, circa 1860s.

win's observations on pigeons and plants to social reform was hopelessly misplaced and often did more harm than good—especially for the most marginalized members of a community. On July 5, 1866, after reading part of Spencer's *Principles of Biology*, Darwin wrote Alfred Wallace, "I suppose you have read the last number of H. Spencer; I have been struck with astonishment at the prodigality of Original thought in it; but how unfortunate it is that it seems scarcely ever possible to discriminate between the direct effect of external influences & 'the survival of the fittest.'"[23] Nearly three decades later, in 1893, Huxley, too, voiced his concerns about social Darwinism's catchphrase: "'Fittest' has a connotation of 'best' and about 'best' there hangs a moral flavor."[24]

Far more damaging to Darwin's legacy was the work of his half-cousin, Francis Galton. In 1869, Galton insisted that brilliance was an inherited trait and included the illustrious Darwin family as an example of his thesis.[25] When writing *The Descent of Man* (1871), Darwin enthusiastically endorsed his cousin's claims that "special tastes and habits, general intelli-

Francis Galton.

gence, courage, bad and good temper, &c., are certainly transmitted," just as "insanity and deteriorated mental powers likewise run in the same families."[26] Darwin also supported Galton's assertion, "If the prudent avoid marriage and the reckless marry, the inferior members tend to supplant the better members of society . . . and the most able should not be prevented by laws or customs from succeeding best and rearing the largest number of offspring."[27]

Galton's pseudoscientific framework of heredity, which he published in 1883—one year after Darwin's death—was perfectly timed for an era stunted by fear, patriarchy, racism, and social Darwinism. Galton apishly imitated his cousin by inventing a phrase to characterize his theory: *eugenics*, from the Greek root, *εὐγενής* or *eugenes*, to mean "good in stock or hereditarily endowed with noble qualities." Galton proposed to improve the public's health and welfare by "giv[ing] to the more suitable races . . . a better chance of prevailing speedily over the less suitable."[28]

Galton's theories soon spread like wildfire throughout Great Britain,

Europe, and North America. Eugenics offered those in positions of power an authoritative, scientific language to substantiate their biases against those they deemed dangerous. The solution was to quarantine, cordon off, and prevent "undesirables," or the unfit, from contaminating, or replacing, the "superior," white, native-born citizens. Such faulty conclusions extended into twentieth-century immigration restriction laws against the entry of so-called unassimilable people, forced sterilization of the mentally and physically disabled, harsh public health regulations, anti-miscegenation laws, and, in Nazi Germany especially, monstrous genocide. Misapplying science at the expense of those deemed inferior, powerful people traded on the currency the world told them was theirs to spend.[29]

Like most men of his time and station, Darwin bandied about his share of racism. His death in 1882, however, effectively halted his complicity in the rise of social Darwinism and eugenics; nor should he be blamed too harshly for those who misapplied his ideas into their overtly racist theories. Nonetheless, we must acknowledge the throughline that extends from Darwin's *Origin* and *The Descent of Man* to Spencer's social Darwinism and Galton's eugenics.[30] Darwin's theory discussed the struggle for life embodied by evolving species in the natural world. The latter two theories were characteristic of a human-made world where those in power abused science to justify and protect their positions of entitlement.[31]

Tossing Darwinian theory into the dustbin of the eugenics movement would be tantamount to an enforced shipwreck of the HMS *Beagle* before it returned to England in 1836. As Darwin wrote in *The Descent of Man*, "False facts are highly injurious to the progress of science, for they often long endure; but false views, if supported by some evidence, do little harm, as everyone takes a salutary pleasure in proving their falseness; and when this is done, one path towards error is closed and the road to truth is often at the same time opened."[32]

What we ought to learn from this knotty history is that there are almost always unintended consequences to an idea, theory, or doctrine—whether generated by scientists, philosophers, theologians, or a higher authority.

It is easy to ridicule ideas of the past; it remains just as easy to spin science and social policies in the present. The controversy over Darwinian theory—and its subsequent uses and abuses—reminds us that we must be ever on guard for the perversion of science or, for that matter, religious faith. Our collective knowledge of the world is always evolving. In 2008, nearly 150 years after *Origin* was first published, for example, a spokesman for the Church of England wrote an "apology of sorts" to Darwin "for misunderstanding you and, by getting our first reaction wrong, encouraging others to misunderstand you still."[33] What we choose to believe today, we may well reject tomorrow.

ʃ

MANY ARE AMAZED to learn that the word *evolve* appears only once in *Origin of Species*. It is the last word on the last page of the book. Darwin much preferred the term *descent with modification* to describe evolution. A fuller appreciation of his concepts requires extensive work in the laboratory, in the scholarly cocoons of archives, and better still, in nature. Understanding the quirky author of *Origins*, however, demands a pilgrimage to his home in what is now called Downe village. A little more than fourteen miles southeast of London's bustle and grime, past St. Mary's Church, and up a hill is Darwin's thinking path, the sand-walk, his experimental garden, and, finally, Down House.[34]

The trek's value becomes clear upon entering the maestro's study—one of the most productive spaces in the history of scientific inquiry. Darwin's rectangular Pembroke writing table is covered with papers, letters, and reports. Off to one side, at the ready, are his microscope and dissecting tools. Filling various nooks and crannies about the room is a parade of biological specimens. Complimenting the ensemble is an overstuffed armchair; a drum table loaded with vials of chemicals, staining dyes, and reagents; a fireplace framed by a carved, marble mantlepiece; and, tucked in the corner behind a folding screen, the metal wash basin Darwin used for his water-douche therapy. On the walls are portraits of the scientists and family members he most admired and loved. Lined up like soldiers on the groaning shelves of his bookcases are scientific tomes, manuscripts, and notebooks. Elsewhere were biographies, memoirs, and a slew of novels by Dickens, Cervantes, Eliot, Trollope, and Austen. Darwin's breadth of reading should not surprise anyone who has profited from his felicitous sentences. As with most distinguished authors, he learned how to write by reading other gifted writers.

Standing in Darwin's cluttered sanctum sanctorum—more than a century and a half after *Origin* became a scientific best seller—one can almost imagine him scribbling out the soaring language of his book's closing paragraphs. In these lines, he noted how "authors of the highest eminence seem to be fully satisfied with the view that each species had been independently created." The restless Darwin sought a different path. He dared to apply "what we know of the laws impressed on matter by the Creator." Darwin viewed "all beings not as special creations but as lineal descendants of some few beings which lived" eons ago, gradually progressing, by means of natural selection, "towards perfection." Charles Darwin's intellectual and global journey allowed him to define nature as "a tangled bank, clothed with many plants of many kinds, with birds singing on the bushes, with various insects flitting about, and with worms crawling, through the damp earth." He changed our understanding of the interactions of all living beings by describing how "these elaborately constructed forms, so different from one another, and dependent on each other in so complex a manner, have all been produced by laws acting around us."[35] We have all benefited from Darwin's perspectives—far from perfect, let alone a creed or eternal text—on the interconnectedness of living beings, botany, geology, the microbial world, the ecological environment, and climate, whether we acknowledge it or not.

Crowning Darwin's explanation of the "war of nature" is one of the most consequential sentences in the entire canon of science, if not Western literature. Darwin wrote these lines while in the thick of several personal wars. That he overcame such existential battles makes the final sentence of *Origin* nothing short of glorious:

> There is grandeur in this view of life, with its several powers, having been originally breathed into a few forms or into one; and that, whilst this planet has gone cycling on according to the fixed law of gravity, from so simple a beginning endless forms most beautiful and most wonderful have been, and are being, evolved.[36]

Today, an appreciation for Darwin's "tangled bank" seems more urgent than ever, given the collective harm we have committed against so many of our planet's inhabitants, the environment, and its climate.[37] After all, the word *evolved* is derived from the Latin root *evolutio*, which describes the unrolling of a long scroll or ancient book, the telling of a complex story without end. No mere afternoon of scientific battle, *evolutio* remains the perfect descriptor for Darwin's—and our—quest to understand nature's "endless forms, most beautiful and most wonderful." Hopefully, we will learn to pay closer attention to the sickly squire's words and begin repairing the injuries we have exacted on the world we all share.

# ACKNOWLEDGMENTS

During the Lent, Easter, and Michaelmas terms of 2022, I was a Visiting Fellow at Clare Hall, Cambridge University. There, I was privileged to reside a few blocks from the Cambridge University Library and its rich Charles Darwin collection. The library's staff, members of the History and Philosophy of Science Department, and the Cambridge University Press have heroically archived, edited, published, and digitized more than fifteen thousand letters written from and to Darwin during his life, as well as his collection of manuscripts, papers, and articles. This monumental effort makes Darwin's work accessible to everyone who seeks to study him. In this book, I quote extensively from this collection, with the permission of the Syndics of the University of Cambridge. Quotations from the Thomas H. Huxley papers are through the kind permission of the Archives of Imperial College of Science, Technology and Medicine, London; and quotes from the Samuel Wilberforce and British Association for the Advancement of Science papers are courtesy of the Bodleian Library Special Collections at Oxford University. I am also indebted to the Darwin scholars I cite in this book; they, and many others, enriched my reading of Darwin.

As with any book, there are many people to thank. I especially benefited from the invaluable and generous help of Professor James Secord (Professor Emeritus of History and Philosophy of Science, Director of the Cambridge University Darwin Correspondence Project, and Fellow, Christ's College, Cambridge); Frank Bowles and Katrina Dean (Keepers of Manuscripts, the Charles Darwin Collection at Cambridge University Library); Samuel Sales, Lucy McCann, and Oliver House (Archivists of the Bodleian Library Special Collections at Oxford University); Anne Barnett and Lucy Shepherd (Curators of the Imperial College of Science, Technology and Medicine, London Archives); Stewart Gilles (Reference Team Leader, British Library News Reference Service); Matthew Barton (Archivist, University of Oxford Museum of Natural History); Dr. J. Alexander Navarro and Heidi Muller (University of Michigan Center for the History of Medicine); and the staffs of the British Museum of Natural History, the British National Portrait Gallery, the British Museum, the National Library of Scotland, the National Library of New Zealand, and the University of Michigan Libraries.

I am also grateful to Professor Catherine DeAngelis and the late Professor James Harris (Department of Pediatrics, the Johns Hopkins University and Hospital); Professor Michael Schoenfeldt (Department of English Literature and Language, the University of Michigan, and Life Fellow, Clare Hall, Cambridge University); Sir Richard Evans (Regius Professor Emeritus of Modern History, Honorary Fellow, Gonville and Caius College, Cambridge); Professor Gregory and Siobhan Toner (Department of Celtic Studies, Queen's University, Belfast, and Life Fellow, Clare Hall, Cambridge University); Professor Geneviève Cartier (Faculty of Law, Université de Sherbrooke, and Life Fellow, Clare Hall, Cambridge University); Professor Marte Spangen (Departments of Archeology, History and Religious Studies, University of Tromsø, and Life Fellow, Clare Hall, Cambridge University); Rabbi Gesa Ederberg, of the Neue Synogogue in Berlin, and Dr. Martin Cetron, former Director, Global Migration and Quarantine, US Centers for Disease Control and Prevention.

I am indebted to my literary agents, Glen Hartley and Lynn Chu, of Writer's Representatives, Inc., who have long championed and encouraged my work; and to my wonderfully astute editor, John Glusman, Editor-in-Chief and Vice President of W. W. Norton and Company, assistant editor Helen Thomides, and copy editor Janet Greenblatt.

On the home front, I thank Sheldon and Geraldine Markel and Samantha Markel for their unflagging love and support. My daughter Samantha, especially, encouraged me to go on my Cambridge adventure in the age of COVID.

# NOTES

## Preface

1. Among the best sources describing Charles Darwin's life, aside from the thirty-volume *Correspondence of Charles Darwin*, are: Janet Browne, *Charles Darwin: Voyaging. Volume I of a Biography* (London: Jonathan Cape; New York: Knopf, 1995); Janet Browne, *Charles Darwin: The Power of Place. Volume II of a Biography* (London: Jonathan Cape; New York: Knopf, 2002) as well as Adrian Desmond and James Moore, *Darwin: The Life of a Tormented Evolutionist* (New York and London: W. W. Norton, 1991; Francis Darwin, ed. *The Life and Letters of Charles Darwin, in Two Volumes* (London: John Murray, 1887; New York: D. Appleton and Co., 1889); and Jacques Barzun. *Darwin, Marx, and Wagner: Critique of a Heritage* (New York: Doubleday Anchor Books, 1958). Another invaluable source of Darwin publications, biographies, and manuscripts is Darwin Online (Darwin-online.org.uk). Throughout this book, I have attempted to use the correspondence as the first choice of reference, but I have benefited greatly by the Browne, Desmond and Moore, and Francis Darwin biographies as well as the other historians, whose important work I cite throughout this book.
2. Charles Darwin to Charles Lyell. June 25, 1860. (Classmark: American Philosophical Society (Mss.B. D25.220)/ Letter # DCP-LETT-2843). In Frederick Burkhardt et al., eds., *The Correspondence of Charles Darwin*, Vol. 8 (Cambridge, UK: Cambridge University Press, in 30 volumes, 1985–2023), 265–266.
3. David N. Livingstone, "Myth 17. That Huxley Defeated Wilberforce in Their Debate over Evolution and Religion," in *Galileo Goes to Jail and Other Myths about Science and Religion*, ed. Ronald L. Numbers (Cambridge, MA: Harvard University Press, 2009), 152–60.
4. Howard Markel, *The Secret of Life: Rosalind Franklin, James Watson, Francis Crick, and the Discovery of DNA's Double Helix* (New York: W.W. Norton, 2021), 6.

5. Kostas Kampouakis, *Understanding Evolution*, 2nd ed. (Cambridge, UK: Cambridge University Press, 2020), 87–89.
6. Leonard Huxley, *Life and Letters of Thomas H. Huxley*, Vol. I, 189.

## Introduction: A Temple of Science

1. Oscar Wilde, "The Soul of Man under Socialism," *Fortnightly Review* 49, no. 290 (February 1891): 292–319, quote is from p. 292.
2. Vyvyan Holland, *Son of Oscar Wilde* (London: Penguin Books, 1957), from the Preface.
3. Charles Darwin, *On the Origin of Species by Means of Natural Selection, or the Preservation of Favoured Races in the Struggle for Life* (London: John Murray, 1859). The phrase "that mystery of mysteries" was coined by the Cambridge scientist, natural philosopher, and astronomer Sir John Herschel in a letter he wrote to Charles Lyell on February 20, 1836. See W. F. Cannon, "The Impact of Uniformitarianism: Two Letters from John Herschel to Charles Lyell, 1836–1837," *Proceedings of the American Philosophical Society* 105, no.3 (1961): 301–314. The full quote is "Of course I allude to that mystery of mysteries the replacement of extinct species by others. Many will doubtless think your speculations too bold—but it is as well to face the difficulty at once."
4. John Holmes and Paul Smith, "Visions of Nature: Reviving Ruskin's Legacy at the Oxford University Museum," *Journal of Art Historiography*, no. 22 (June 2020).
5. Charles Daubeny, "Address Delivered as President of the British Association for the Advancement of Science, to the Meeting at Cheltenham, 1856," in *Report of the 26th Meeting of the British Association for the Advancement of Science, Held at Cheltenham in August 1856* (London: John Murray, 1857), lxxi; "Report of Societies," *British Association Medical Journal* 4, no. 190 (1856): 727–729; see also Janine Rogers and John Holmes, "Monkey Business: The Victorian Natural History Museum, Evolution, and the Medieval Manuscript," *Romanticism on the Net*, no. 70 (2018): 1–27.
6. Carla Yanni, *Nature's Museums: Victorian Science and the Architecture of Display* (New York: Princeton Architectural Press, 2005), 62–90.
7. This concept, known as classical cell theory, helped usher in the biological revolution. It proscribed that all organisms are composed of cells—the basic organizational unit of life—and that all new cells come from preexisting cells. The first portion of this concept was espoused by botanist Matthias Schleiden and zoologist Theodor Schwann in 1838. In 1858, Rudolf Virchow elucidated the latter point of the theory: "*Omnis cellula e cellula* (all cells come from cells)." John Holmes, *Temple of Science: The Pre-Raphaelites and Oxford University Museum of Natural History* (Oxford, UK: Bodleian Library and the Oxford University Museum of Natural History, 2020), 62.
8. Darwin, *On the Origin of Species*, 1st ed., 184.
9. Holmes, *Temple of Science*, 146; Henry W. Acland, *The Unveiling of the Stature of Sydenham in the Oxford Museum, August 9, 1894, by the Marquess of Salisbury, with an Address by Sir Henry W. Acland* (Oxford, UK: Horace Hart, Printer to the University, 1894); Henry W. Acland and John Phillips, *The Oxford Museum*, 4th ed. (Oxford, UK: James Parker and Co., 1967).

10. *Report of the Thirtieth Meeting of the British Association for the Advancement of Science; Held at Oxford in June and July 1860* (London: John Murray, 1861), 115–116.
11. Adrian Desmond and James Moore, *Darwin: The Life of a Tormented Evolutionist* (New York: W. W. Norton, 1991), 451.
12. Richard Owen, "On the Characters, Principles of Division, and Primary Groups of the Class Mammalia," *Zoological Journal of the Linnean Society* 2, no. 5 (1857): 1–37. See pp. 14 and 20.
13. Desmond and Moore, *Darwin: The Life of a Tormented Evolutionist*, 451–453; Charles Darwin to Joseph D. Hooker, July 5, 1857 (Cambridge University Library [hereafter, CUL] Classmark: DAR 114: 203/Letter # DCP-LETT-2177). In Frederick Burkhardt et al., eds., *The Correspondence of Charles Darwin*, Vol. 6 (Cambridge, UK: Cambridge University Press, in 30 volumes, 1985–2023), 419–420.
14. "The British Association at Oxford," *Guardian*, July 4,1860, 593. The Mr. James is George Payne Rainsford James, 1799–1860, whose best-known work was *Richelieu: A Tale of France* (1829). Cumole is the dated chemical term for the compound $C_{18}H_{12}$,—a volatile hydrocarbon derived from cuminic acid. See C. M. Warren, "Researches on the Volatile Hydrocarbons," *Memoirs of the American Academy of Arts and Sciences* 9, no. 1 (1867): 135–176.
15. Obituary of John Richard Green, *Nature* 27, no. 462 (1883).
16. Leslie Stephen, ed., *Letters of John Richard Green* (New York and London: Macmillan, 1901), 44.
17. William Paley, *Natural Theology; Or Evidences of the Existence and Attributes of the Deity from the Appearances of Nature* (London: R. Faulder, 1802).
18. Charles Darwin to John Lubbock, November 22, 1859. (CUL. Classmark: DAR 263: 17 (EH 88206466)/Letter #: DCP-LETT-2532). In Burkhardt et al., *Correspondence*, Vol. 7, 388.
19. Charles Lyell (edited by his sister-in-law, Mrs. Katherine M. Lyell), *Life, Letters and Journals of Sir Charles Lyell, Bart*, Vol. 2 (London: John Murray, 1881) 335.
20. Charles Darwin to Thomas H. Huxley, July 3, 1860. (CUL. Classmark: Imperial College of Science, Technology and Medicine, Archives (Huxley 5: 121)/ Letter #: DCP-LETT-2854); In Burkhardt et al., *Correspondence*, Vol. 8, 277; David N. Livingstone, "Myth 17: That Huxley Defeated Wilberforce in Their Debate over Evolution and Religion," in Ronald L. Numbers, ed., *Galileo Goes to Jail and Other Myths about Science and Religion* (Cambridge, MA: Harvard University Press, 2009), 152–160; see also Howard Markel, "How Galileo's Groundbreaking Works Got Banned," *PBS NewsHour*, February 15, 2022.
21. James C. Ungureanu, "A Yankee at Oxford: John William Draper at the British Association for the Advancement of Science at Oxford, 30 June 1860," *Notes and Records of the Royal Society of London* 70 (2015): 135–150.

## Part I: Down

1. Charles Darwin to Joseph D. Hooker, June 2, 1857. (CUL. Classmark: DAR 114:119/ Letter # DCP-LETT-2099). In Frederick Burkhardt et al., eds., *The Correspondence*

*of Charles Darwin*, Vol. 6 (Cambridge, UK: Cambridge University Press, in 30 volumes, 1985–2023), 403–4.

## 1. The Letter

1. Charles Darwin to A. R. Wallace, December 22, 1857. (CUL. Classmark: The British Library-Add MS 46434/Letter # DCP-LETT-2192). In Frederick Burkhardt et al., eds., *The Correspondence of Charles Darwin*, Vol. 6 (Cambridge, UK: Cambridge University Press, in 30 volumes, 1985–2023), 514–15.
2. Charles Darwin to W. D. Fox, February 8, 1857. (CUL. Classmark: Christ's College Library, Cambridge. MS 53 Fox 110/Letter # DCP-LETT-2049). In Burkhardt et al., *Correspondence*, Vol. 6, 334–36.
3. William Lewins, *Her Majesty's Mails* (London: Sampson Low, Son, and Marston, 1865), 278.
4. The suburbs of London were kept to six deliveries per day at this time. Rowland Hill, "On the Post-Office," *Notices of the Proceedings of the Royal Institution of Great Britain* 3 (1862): 457–466.
5. This figure is for 1859. J. C. Hemmeon, *The History of the British Post Office* (Cambridge, MA: Harvard University Press, 1912), Harvard Economic Studies, Vol. 7, 71.
6. Janet Browne, *Charles Darwin: The Power of Place. Volume II of a Biography* (London: Jonathan Cape; New York: Knopf, 2002), 13; Duncan Campbell-Smith, *Masters of the Post: The Authorized History of the Royal Mail* (London: Allen Lane, 2011), 113–169.
7. *Darwin in Conversation: The Endlessly Curious Life and Letters of Charles Darwin. An Exhibition Guide* (Cambridge, UK: Cambridge University Library, 2022).
8. Charles Darwin to Asa Gray, January 22, 1862. (CUL. Classmark: Gray Herbarium Harvard University, [74]/Letter # DCP-LETT-3404). In Burkhardt et al., *Correspondence*, Vol. 10, 40–42.
9. Alfred Russel Wallace, *My Life: A Record of Events and Opinions*, Vol. 2 (London: Chapman and Hall, 1905), 1–22, 23–50.
10. See Alfred R. Wallace, *Palm Trees of the Amazon and Their Uses* (London: John van Voorst, 1853); and Alfred R. Wallace, *Narrative of Travels on the Amazon and Rio Negro, with an Account of the Native Tribes, and Observations on the Climate, Geology, and Natural History of the Amazon Valley* (London: Reeve and Co., 1853).
11. Wallace, *My Life: A Record of Events and Opinions*, Vol. 2, 1–22, 23–50.
12. Alfred Russel Wallace, *The World of Life: A Manifestation of Creative Power, Directive Mind, and Ultimate Purpose* (London: Chapman and Hall, 1910); Wallace, *My Life: A Record of Events and Opinions*; Charles H. Smith, "Wallace, Darwin, and Ternate, 1858," *Notes and Records of the Royal Society of London* 68 (2014): 165–70.
13. Alfred Russel Wallace, "On the Tendency of Varieties to Depart Indefinitely from the Original Type." Originally published in *Journal of the Proceedings of the Linnean Society (Zoology)* 3 (1859): 45–62; also in Burkhardt et al., *Correspondence*, Vol. 7, Appendix IV, 512–20. This quote is from p. 512. See also Wallace's autobiography, *My Life: A Record of Events and Opinions*, Vol. 1, 358–63.
14. Alfred Russel Wallace, "The Dawn of a Great Discovery: My Relations with Darwin

in Reference to the Theory of Natural Selection," *Black and White* 25 (January 1903): 78; Wallace, "On the Tendency of Varieties to Depart Indefinitely, from the Original Type," *Journal of the Proceedings of the Linnean Society* 3: 45–62. In Burkhardt et al., *Correspondence*, Vol. 7, Appendix IV, 512–20; quote is from pp. 512–13.

15. Wallace, "On the Tendency of Varieties to Depart Indefinitely," *Journal of the Proceedings of the Linnean Society* 3: 45–62; in Burkhardt et al., *Correspondence*, Vol. 7, Appendix IV, 512–20; quote is from pp. 512–13.
16. Wallace, "On the Tendency of Varieties to Depart Indefinitely," *Journal of the Proceedings of the Linnean Society* 3: 45–62; in Burkhardt et al., *Correspondence*, Vol. 7, Appendix IV, 510–20.
17. Description of the portrait of Sir Charles Lyell, 1st Baronet. National Portrait Gallery Archives, London, United Kingdom.
18. Charles Lyell, *Principles of Geology: An Attempt to Explain the Former Changes of the Earth's Surface by Reference to Causes Now in Operation* (London: John Murray; Vol. 1, 1830; Vol. 2, 1832; Vol. 3, 1833; Vol. 4, 1837); see also "Review of Lyell's *Principles of Geology*," *Quarterly Review* 43 (October 1830): 411–469.
19. James Hutton, *Theory of the Earth, with Proof and Illustrations* (Edinburgh: William Creech, 1795); William Whewell, "Review: Lyell's *Principles of geology*," 47 (1832): 103–32, see p. 126.
20. Roy Porter, "Charles Lyell and the Principles of the History of Geology," *British Journal for the History of Science* 9, no. 2 (July 1976): 91–103, quote is on pp. 93–94; K. M. Lyell, ed., *Life, Letters and Journals of Sir Charles Lyell, Bart*, Vol. 1 (London: John Murray, 1881), 168.
21. John S. Henslow to Charles Darwin, August 24, 1831. (CUP. Classmark: DAR 97, series 2: 4–5/DCP-LETT-105). In Burkhardt et al, *Correspondence*, Vol. 1, 128–129.
22. Robert W. Darwin to Josiah Wedgwood II, September 1, 1831. (Classmark: Victoria and Albert Museum/ Wedgwood Collection. MS W/M 96/ Letter # DCP-LETT-111). In Burkhardt et al., *Correspondence*, Vol. 1, 132.
23. Francis Darwin, ed., *The Life and Letters of Charles Darwin*, Vol. 1, 50–51; transcribed from Charles Darwin, "Recollections of My Mind and Character." [Autobiography, 1876–4.1882] (CUL-Classmark: DAR 26.1–121). Darwin Papers, Special Collections, Cambridge University Library; Stanley Edgar Hyman, "A Darwin Sidelight: The Shape of the Young Man's Nose," *Atlantic Monthly*, November 1967.
24. Nora Henslow, ed., *Darwin and Henslow: The Growth of an Idea, Letters 1831–1860* (London: Bentham-Moxon Trust/John Murray, 1967), 15–16, 136–37, 139–40.
25. Darwin ordered the second volume, which he received in Montevideo in November 1832, and he read the third volume midway through his voyage, in May 1834, just after exploring Santa Cruz. James A. Secord, *Introduction. Lyell's Principles of Geology* (London: Penguin Classics, 1997), ix–xliii; Humphrey Carpenter, *The Seven Lives of John Murray: The Story of a Publishing Dynasty* (London: John Murray, 2009), 190–192.
26. Charles Darwin to Leonard Horner, August 29. 1844. (CUL. Classmark: American Philosophical Society, [Mss. B.D25.38/Letter # DCP-LETT-771). In Burkhardt et al., *Correspondence*, Vol. 3, 54–55.

27. A. C. Steward, "Sir Joseph Hooker and Charles Darwin: The History of a Forty Years' Friendship," *New Phytologist* 11, no. 5/6 (May–June 1912): 195–206; see also Leonard Huxley, *Life and Letters of Sir Joseph Dalton Hooker* (London: John Murray, 1918).
28. J. D. Hooker, "Reminiscences of Darwin," *Nature* 60 (1899): 187–88, quote is from p. 187; Adrian Desmond and James Moore, *Darwin: The Life of a Tormented Evolutionist* (New York: W.W. Norton, 1991), 314.
29. Charles Darwin, *Journal of Researches into the Natural History and Geology of the Countries Visited during the Voyage of the H.M.S. Beagle Round the World, under the Command of Captain FitzRoy, R.A.*, 2nd ed. (London: John Murray, 1845), Corrected with Additions/Colonial and Home Library, Vol. 12; the first edition was published by Coburn in London in 1839.
30. J. D. Hooker, *Flora Antarctica: The Botany of the Antarctic Voyage of H.M. Discovery Ships Erebus and Terror in the Years 1839–1843, under the Command of Captain Sir James Clark Ross*, 3 volumes: 1844–General, 1853–New Zealand, 1859–Tasmania (London: Reeve Brothers, 1844, 1853, 1859).
31. Charles Darwin to William Hooker, March 12, 1843. (CUL. Classmark: Royal Botanic Gardens, Kew/Director's Correspondence: S. American letters, 1838–1844, 69:40/ Letter # DCP-LETT-664). In Burkhardt et al., *Correspondence*, Vol. 2, 252.
32. Charles Darwin to Joseph D. Hooker, January 11, 1844. (CUL. Classmark: DAR 114:3/Letter # DCP-LETT-729). In Burkhardt et al., *Correspondence*, Vol. 3, 1–3.
33. Joseph D. Hooker to Charles Darwin, January 29, 1844. (CUL. Classmark: DAR 100:5–7/Letter # DCP-LETT-734). In Burkhardt et al., *Correspondence*, Vol. 3, 5–8.
34. Charles Darwin to Joseph D. Hooker, July 2, 1860. (CUL. Classmark: DAR 115:64/ Letter # DCP-LETT-2853). In Burkhardt et al., *Correspondence*, Vol. 8, 272–73; see also L. Huxley, *Life and Letters of Sir Joseph Dalton Hooker*, Vol. 1, 486.
35. Alfred R. Wallace, "On the Law Which Has Regulated the Introduction of New Species," *Annals and Magazine of Natural History 2nd ser.* 16 (1855): 184–196.
36. Letter from Charles Lyell to Charles Darwin, May 1–2, 1856. (CUL. Classmark: DAR 205.3: 282/Letter # DCP-LETT-1862); Charles Darwin to Charles Lyell, May 3, 1856. (CUL. Classmark: American Philosophical Society: Mss. B.D25.127/Letter # DCP-LETT-1866). In Burkhardt et al., *Correspondence*, Vol. 6, 99–101; see also Roy Davies, "1 July 1858: What Wallace Knew; What Lyell Thought He Knew; What Both He and Hooker Took on Trust; and What Charles Darwin Never Told Them," *Biological Journal of the Linnean Society* 109, no. 3 (2013): 725–36.
37. Charles Darwin to Alfred Wallace, May 1, 1857. (CUL. Classmark: The British Library [Add MS 46434/Letter # DCP-LETT-2086). In Burkhardt et al., *Correspondence*, Vol. 6, 387–88.
38. Charles Darwin to Charles Lyell, June 25, 1858. (CUL. Classmark: American Philosophical Society, Mss. B. D25.153/ DCP-LETT-2294). In Burkhardt et al., *Correspondence*, Vol. 7, 117–18.
39. Anthony Wohl, *Endangered Lives: Public Health in Victorian Britain*. (London; J. M. Dent, 1983); Populations Past—Atlas of Victorian and Edwardian Population. Infant and Early Childhood Mortality Rates. University of Cambridge, accessed March 11,

2022 at https://www.populationspast.org/about/ AND https://www.populationspast.org/ecmr/1861/#8/51.034/0.893/bartholomew.

40. Tim M. Berra, *Darwin and His Children: His Other Legacy* (New York: Oxford University Press, 2013).
41. Charles Darwin to Joseph D. Hooker, June 23, 1858. (CUL. Classmark: DAR 114:238/ Letter # DCP-LETT-2290). In Burkhardt et al., *Correspondence*, Vol. 7, 116.
42. Another potential diagnosis might be acute epiglottitis, an infection of the throat caused by the bacterium *Haemophilus influenzae*. See Howard Markel, "Long Ago, Against Diphtheria, the Heroes Were Horses," *New York Times*, July 10, 2007; Howard Markel, "December 14, 1799: The Excruciating Final Hours of President George Washington," *PBS NewsHour*, December 14, 2014.
43. See, for example, Charles Darwin to Asa Gray, October 31, 1860. (CUL. Classmark: Archives of the Gray Herbarium, Harvard University, [45 and 24a]/Letter # DCP-LETT-2969). In Burkhardt et al., *Correspondence*, Vol. 8, 451.
44. Charles Darwin, *The Descent of Man and Selection in Relation to Sex* (London: John Murray, 1871).
45. Charles Darwin, *The Autobiography of Charles Darwin, 1809–1882*, ed. Nora Barlow (London: Collins, 1958).
46. Henrietta Litchfield, ed., *Emma Darwin: A Century of Family Letters, 1792–1896* (London: John Murray, 1915).
47. Charles Darwin to W. D. Fox, June 24, 1858. (CUL. Classmark: Christ's College Library, Cambridge [MS 54 Fox 114]/Letter # DCP-LETT-2293). In Burkhardt et al., *Correspondence*, Vol. 7, 116. Emma dated the onset of Charles's illness in her diary as beginning on June 23 and Henrietta's as beginning on June 18. See Emma Darwin's Diary for 1858, Cambridge University Library, Special Collections, [CUL. Classmark: DAR 242.22]; transcribed by Christine Chua and edited by John van Whye. Available online at http://darwin-online.org.uk.
48. Howard Markel, "On John Snow," in *Literatim: Essays at the Intersection of Medicine and Culture* (New York: Oxford University Press, 2019), 75–78.
49. J. L. Down, "Observations on an Ethnic Classification of Idiots," *London Hospital Clinical Lectures and Reports* 3 (1866): 259–62.
50. A. Mégarbané, A. Ravel, C. Mircher, et al., "The 50th Anniversary of the Discovery of Trisomy 21: The Past, Present, and Future of Research and Treatment of Down Syndrome," *Genetics in Medicine* 11 (2009): 611–16.
51. Howard Markel and Frank A. Oski, *The H. L. Mencken Baby Book* (Philadelphia: Hanley and Belfus, 1990), 174–84.
52. Thomas Sydenham, "Febris Scarlatina," *Opera Omnia Medica* (Padua: J. Manfré, 1714), 299–301.
53. Charles Darwin to Charles Lyell, June 26, 1858. (CUL. Classmark: American Philosophical Society, Mss. B.D25.154/DCP-LETT-2295). In Burkhardt et al., *Correspondence*, Vol. 7, 117–18.
54. Howard Markel, "Cases: Can That Strange Rash Really Be Scarlet Fever?" *New York Times*, July 25, 2006, F5.

55. Charles Darwin, "The Death of Charles Warren Darwin. July 2, 1858." (CUL. Classmark: DAR 210.13.42). In Burkhardt et al., *Correspondence*, Vol. 7, Appendix V, 521.
56. Litchfield, *Emma Darwin: A Century of Family Letters*, Vol. 2, 162.
57. Randal Keynes, *Annie's Box: Charles Darwin, His Daughter, and Human Evolution* (London: Fourth Estate/HarperCollins, 2002), 83–85; Berra, *Darwin and His Children: His Other Legacy.*
58. Alfred Wrigley to Charles Darwin, January 2 1868. (CUL. Classmark: DAR 181:180/ DCP-LETT-5773). In Burkhardt et al., *Correspondence*, Vol. 16, 6.
59. Charles Darwin, "Memoir of Anne Darwin, 30 April 1851." (CUL. Classmark: DAR 210.13.40). In Burkhardt et al., *Correspondence*, Vol. 5, 542; Charles Darwin to Emma Darwin. April 20, 1851. (CUL. Classmark: DAR 210.13:18/ Letter # DCP-LETT-1406). In Burkhardt et al., *Correspondence*, Vol. 5, 18–19.
60. Keynes, *Annie's Box*, 177, 79.
61. Keynes, *Annie's Box*, 147–79, 180–98; Litchfield, *Emma Darwin: A Century of Family Letters*, Vol. 2, 137.
62. Charles Darwin to W. D. Fox, April 29, 1859. (CUL. Classmark: Christ's College Library, Cambridge. MS 53,Fox 79/ DCP-LETT-425). In Burkhardt et al., *Correspondence*, Vol. 5, 32.
63. Darwin, "Memoir of Anne Darwin, 30 April 1851."
64. Darwin, "Memoir of Anne Darwin, 30 April 1851."
65. Keynes, *Annie's Box*, 183, 216.
66. Charles Darwin to Francis Galton, May 28, 1873. (CUL. Classmark: University College, London Library Services, Special Collections. [Galton/1/1/9/5/7/15); Pearson 1914–30,2: 178/ Letter #: DCP-LETT-1924). In Burkhardt et al., *Correspondence*, Vol. 21, 234–35; Dov Ospovat, "God and Natural Selection: The Darwinian Idea of Design," *Journal of the History of Biology* 13 (1980): 169–94.
67. Charles Darwin to Joseph D. Hooker, June 29, 1858. (CUL. Classmark: DAR 114: 240/Letter #: DCP-LETT-2298). In Burkhardt et al., *Correspondence*, Vol. 7, 121.

## 2. First

1. Letter from Alfred Russel Wallace to Henry Walter Bates, January 4, 1858. In Alfred Russel Wallace, *My Life: A Record of Events and Opinions*, Vol. 1 (New York: Dodd, Mead, and Co., 1905), 358.
2. Howard Markel, "'Who's on First?' Medical Discoveries and Scientific Priority," *New England Journal of Medicine* 351 (2004): 2792–94.
3. Charles Lyell, *The Geological Evidences of the Antiquity of Man with Remarks on Theories of the Origin of Species by Variation*, 3rd ed., rev. (London: John Murray, 1863), 408.
4. Charles Darwin, "Sketch of the Tree of Life." In "Notebook B on the Transmutation of Species, 1837." (CUL. Classmark: DAR 121, p. 36), Charles Darwin Papers, Cambridge University Library; see also 1837, Notebook C, (CUL. Classmark DAR 122); Notebook D—which contains a formulation of his ideas on natural selection— (CUL. Classmark: DAR 123); Notebook E (CUL. Classmark: DAR 124), Charles Darwin Papers, Special Collections, Cambridge University Library; see also Jonathan Hodge, "The

Notebook Programmes and Projects of Darwin's London Years," in *The Cambridge Companion to Darwin*, ed. Jonathan Hodge and Gregory Radick (Cambridge, UK: Cambridge University Press, 2009), 44–72. The italics are the author's (H.M.).

5. These manuscripts are filed in the CUL as "1842 Pencil Sketch; 1844 Essay Part 1, Draft A; and 1857 Outline of Species Theory" (MSS. DAR 6); 1844 Essay, fair copy: (MSS. DAR 113); Natural Selection, summary of whole book-plan, table of contents, Chapter 3: (MSS. DAR 8); Natural Selection, Chapter 4: (MSS.DAR 9). Darwin Manuscript Collection, Special Collections and Archives, Cambridge University Library, Cambridge, UK.
6. Lyell, *The Geological Evidences of the Antiquity of Man*, 408.
7. Charles Darwin to Joseph D. Hooker, January 11, 1844. (CUL. Classmark: DAR 107: 66–7; Letter #: CDP-LETT-1496). In Frederick Burkhardt et al., eds., *The Correspondence of Charles Darwin*, Vol. 3 (Cambridge, UK: Cambridge University Press, in 30 volumes, 1985–2023), 1–3.
8. Alfred Russel Wallace, "Attempts at a Natural Arrangement of Birds.," *Annals and Magazine of Natural History, 2nd ser.* 18 (1856): 193–216; for a superb analysis of this process, see Dov Ospovat, *The Development of Darwin's Theory: Natural History, Natural Theology, and Natural Selection, 1838–1859* (Cambridge, UK: Cambridge University Press, 1981); Roy Davies, "1 July 1858: What Wallace Knew; What Lyell Thought He Knew; What Both He and Hooker Took on Trust; and What Charles Darwin Never Told Them," *Biological Journal of the Linnean Society* 109, no. 3 (2013): 725–36.
9. Charles Darwin to Charles Lyell, June 17, 1860. (CUL. Classmark: American Philosophical Society [Mss. B.D25.217/Letter # DCP-LETT-2833). In Burkhardt et al., *Correspondence*, Vol. 8, 258; see also William Paley, *Natural Theology; Or Evidences of the Existence and Attributes of the Deity from the Appearances of Nature* (London: R. Faulder, 1802).
10. Emma Darwin to Charles Darwin, February 1839. Charles Darwin Collection, Cambridge University Library, Special Collections. (CUL Classmark: DAR 210.8: 14; Letter # DCP-LETT-471). In Burkhardt et al., *Correspondence*, Vol. 2, 171–73.
11. Henrietta Darwin Litchfield, ed., *Emma Darwin: A Century of Family Letters, 1792–1896*, Vol. 2 (London: John Murray, 1915), 146.
12. Francis Darwin, Preliminary Draft (circa 1884) of "Reminiscences of My Father's Everyday Life" CUL-DAR 140.3.1–159; and later published as Chapter 3 in Francis Darwin, ed., *The Life and Letters of Charles Darwin*, Vol. 1, (New York: D. Appleton and Co.), 87–138. This book was first published in London two years earlier (1887) by John Murray.
13. Litchfield, *Emma Darwin: A Century of Family Letters*, Vol. 2, 118, 146, 162.
14. Conway Zirkle, "Natural Selection before the 'Origin of Species'," *Proceedings of the American Philosophical Society*, 84, no. 1 (April 1941): 71–123; James A. Secord, *Victorian Sensation: The Extraordinary Publication, Reception, and Secret Authorship of Vestiges of the Natural History of Creation* (Chicago: University of Chicago Press, 2000).
15. Erasmus Darwin, *Zoonomia, Or the Laws of Organic Life*, Vol. 1 (London: J. Johnson of St. Paul's Church-Yard, 1794), 503; Hesketh Pearson, *Doctor Darwin* (London: Penguin Books, 1943).

16. Erasmus Darwin, *The Botanic Garden: Part I: The Economy of Vegetation; Part II: The Loves of Plants* (London: J. Johnson of St. Paul's Church-Yard, 1791); Erasmus Darwin, *The Temple of Nature or, The Origin of Society, A Poem with Philosophical Notes* (London: J. Johnson of St. Paul's Church-Yard, 1803).
17. J. B. Lamarck, *Zoological Philosophy, or Exposition with Regard to the Natural History of Animals*, trans. Hugh Elliot (London: Macmillan, 1914).
18. Anonymous [Robert Chambers], *Vestiges of the Natural History of Creation* (London: John Churchill, 1844); the book went through several printings and editions, and its impact and history are described in Secord, *Victorian Sensation*, 9–10, 165, 406, 429–31, 436. I am very grateful to Professor Secord for generously sharing his ideas and knowledge on Darwin and *Vestiges* with me.
19. Darwin's notes on Sedgwick, 1859. In Mario A. Di Gregorio, with the assistance of N. W. Gill, *Charles Darwin's Marginalia, Vol. 1* (New York: Garland Publishing, 1990), 750; Secord. *Victorian Sensation*, 429–31.
20. The "fear and trembling" quote is from Charles Darwin to Charles Lyell, October 8, 1845. (CUL. Classmark: American Philosophical Society, Mss. B.D24.46/ DCP-LETT-919). In Burkhardt et al., *Correspondence*, Vol. 7, 412–13. The "flea-infested" quote is from Secord, *Victorian Sensation*, 429.
21. Secord, *Victorian Sensation*, 429–31.
22. Secord, *Victorian Sensation*, 428–29.
23. Charles Darwin to Joseph D. Hooker, January 7, 1845. (CUL. Classmark: DAR 114:25/ Letter #: DCP-LETT-814). In Burkhardt et al., *Correspondence*, Vol. 3, 108.
24. Herschel also made several important contributions to mathematics, photography, and meteorology and even produced a translation of Homer's *Iliad*. Steven Ruskin, *John Herschel's Cape Voyage: Private Science, Public Imagination, and the Ambitions of Empire* (London: Taylor and Francis, 2018); John Herschel, *Herschel at the Cape: Diaries and Correspondence, 1834–1838*, ed. David Evans (Austin, TX: University of Texas Press, 1995).
25. *Cambridge Chronicle*, June 21, 1845, 1.
26. John Herschel, "Presidential Address," *British Association for the Advancement of Science Annual Report, 1845* (London: John Murray, 1846), xxvii–xxviii; Secord. *Victorian Sensation*, 406–407; Herschel's weak delivery and illness is described in the *Cambridge Chronicle*, June 21, 1845, 2.
27. Secord, *Victorian Sensation*, 409; Wilberforce's comments are quoted in "Scientific Meeting at Cambridge," *Cambridge Chronicle*, June 28, 1845, 2.
28. Samuel Wilberforce, *Pride a Hindrance to True Knowledge: A Sermon Preached in the Church of St. Mary the Virgin, Oxford, before the University, on Sunday 27 June 1847* (London: Rivington, 1847); Secord, *Victorian Sensation*, 436.
29. For a superb description of the VISTA acronym, see "How Does Natural Selection Work?" American Museum of Natural History, https://www.amnh.org/exhibitions/darwin/evolution-today/natural-selection-vista.
30. Charles Darwin, *The Autobiography of Charles Darwin, 1809–1882*, ed. Nora Barlow (New York: W. W. Norton, 1969), 76.
31. Charles Darwin to Charles Lyell, June 18, 1858. (CUL. Classmark: American Philo-

sophical Society, Mss.BD25.152/Letter # DCP-LETT-2285). In Burkhardt et al., *Correspondence*, Vol. 7, 107.

32. Charles Darwin to Charles Lyell, June 25, 1858. (CUL. Classmark: American Philosophical Society [Mss. B.D25.153/ Letter # DCP-LETT-2294). In Burkhardt et al., *Correspondence*, Vol. 7, 117–18.
33. Charles Darwin to Charles Lyell, June 26, 1858. (CUL. Classmark: American Philosophical Society [Mss. B.D25.154/ Letter # DCP-LETT-2295). In Burkhardt et al., *Correspondence*, Vol. 7, 119.
34. Charles Darwin to Joseph Hooker, June 29, 1858. (CUL. Classmark: DAR 114: 239/ Letter # DCP-LETT-2297). In Burkhardt et al., *Correspondence*, Vol. 7, 121.
35. This was the second letter Darwin wrote to Hooker on June 29 and it is dated "Tuesday Night." The bold and italicized words are Darwin's. Charles Darwin to Joseph D. Hooker. June 29, 1858. (CUL. Classmark: DAR 114: 240/Letter # DCP-LETT-2298). In Burkhardt et al., *Correspondence*, Vol. 7, 121–22.
36. Charles Darwin to Joseph D. Hooker, June 29, 1858. In Burkhardt et al., *Correspondence*, Vol. 7, 121–22.

## 3. Survival of the Fittest

1. Charles Darwin to J. S. Henslow, August 4, 1858. (CUL. Classmark: DAR 93: A53–5/ Letter #: DCP-LETT-2320). In Frederick Burkhardt et al., eds., *The Correspondence of Charles Darwin*, Vol. 7 (Cambridge, UK: Cambridge University Press, in 30 volumes, 1985–2023), 146.
2. Charles Lyell and Joseph Hooker to J. J. Bennett, Secretary of the Linnean Society, June 30, 1858. (CUL. Classmark: Original has not been found; Letter # DCP-LETT-2299). In Burkhardt et al., *Correspondence*, Vol. 7, 122–24; this letter was formally published in the *Journal of the Proceedings of the Linnean Society (Zoology)* 3 (1858): 45–46.
3. Alfred Russel Wallace, "On the Law Which Has Regulated the Introduction of New Species," *Annals and Magazine of Natural History 2nd ser.* 16 (1855): 184–96; Charles Darwin to Alfred Wallace. May 1, 1857. In Burkhardt et al., *Correspondence*, Vol. 6, 387–88.
4. These manuscripts are filed as "1842 Pencil Sketch; 1844 Essay Part 1, Draft A; and 1857 Outline of Species Theory" (DAR 6); 1844 Essay, fair copy: (DAR 113); Natural Selection, summary of whole book-plan, table of contents, Chapter 3: (DAR 8); Natural Selection, Chapter 4: (DAR 9). Darwin Manuscript Collection, Special Collections and Archives, Cambridge University Library, Cambridge, UK.
5. Charles Lyell and Joseph Hooker to J. J. Bennett, Secretary of the Linnean Society, June 30, 1858. (CUL. Classmark: Original has not been found; Letter # DCP-LETT-2299). In Burkhardt et al., *Correspondence*, Vol. 7, 122–24; quote is from p. 123. The date of 1839 was also likely incorrect given that Darwin's early sketch was written in 1842 and copied in 1844.
6. Charles Lyell and Joseph Hooker to J. J. Bennett, Secretary of the Linnean Society, June 30, 1858. (CUL. Classmark: Original has not been found; Letter # DCP-LETT-2299). In Burkhardt et al., *Correspondence*, Vol. 7, 122–24; quote is from p. 123.

7. Charles Lyell and Joseph Hooker to J. J. Bennett, Secretary of the Linnean Society, June 30, 1858. (CUL. Classmark: Original has not been found; Letter # DCP-LETT-2299). In Burkhardt et al., *Correspondence*, Vol. 7, 122–24; quote is on pp. 123; see also, Dov Ospovat, *The Development of Darwin's Theory: Natural History, Natural Theology, and Natural Selection, 1838–1859* (Cambridge, UK: Cambridge University Press, 1981); Roy Davies, "1 July 1858: What Wallace Knew; What Lyell Thought He Knew; What Both He and Hooker Took on Trust; and What Charles Darwin Never Told Them," *Biological Journal of the Linnean Society* 109, no. 3 (2013): 725–36.
8. Charles Lyell and Joseph Hooker to J. J. Bennett, Secretary of the Linnean Society, June 30, 1858. (CUL. Classmark: Original has not been found; Letter # DCP-LETT-2299). In Burkhardt et al., *Correspondence*, Vol. 7, 122–24; quote is from p. 123.
9. Charles Lyell and Joseph Hooker to J. J. Bennett, Secretary of the Linnean Society, June 30, 1858. (CUL. Classmark: Original has not been found; Letter # DCP-LETT-2299). In Burkhardt et al., *Correspondence*, Vol. 7, 122–24; quote is on pp. 123–124; the texts of all the relevant published material, including Wallace's essay, are provided in *Correspondence*, Vol. 7, Appendixes III and IV.
10. D. J. Mabberley, *Jupiter Botanicus. Robert Brown of the British Museum* (London: J. Cramer, 1985).
11. Derek Partridge, "The Famous Linnean Society Meeting: From Old Errors to New Insights," *Biological Journal of the Linnean Society* 137, no. 3 (2022): 556–67; Derek Partridge, "Further Details Concerning the Darwin–Wallace Presentation to the Linnean Society in 1858, Including Its Submission on 1 July, Not 30 June," *Journal of Natural History* 50, nos. 15/16 (2016): 1035–44; J .W. T. Moody, "The Reading of the Darwin and Wallace Papers: An Historical 'Non-Event'," *Journal of the Society for the Bibliography of Natural Hist*ory 6, no. 6 (1971): 474–76.
12. Nicholas Savage, *Burlington House: Home of the Royal Academy of Arts* (London: Royal Academy of Arts, 2018), 157–58.
13. In 1871, the Linnean Society moved to a northwest addition of the Burlington House. F. H. W. Sheppard, ed., Chapter 23, "Burlington House," in *Survey of London: Volumes 31 and 32, St James Westminster, Part 2* (London: London County Council, 1963), *British History Online*; see also Savage, *Burlington House*, 157–158.
14. "Quinsy" is the antiquated term for a peritonsillar abscess. Charles Darwin to Joseph D. Hooker, July 5, 1858. (CUL. Classmark: DAR 114: 241/ Letter #: DCP-LETT-2303). In Burkhardt et al., *Correspondence*, Vol. 7, 127.
15. Charles Darwin to Asa Gray, July 4, 1858. (CUL. Classmark: Archives of the Gray Herbarium, Harvard University [20]; DCP-LETT-2302). In Burkhardt et al., *Correspondence*, Vol. 7, 125–26; the 1857 letter is: Charles Darwin to Asa Gray, September 5, 1857. (CUL. Classmark; Archives of the Gray Herbarium, Harvard, University, 48/ DCP-LETT-2136). In Burkhardt et al., *Correspondence*, Vol. 6, 445–50.
16. Charles Darwin to Joseph D. Hooker, July 5, 1858. (CUL. Classmark: DAR 114: 241/ Letter #: DCP-LETT-2303). In Burkhardt et al., *Correspondence*, Vol. 7, 127.
17. Charles Darwin to J. D. Hooker, July 5, 1858. (CUL. Classmark: DAR 114:241/ DCP-LETT-2303). In Burkhardt et al., *Correspondence*, Vol. 7, 127–28.

18. Charles Darwin to J. D. Hooker, December 18, 1861. (CUL. Classmark: DAR 115: 137/ Letter #: DCP-LETT-3346). In Burkhardt et al., *Correspondence*, Vol. 9, 373.
19. The concept of protective sequestration was first described in Howard Markel et al., "Nonpharmaceutical Influenza Mitigation Strategies, U.S. Communities, 1918–1920 Pandemic," *Emerging Infectious Diseases* 12, no 12 (2006): 1961–64.
20. Charles Darwin to J. D. Hooker, July 13, 1858. (CUL. Classmark: DAR 114:242/ Letter # DCP-LETT-2306). In Burkhardt et al., *Correspondence*, Vol. 7, 129. Bold emphasis is Darwin's.
21. Charles Darwin to J. D. Hooker, July 13, 1858. (CUL. Classmark: DAR 114:242/ Letter # DCP-LETT-2306). In Burkhardt et al., *Correspondence*, Vol. 7, 129–31.
22. Janet Browne, *Charles Darwin: The Power of Place. Volume II of a Biography* (London: Jonathan Cape; New York: Knopf, 2002), 47.
23. Charles Darwin to J. D. Hooker, July 13, 1858. (CUL. Classmark: DAR 114:242/ Letter # DCP-LETT-2306). In Burkhardt et al., *Correspondence*, Vol. 7, 129–31. Darwin continued to observe ants once he arrived at the Isle of Wight; for his notes on these observations, see: CUL. Classmark: DAR 205.11 (2): 94 in the Charles Darwin Collections, Cambridge University Library. Bold and italics are Darwin's.
24. Charles Darwin to J. D. Hooker, August 10, 1858. (CUL. Classmark: DAR 114:245/ DCP-LETT-2318). In Burkhardt et al., *Correspondence*, Vol. 7, 148.
25. Charles Darwin to Charles Lyell, July 18, 1858. (CUL. Classmark: American Philosophical Society [Mss. B.D25.155/Letter # DCP-LETT-2309). In Burkhardt et al., *Correspondence*, Vol. 7, 137.
26. Charles Darwin to Joseph D. Hooker, July 21, 1858. (CUL. Classmark: DAR 114:244/ Letter # DCP-LETT-2311). In Burkhardt et al., *Correspondence*, Vol. 7, 139. The italic emphasis is Darwin's.
27. Charles Darwin, Esq., FRS, FLS, FGS, and Alfred Wallace, Esq., "On the Tendency of Species to Form Varieties; and on the Perpetuation of Varieties and Species by Natural Means of Selection," *Journal of the Proceedings of the Linnean Society (Zoology)* 3, no. 9 (1858): 45–62.
28. Alfred R. Wallace to Joseph D. Hooker, October 6, 1858. (CUL. Classmark: Linnean Society of London-Quentin Keynes Collection/ Letter # DCP-LETT-2337). In Burkhardt et al., *Correspondence*, Vol. 7, 167.
29. Alfred R. Wallace, *Darwinism: An Exposition of the Theory of Natural Selection, with Some of Its Applications* (London: Macmillan, 1889), viii.
30. Howard Markel, "'Who's on First?' Medical Discoveries and Scientific Priority," *New England Journal of Medicine* 351 (2004): 2792–94.
31. Anniversary Meeting Address of the President of the Linnean Society, Thomas Bell, Esquire, delivered on 24 May 1859, *Journal of the Proceedings of the Linnean Society (Zoology)* (London: Longman, Green, Longmans and Roberts and Williams and Norgate, 1860), Vol. 4, 2–15; quote is from pp. 3–4. Oersted's name is spelled Ørsted in Danish.
32. Frederick Burkhardt, "Darwin and the Biological Establishment," *Biological Journal of the Linnean Society* 17, no. 1 (1982): 39–44; Anonymous [George Bentham],

*Proceedings of the American Academy of Arts and Sciences* 20 (1884): 527–38. Bentham was coauthor, with Hooker, on *Genera Plantarum*, an important system of plant taxonomy that, despite a post-*Origins* publication date, did not take evolution into account. G. Bentham and J. Hooker, *Genera Plantarum: Ad Exemplaria Imprimis in Herbariis Kewensibus Servata Definita* (London: L. Reeve and Co., 1862–1883).

33. This undated letter from J. D. Hooker to Francis Darwin appears in the British edition of Francis Darwin, ed., *The Life and Letters of Charles Darwin*, Vol. 2 (London: John Murray, 1887–1888), 126.
34. Charles Darwin to Charles Lyell, July 18, 1858. (CUL. Classmark: American Philosophical Society [Mss. B. D25.155/ DCP-LETT-2309). In Burkhardt et al., *Correspondence*, Vol. 7, 137.

## Part II: The Book

1. George Eliot to Barbara Bodichon, December 5, 1859. In Gordon S. Haigh, ed., *The George Eliot Letters,* Vol. 3 (New Haven, CT: Yale University Press, 1954–1956), 227.

### 4. The Devil's Chaplain

1. Charles Darwin to Joseph D. Hooker, July 13, 1856. (CUL. Classmark: DAR 114:169/ Letter # DCP-LETT-1924). In Frederick Burkhardt et al., eds., *The Correspondence of Charles Darwin*, Vol. 6 (Cambridge, UK: Cambridge University Press, in 30 volumes, 1985–2023), 178–79.
2. Charles Darwin to Joseph D. Hooker, July 5, 1858. (CUL. Classmark: DAR 114:246/ Letter # DCP-LETT-2313). In Burkhardt et al., *Correspondence*, Vol. 7, 127–28.
3. Hyacinth Symonds Hooker and Leonard Huxley, eds., *Life and Letters of Sir Joseph Dalton Hooker*, Vol. 1 (London: John Murray, 1918), 405–12; quote is on p. 409.
4. Janet Browne, *Charles Darwin: The Power of Place. Volume II of a Biography* (London: Jonathan Cape; New York: Knopf, 2002), 47.
5. *Darwin in Conversation: The Endlessly Curious Life and Letters of Charles Darwin. An Exhibition Guide* (Cambridge, UK: Cambridge University Library, 2022).
6. Charles Darwin, *Journal of Researches into the Natural History and Geology of the Countries Visited during the Voyage of the H.M.S. Beagle Round the World, under the Command of Captain FitzRoy, R.A.* (London: John Murray, 1845; 2nd ed., Corrected with Additions/Colonial and Home Library, Vol. 12); the first edition was published by Henry Coburn in London in 1839 under the aegis of Darwin's commanding officer Robert FitzRoy. Darwin sold the copyright to Murray for £150 and edited out about 12,000 words for a book that still contained over 213,000 words. It was reprinted several times, in 1852, 1860, 1870, and 1888, by which time it had sold over eighteen thousand copies. Other editions were published by Murray in 1890, 1901 and 1913. It has remained in print ever since.
7. Gillian Beer, *Darwin's Plots. Evolutionary Narratives in Darwin, George Eliot, and Nineteenth-Century Fiction*, 3rd ed. (Cambridge, UK: Cambridge University Press, 2009); see also, David Quammen, *The Reluctant Mr. Darwin: An Intimate Portrait of*

*Charles Darwin and the Making of His Theory of Evolution* (New York: W. W. Norton/Atlas Books, 2006), 181–204; and Browne, *Charles Darwin: The Power of Place,* 47–73.

8. Beer, *Darwin's Plots,* 1–17.
9. Darwin, *On the Origin of Species* (1st ed.), 62.
10. Charles Darwin to J. D. Hooker, October 6, 1858. (CUL. Classmark: DAR 114:248/ Letter # DCP-LETT-2335). In Burkhardt et al., *Correspondence,* Vol. 7, 164–65.
11. Charles Darwin to J. D. Hooker, December 24, 1858. (CUL. Classmark: DAR 114:257/ Letter # DCP-LETT-2384). In Burkhardt et al., *Correspondence,* Vol. 7, 220–22.
12. The estimated length of the manuscript is between 550 and 580 pages. See R. Keynes and D. Kohn, "*Origin of Species* 1859: Surviving Manuscript Leaves, July 1858–March 1859," cited in Adam M. Goldstein, "The *Origin* Manuscripts at the 'The Darwin Manuscripts Project,'" *Evolution: Education and Outreach* 3 (2010): 121–27.
13. Charles Darwin to Alfred Wallace, January 25, 1859. (CUL. Classmark: The British Library [Add MS 46434]/Letter # DCP-LETT-2405). In Burkhardt et al., *Correspondence,* Vol. 7, 240–41. The letter Wallace wrote to Darwin is lost, but it appears that Darwin received it in mid to late January of 1859 and that it was written in October. See Charles Darwin to J. D. Hooker. January 23, 1859. (CUL. Classmark: DAR 115: 3/ Letter # DCP-LETT-2403). In Burkhardt et al., *Correspondence,* Vol. 7, 238.
14. Col. Robert Gordon to John Murray, October 21, 1769. John Murray Archives, National Library of Scotland, Edinburgh. Acc. No. 12604/1464.
15. Humphrey Carpenter, *The Seven Lives of John Murray: The Story of a Publishing Dynasty* (London: John Murray, 2009), 7, 67–148, 149–217, 235–36.
16. Darwin, *Journal of Researches into the Natural History and Geology of the Countries Visited.*
17. John Murray IV, *John Murray III: A Brief Memoir* (London: John Murray, 1919), 13–25.
18. Murray preferred P. G. Tait's 1876 warning to Lyell, Darwin, et al.: "So much the worse for geology, since physical conditions render it impossible to allow her more than 10 of 15 millions of years." See "Verifier" (John Murray III), *Skepticism in Geology and the Reasons for It* (London: John Murray, 1877), vii.
19. John Murray IV, *John Murray III: A Brief Memoir,* 25–27.
20. By 1861, the gorilla was not entirely unknown. In 1847, an American missionary named Thomas Savage—yes, that was his real name—was exploring the jungles of Africa where he acquired the bones and skull of a heretofore undescribed and especially large species of apes. Savage named them *Troglodytes gorilla,* from the Greek for "wild, hairy beings." See P. B. Du Chaillu, *Explorations and Adventures in Equatorial Africa; with Accounts of the Manners and Customs of the People and of the Chase of the Gorilla, Crocodile, Leopard, Elephant, Hippopotamus, and Other Animals* (London: John Murray, 1861), 61; see also Merian C. Cooper, Ernest B. Schoedsack, and David O. Selznick, producers, screenplay by James Creelman and Ruth Rose, *King Kong* (Hollywood: RKO Radio Pictures, 1933).
21. David McClay, *Dear Mr. Murray: Letters to a Gentleman Publisher* (London: John Murray, 2018), ix–x, 1–2.

22. Carpenter, *The Seven Lives of John Murray*, 204–205.
23. Thomas Huxley met Du Chaillu just before a return to Africa to catch a live gorilla for Richard Owen. Unimpressed, he described the wild game hunter as "nothing but inexplicable confusion." Adrian Desmond, *Huxley: From Devil's Disciple to Evolution's High Priest* (London; Penguin Books, 1997), 296–97; see also Alison Bashford, *The Huxleys: An Intimate History of Evolution* (Chicago: University of Chicago Press, 2022), 208–13.
24. Charles Darwin to Charles Lyell, March 28, 1859. (CUL. Classmark: American Philosophical Society [Mss. B.D25.163/Letter # DCP-LETT-2437). In Burkhardt et al., *Correspondence*, Vol. 7, 269–70.
25. Charles Darwin to Charles Lyell, March 30, 1859. (CUL. Classmark: American Philosophical Society [Mss. B.D25.164/Letter # DCP-LETT-2439). In Burkhardt et al, *Correspondence*, Vol. 7, 272; Charles Darwin to John Murray, March 31, 1859. (CUL. Classmark: National Library of Scotland/John Murray Archive. Ms. 42153 ff. 12–13/ Letter # DCP-LETT-2441). In Burkhardt et al., *Correspondence*, Vol. 7, 273.
26. Those terms were as follows: "As soon as I can ascertain the cost of its production I will make you an offer amounting as nearly as I can ascertain to 2/3 [pounds and shillings] of the net proceeds of the edition—payable by note of hand at six months from the day of publication, I shall be quite willing to give the author 12 copies for himself & as many more as he may require at the trade price." See John Murray to Charles Darwin, April 1, 1859. (CUL. Classmark: National Library of Scotland/John Murray Archive. Ms. 41913, p. 32/Letter # DCP-LETT-2443). In Burkhardt et al., *Correspondence*, Vol. 7, 275.
27. Charles Darwin to John Murray, April 2, 1859. (CUL. Classmark: National Library of Scotland/John Murray Archive. Ms. 42153 ff. 18–19/ Letter # DCP-LETT-2445). In Burkhardt et al., *Correspondence*, Vol. 7, 277. The bold is Darwin's.
28. Charles Darwin to J. D. Hooker, April 2, 1859. (CUL. Classmark: DAR 115: 9/Letter #: DCP-LETT-2446). In Burkhardt et al., *Correspondence*, Vol. 7, 275–76. The bold is Darwin's.
29. Charles Darwin to J. D. Hooker, April 7, 1859. CUL. Classmark: DAR 115: 10/Letter #: DCP-LETT-2450). In Burkhardt et al., *Correspondence*, Vol. 7, 280–81.
30. J. D. Hooker to Charles Darwin, April 8–11, 1859. (CUL. Classmark: DAR 100: 127/ Letter #: DCP-LETT-2444). In Burkhardt et al., *Correspondence*, Vol. 7, 281–82.
31. Whitwell Elwin to John Murray, May 3, 1859. (CUL. Classmark: National Library of Scotland/John Murray Archive, Ms. 42197)/ Letter # DCP-LETT-2457A). In Burkhardt et al., *Correspondence*, Vol. 7, 288–90. Italics are Elwin's.
32. Whitwell Elwin to John Murray, May 3, 1859. (CUL. Classmark: National Library of Scotland/John Murray Archive, Ms. 42197)/ Letter # DCP-LETT-2457A). In Burkhardt et al., *Correspondence*, Vol. 7, 288–90. Italics are Elwin's.
33. Charles Darwin, *On the Origin of Species by Means of Natural Selection, or the Preservation of Favoured Races in the Struggle for Life* (London: John Murray, 1859), 29.
34. Charles Darwin to John Murray, May 6, 1859. CUL. Classmark: National Library of Scotland/John Murray Archive, Ms. 42152, ff. 57–57A)/ Letter # DCP-LETT-2459). In Burkhardt et al., *Correspondence*, Vol. 7, 295–96.

35. Charles Darwin to John Murray, May 10, 1859. CUL. Classmark: National Library of Scotland/John Murray Archive, Ms. 42152, ff. 56–57)/ Letter # DCP-LETT-2460). In Burkhardt et al., *Correspondence*, Vol. 7, 296.
36. Charles Darwin to J. D. Hooker, May 11, 1859. (CUL. Classmark: DAR 115:15/Letter # DCP-LETT-2461). In Burkhardt et al., *Correspondence*, Vol. 7, 296–97.
37. Charles Darwin to John Murray, May 14, 1859. (CUL. Classmark: National Library of Scotland/John Murray Archive, Ms. 42152, ff. 40–40A)/ Letter # DCP-LETT-2462). In Burkhardt et al., *Correspondence*, Vol. 7, 298–99. Italics and bold are Darwin's.
38. Charles Darwin to John Murray, May 14, 1859. (CUL. Classmark: John Wilson, (dealer)/ Letter # DCP-LETT-2463). In Burkhardt et al., *Correspondence*, Vol. 7, 298–99.
39. Charles Darwin to Joseph D. Hooker, May 18, 1859. (CUL. Classmark: DAR 115: 16/ Letter # DCP-LETT-2463). In Burkhardt et al., *Correspondence*, Vol. 7, 299.
40. Charles Darwin to Joseph D. Hooker, May 26, 1859. (CUL. Classmark: DAR 115: 17/ Letter # DCP-LETT-2464). In Burkhardt et al., *Correspondence*, Vol. 7, 300.
41. A superb account of this period of Darwin's life can be found in Janet Browne's definitive biography, *Charles Darwin: The Power of Place* (New York: Knopf, 2002), 53–81; see also Adrian Desmond and James Moore, *Darwin: The Life of a Tormented Evolutionist* (New York: W. W. Norton, 1991), 470–77.
42. Charles Darwin to Joseph D. Hooker, April 12, 1859. (CUL. Classmark: DAR 115: 12/ Letter # DCP-LETT-2453). In Burkhardt et al., *Correspondence*, Vol. 7, 284–85. The bold is Darwin's.
43. Leonard Huxley, *Life and Letters of Sir Joseph Dalton Hooker, O.M., G.C.S.I., Based on Materials Collected and Arranged by Lady Hooker*, Vol. 1 (London: John Murray, 1918), 496.
44. Rachel Cohen, "Can You Forgive Him?" *New Yorker*. November 8, 2004, 48–65.
45. Charles Darwin to Charles Lyell, September 20, 1859. (CUL. Classmark: American Philosophical Society/Mss.B. D25.169/ Letter # DCP-LETT-2492). In Burkhardt et al., *Correspondence*, Vol. 7, 333–35.
46. *Athenaeum*, September 24, 1859, 403–404; quote is from p. 403; Charles Lyell, Introductory Address by the President of the Geology Section; and Charles Lyell, "On the Occurrence of Works of Human Art in Post-Pliocene Deposits," *Report of the Twenty-Ninth Meeting of the British Association for the Advancement of Science, held at Aberdeen, 14–21 September 1859* (London: John Murray, 1860), Notices and Abstracts, 93–95.
47. *Athenaeum*, September 24, 1859, 403–404; quote is on p. 404.
48. Charles Darwin to Charles Lyell, September 25, 1859. (CUL. Classmark: American Philosophical Society/Mss.B. D25.170/ Letter # DCP-LETT-2494). In Burkhardt et al., *Correspondence*, Vol. 7, 336–37.
49. *Athenaeum*, September 24, 1859, 403–404.
50. Elizabeth Longford, *Victoria R.I.* (London: Weidenfeld and Nicholson, 1964), 286–87.
51. Appendix II, Chronology, 1858–1859. In Burkhardt et al., *Correspondence*, Vol. 7, 504.
52. John Murray to Charles Darwin, November 2, 1859. (CUL. Classmark: National

Library of Scotland. John Murray Archive/Ms. 4193, pp. 53–54/ Letter # DCP-LETT-2513A). In Burkhardt et al., *Correspondence*, Vol. 7, 364–65.

53. Charles Darwin to W. D. Fox, September 23, 1859. (CUL. Classmark: Christ's College Library, Cambridge, MS 53 Fox 122/ DCP-LETT-2493). In Burkhardt et al., *Correspondence*, Vol. 7, 335–36.
54. Charles Darwin to Charles Lyell, March 28, 1859. (CUL. Classmark: American Philosophical Society [Mss.B.D25.163]/Letter # DCP-LETT-2437). In Burkhardt et al., *Correspondence*, Vol. 7, 269–70.
55. Koen B. Tanghe, "On the Origin of Species: The Story of Darwin's Title," *Notes and Records of the Royal Society of London* 73 (2019): 83–100.
56. William Whewell, Bridgewater Treatise No. III, in *The Bridgewater Treatises: On the Power, Wisdom, and Goodness of God, as Manifested in the Creation* (London: William Pickering, 1833–1839). As a sign of Whewell's inner conflict on science and religion, one need only read his anonymously published 1853 essay "Of the Plurality of Two Worlds," in which he argued that human life existed only on earth, thanks to God's special relationship with his greatest creation, and railed against those who tried to usurp Judeo-Christian doctrines with unproven scientific theories. See William Whewell, *On the Plurality of Two Worlds*, ed. M. Ruse (Chicago: University of Chicago Press, 2001; facsimile of 1853 edition, with additions); John Hedley Brook, "Natural Theology and the Plurality of Worlds: Observations on the Brewster-Whewell Debate.," *Annals of Science* 34, no. 3 (1977): 221–86. See also Francis Bacon, *Of the Advancement and Proficience of Learning, or the Partitions of Sciences, IX Bookes* (Oxford, UK: Printed by Leonard Lichfield, printer for the University of Oxford, 1640, originally published in Latin).
57. Charles Darwin, opening quotations, *On the Origin of Species*, 1. In the second edition and henceforth, he added another quotation, from Joseph Butler's *Analogy of Revealed Religion*: "The only distinct meaning of the word 'natural' is *stated, fixed*, or *settled*; since what is natural as much requires and presupposes an intelligent agent to render it so, *i.e.* to effect it continually or at stated times, as what is supernatural or miraculous does to effect it for once."
58. In 1859, Darwin went to Moor Park on February 5–19, which "did not do me so much good as usual," May 21–28, July 26, and then to the Ilkley hydropathic institute for a nine-week stay beginning on October 4 and lasting into early December. Appendix II, Chronology, 1858–1859. He visited Malvern again in 1863. In Burkhardt et al., *Correspondence*, Vol. 7, 503–506.
59. Mike Dixon and Gregory Radick, *Darwin in Ilkley* (Stroud, UK: History Press, 2009) 51.
60. Charles Darwin to his son, William E. Darwin, October 14, 1859. (CUL. Classmark: 210.6: 49/ Letter # DCP-LETT-2498). In Burkhardt et al., *Correspondence*, Vol. 7, 348–49.
61. Dixon and Radick, *Darwin in Ilkley*, 88.
62. William was, by now, an undergraduate at Christ's College, Cambridge University, and George was a boarding student at the Clapham Grammar School. See Charles Darwin to Lady Elizabeth Drysdale, October 22 or 29, 1859. (CUL. Classmark: John

Wilson Autographs, Catalogue 88; Clive Farahar and Sophie Dupre, Catalogue 55; B&L Rootenberg, May 1991; Letter # DCP-LETT-2498A). In Burkhardt, et al., *Correspondence,* Vol. 13, Supplement, 416, and footnote 4, p. 417; Dixon and Radick, *Darwin in Ilkley,* 89; William Irvine, *Apes, Angels, and Victorians: The Story of Darwin, Huxley, and Evolution* (New York: Time-Life Books, 1955), 145–46.

63. Charles Darwin to J. D. Hooker, October 27 or November 3, 1859. (CUL. Classmark: 115: 25/Letter # DCP-LETT-2512). In Burkhardt et al., *Correspondence,* Vol. 7, 361.
64. James M. Gully, *The Water Cure in Chronic Disease,* 9th ed. (London: Simpkin, Marshall and Co., 1863), 444–48; Dixon and Radick, *Darwin in Ilkley,* 92–95.
65. Charles Darwin to A. R. Wallace, November 13, 1859. (CUL. Classmark; The British Library, Add MS 46434/DCP-LETT-2529). In Burkhardt et al., *Correspondence,* Vol. 7, 375–76.
66. Henrietta Litchfield, ed., *Emma Darwin: A Century of Family Letters, 1792–1896,* Vol. 2 (London: John Murray, 1915), 172.

## 5. Best Seller

1. Incomplete letter without address. Charles Darwin to ? After November 2, 1859. (CUL. Classmark: Michael S. Hollander, dealers/ Letter # DCP-LETT-2451). In Frederick Burkhardt et al., eds., *The Correspondence of Charles Darwin,* Vol. 7 (Cambridge, UK: Cambridge University Press, in 30 volumes, 1985–2023), 365.
2. R. B. Freeman, *The Works of Charles Darwin: An Annotated Bibliographical Handlist,* 2nd ed. (Hamden, CT: Archon Books, 1977), 75.
3. Darwin was not using the word "race" in the form it is used today. In Darwin's era, this was a crude term for the different species of mammals, birds, and insects, as well as a racist rubric for delineating the distinctions of various human ethnicities, nationalities, skin color, and even religions. For an excellent discussion of the book, see David Quammen, *The Reluctant Mr. Darwin: An Intimate Portrait of Charles Darwin and the Making of His Theory of Evolution* (New York: W. W. Norton/Atlas Books, 2006), 171–81; for recent prices of first editions of *Origin,* see Peter Harrington of London, Booksellers Catalogue, 2022. Accessed on November 16, 2022, at https://www.peterharrington.co.uk/on-the-origin-of-species-by-means-of-natural-selection-146564.html.
4. John Murray to Charles Darwin, November 2, 1859. (CUL. Classmark: National Library of Scotland, John Murray Archive, Ms. 42153, ff. 53–54/ Letter # DCP-LETT-2513A). In Burkhardt et al., *Correspondence,* Vol. 7, 364–65.
5. Letter from Charles Darwin to John Murray, November 3, 1859. (CUL. Classmark: National Library of Scotland, John Murray Archive, Ms. 42152, f. 49/ Letter # DCP-LETT-2514). In Burkhardt et al., *Correspondence,* Vol. 7, 365–66. The italics are Darwin's.
6. Charles Darwin, *On the Origin of Species by Means of Natural Selection, or the Preservation of Favoured Races in the Struggle for Life* (London: John Murray, 1859), 1–2; John Herschel is the great philosopher referring to the mystery of mysteries and whom Darwin obliquely quotes using his phrase. See W. F. Cannon, "The Impact

of Uniformitarianism: Two Letters from John Herschel to Charles Lyell, 1836–1837," *Proceedings of the American Philosophical Society* 105 (1961): 301–14.

7. Isaac Newton, *Philosophiæ Naturalis Principia Mathematica* (London: S. Pepys, 1687). Newton presented the first part of this book to the Royal Society in 1687. As the story goes, Robert Hooke stormed out of the room after claiming that Newton took the inverse square law of gravitation from him! In the wake of Newton's landmark work, several physicists and chemists of the seventeenth century and beyond developed and proved their own "laws."
8. Thomas Robert Malthus, *An Essay on the Principle of Population: As It Affects the Future Improvement of Society with Remarks on the Speculations of Mr. Godwin, M. Condorcet, and Other Writers* (London: J. Johnson, 1798.) This edition was published anonymously; Malthus's name appears on the 1803, or second, edition.
9. Darwin first read Malthus in 1838 and again, more closely, in 1847. Charles Darwin to Ernst Haeckel, sometime after August 10–October 8, 1864. (CUL. Classmark: Ernst-Haeckel-Haus/Bestand A-Abt. 1:1–52/5/ Letter # DCP-LETT-4631). In Burkhardt et al., *Correspondence*, Vol. 12, 301–304; for an account of Malthus's influence on Darwin, see Antonello Vergata, "Images of Darwin: A Historiographic Overview," in *The Darwinian Heritage*, ed. David Kohn (Princeton, NJ: Princeton University Press, 1985), 901–72.
10. Charles Darwin to Otto Zacharias, June 11, 1875. (CUL. Classmark: Zacarias, 1882, p. 80/Letter # DCP-LETT-10013F). In Burkhardt et al., *Correspondence*, Vol. 23, 225.
11. Darwin, *On the Origin of Species*, 1st ed., 489–90.
12. Quammen, *The Reluctant Mr. Darwin*, 175–204; Paul Johnson, *Darwin: Portrait of a Genius* (New York: Penguin Books, 2012), 80, 89–99. Freeman disputes whether all the copies of the first edition sold that day; the origin of this claim can be found in Darwin's daily diary, which stated: "1250 copies printed. The first edition was published on November 24th, and all copies sold first day." See Freeman, *The Works of Charles Darwin*, 75; Charles Darwin, Pocket Diary (Journal), 1838–1881. Charles Darwin Collection. DAR 158. Reference Code: GBR/0012/MS DAR 158:38. (January–December 1859). Cambridge University Library, Cambridge UK. Accessed from the American Museum of Natural History,on August 2, 2022, at https://www.amnh.org/research/darwin-manuscripts/journals-diaries/journal-1838–1881.
13. Phillip V. Allingham, "Charles Dickens's *A Tale of Two Cities* (1859) Illustrated: A Critical Reassessment of Hablot Knight Browne's Accompanying Plates," *Dickens Studies Annual* 33 (2003): 109–57.
14. Charles Darwin. *On the Origin of Species*, 2nd ed. (London: John Murray, 1860), 481.
15. See Charles Darwin, *On the Origin of Species*, 3rd ed., xiii–xix. He revised the historical sketch again for the fourth 1866 edition (pp. xiii–xx), and the sketch appears in the remaining editions as well; "Rewriting Origin—the Later Editions." Darwin Correspondence Project. Darwin Collections, Cambridge University Library, Cambridge, UK. https://www.darwinproject.ac.uk/letters/darwins-works-letters/rewriting-origin-later-editions.
16. A superb analysis of the various editions of *On the Origin of Species* can be found in Quammen, *The Reluctant Mr. Darwin*, 175–204.

17. Charles Darwin to Alfred Wallace, November 13, 1859. (CUL. Classmark: The British Library [Add MS 46434]/ Letter #: DCP-LETT-2529). In Burkhardt et al., *Correspondence*, Vol. 7, 375–76.
18. "Ague" was the nineteenth-century term for malarial fever but could also be a signifier of many other maladies. Erasmus Darwin to Charles Darwin, November 23, 1859. (CUL. Classmark: DAR 98: B14–15/Letter #: DCP-LETT-2545). In Burkhardt et al., *Correspondence*, Vol. 7, 389–90.
19. Janet Browne, *Charles Darwin: The Power of Place. Volume II of a Biography* (London: Jonathan Cape; New York: Knopf, 2002), 101–11.
20. Hugh Chisholm, ed., "Charles Edward Mudie," in *Encyclopedia Britannica*, Vol. 18, 11th ed. (Cambridge, UK: Cambridge University Press, 1911), 955; Lewis Roberts, "Trafficking in Literary Authority: Mudie's Select Library and the Commodification of the Victorian Novel," *Victorian Literature and Culture* 34, no. 1 (2006): 1–25.
21. Emma Darwin to William Erasmus Darwin, undated but likely late November or early December of 1859. Charles Darwin Papers, University of Cambridge Library Special Collections (CUL. Classmark DAR 210.6); see also, footnote 1 for a letter Charles Darwin wrote to John Murray on November 24, 1859. (CUL. Classmark: National Library of Scotland, John Murray Archive, Ms. 42152, ff. 70–71/Letter #: DCP-LETT-2549). In Burkhardt et al., *Correspondence*, Vol. 7, 395.
22. Charles Lyell to Charles Darwin, October 3, 1859. (CUL. Classmark: DAR 98: B1–6/ Letter #: DCP-LETT-2501). In Burkhardt et al., *Correspondence*, Vol. 7, 339–42.
23. Charles Lyell to Charles Darwin, October 22, 1859. (CUL. Classmark: University of Edinburgh, Centre for Research Collections, Lyell Collection, Coll-203/A1/242: 15–24). In Burkhardt et al., *Correspondence*, Vol. 13 (Supplement), 418–19.
24. See Charles Lyell, *The Geological Evidences of the Antiquity of Man, with Remarks on Theories of the Origin of Species by Variation* (London: John Murray, 1863) and *Principles of Geology; or the Modern Changes of the Earth and Its Inhabitants, Considered as Illustrative of Geology*, 10th ed., in 2 volumes (London: John Murray, 1868); Lyell also wondered if dogs might have originated from several species. Further, although he and Darwin had agreed all humans shared a common ancestry, Lyell mentioned the work of others, such as P. S. Pallas, who speculated that various human races "European, Negro, Hottentot, and Australian races . . . sprung from several indigenous stocks or species settled in remote & isolated regions." See Charles Lyell to Charles Darwin, October 22, 1859. (CUL. Classmark: University of Edinburgh, Centre for Research Collections, Lyell Collection, Coll-203/A1/242: 15–24). In Burkhardt et al., *Correspondence*, Vol.13 (Supplement), 418–19.
25. Anonymous. review. "The Antiquity of Man," *The Saturday Review of Politics, Literature, Science and Art* 15, no. 384 (March 7, 1863): 311–12; quote is on p. 311.
26. Charles Lyell to Charles Darwin, March 11, 1863. In Charles Lyell, (edited by his sister-in-law, Mrs. Katherine M. Lyell), *Life, Letters and Journals of Sir Charles Lyell, Bart*, Vol. 2 (London: John Murray, 1881), 362–64. (CUL Letter # DCP-LETT-4035). In Burkhardt et al., *Correspondence*, Vol. 11, 217–19.
27. Joseph Hooker to Charles Darwin, December 12, 1859. (CUL. Classmark: DAR 100: 137–138/ Letter #: DCP-LETT-2579). In Burkhardt et al., *Correspondence*, Vol. 7, 425–

27. To gild the lily further, Hooker repeated this praise in an unsigned essay for the December 1859 issue of *Gardener's Chronicle and Agricultural Gazette*: "It cannot be denied that it would be difficult in the whole range of the literature of science to find a book so exclusively devoted to the development of theoretical inquiries, which at the same time is so full of conscientious care, so fair in argument, and so considerate in tone." Predictably, Darwin had some quibbles even with this laudatory piece until he learned who wrote it and ultimately enthused to Charles Lyell, "I have reread *Gardeners Ch.* with extreme pleasure now that I know Hooker is author."
28. Calvin Trillin, *Quite Enough of Calvin Trillin: Forty Years of Funny Stuff* (New York: Random House, 2011), 87.
29. Anonymous [John Leifchild], "Literature: Review of *On the Origin of Species by Means of Natural Selection, or the Preservation of Favoured Races in the Struggle for Life*," *Athenaeum*, no. 1673 (November 19, 1859 ): 659–60; quotes appear on p. 660.
30. This quote first appears in Darwin's 1837–1838 notebook B, on the transmutation of species, p. 232. Charles Darwin Papers. (CUL Classmark: DAR 121). Special Collections, Cambridge University Library. Darwin scratches out the original word "these" and replaces it with "one" for the phrase "one common ancestor." He also notes that "this book was commenced about July 1837, p. 235 was written in January 1838. Probably ended in beginning of February."
31. Darwin. *On the Origin of Species*, 1st ed., 488.
32. Darwin, *On the Origin of Species*, 1st ed., 484.
33. Charles Darwin, *The Descent of Man, and Selection in Relation to Sex*, Vol. 2, 1st ed. (London: John Murray, 1871), 389.
34. Charles Darwin to J. D. Hooker, November 22, 1859. (CUL. Classmark: DAR 115: 26/Letter #: DCP-LETT-2542). In Burkhardt et al., *Correspondence*, Vol. 7, 387.
35. Charles Darwin to J. D. Hooker, November 22, 1859. (CUL. Classmark: DAR 115: 26/Letter #: DCP-LETT-2542). In Burkhardt et al., *Correspondence*, Vol. 7, 387; see also Charles Darwin to T. H. Huxley, November 25, 1859. In Burkhardt et al., *Correspondence*, Vol. 7, 387–88; Hooker wrote Darwin on November 21, 1859, to tell him Huxley "talked about giving a R. Inst. Friday Evening to your book—but pray say nothing of this—it may come to nothing—he is vastly pleased with it." (CUL. Classmark: DAR 100: 135–6/Letter # DCP-LETT-2539). In Burkhardt et al., *Correspondence*, Vol. 7, 385–84. The lecture was arranged and confirmed by November 25, 1859; Darwin wrote Huxley to tell him "Hurrah for Lecture it is grand!" Charles Darwin to T. H. Huxley, November 25, 1859. (CUL. Classmark: Imperial College of Science, Technology and Medicine Archives, Huxley 5: 74/. Letter #: DCP-LETT-2554). In Burkhardt et al., *Correspondence*, Vol. 7, 399–400.
36. Charles Darwin to Thomas Huxley, December 9, 1859. (CUL. Classmark: DAR 145: 189/ Letter #: DCP-LETT-2574). In Burkhardt et al., *Correspondence*, Vol. 7, 418; see also Charles Darwin to Thomas Huxley, December 25, 1859. (CUL. Classmark: Imperial College of Sceince, Technology and Medicine Archives, Huxley 5: 70/ Letter # DCP-LETT-2603). In Burkhardt et al., *Correspondence*, Vol. 7, 451–52; see also Francis Darwin, ed., *The Life and Letters of Charles Darwin*, Vol. 2 (London: John Murray, 1887), 251; Leonard Huxley, *Life and Letters of Thomas H. Huxley*, Vol. 1

(London: Macmillan, 1900), 171. Holland was a distant cousin on the Wedgwood side of his family.

37. John M. Thomas, *Michael Faraday and the Royal Institution: The Genius of Man and Place* (London: Taylor and Francis, 1991); Colin Russell, *Michael Faraday, Physics and Faith* (Oxford, UK: Oxford University Press, 2000); Joseph Agassi, *Faraday as a Natural Philosopher* (Chicago: University of Chicago Press, 1971).
38. John Howard Smith, *The Perfect Rule of the Christian Religion: A History of Sandemanianism in the Eighteenth Century* (Albany: SUNY Press, 2008).
39. Gwendy Caroe, *The Royal Institution: An Informal History* (London: John Murray, 1985).

## Part III: Friends and Foes

1. Kingsley is quoting here from Romans 3:5 and refers to a passage from Plato's *Republic*, Section 365c. Charles Kingsley to Charles Darwin, November 18, 1859 (CUL. Classmark: DAR 98: B7–8/ Letter #: DCP-LETT-2534). In Frederick Burkhardt et al., eds., *The Correspondence of Charles Darwin*, Vol. 7 (Cambridge, UK: Cambridge University Press, in 30 volumes, 1985–2023), 379–80.
2. [Richard Owen], "Review of Darwin on the Origin of Species" [and other works: Alfred Wallace's 1859 Linnean Society paper, as well as papers by John Hooker, Flourens, Agassiz, Pouchet, the creationist Reverend Baden Powell, and some of Owen's recent work on paleontology and species], *Edinburgh Review* 111, no. 3, Article VIII (April 1860): 487–532.

### 6. Darwin's Bulldog

1. Thomas Huxley to Charles Darwin, November 23, 1859. (CUL. Classmark: DAR 98: B11–12/ Letter #: DCP-LETT-2544). In Frederick Burkhardt et al., eds., *The Correspondence of Charles Darwin*, Vol. 7 (Cambridge, UK: Cambridge University Press, in 30 volumes, 1985–2023), 390–91.
2. William Irvine, *Apes, Angels, and Victorians: The Story of Darwin, Huxley, and Evolution* (New York: McGraw-Hill, 1955), 1.
3. Adrian Desmond, *Huxley: The Devil's Disciple* (New York: Penguin Books, 1998.); see also Paul White, *Thomas Huxley: Making the "Man of Science"* (Cambridge, UK: Cambridge University Press, 2002); Leonard Huxley, *Life and Letters of Thomas H. Huxley* (London: Macmillan, 1900); Alison Bashford, *The Huxleys: An Intimate History of Evolution* (Chicago: University of Chicago Press, 2022).
4. John van Whye, "Why There Was No 'Darwin's Bulldog': Thomas Henry Huxley's Famous Nickname," *Linnean Society of London News*, July 1, 2019.
5. Henry Fairfield Osborn, "A Student's Reminiscences of Huxley," *Biological Lectures Delivered at the Marine Biological Laboratory of Woods Hole, in the Summer Session of 1895*, Vol. 4 (Boston: Ginn and Co., 1896), 29–42; quotes are from pp. 30, 37, and 32.
6. Bulldog, Definition No. 2. "A sheriff's officer (*obsolete*); one of the Proctors' attendants at the Universities of Oxford and Cambridge. *Colloquial.*" In *The Oxford*

*English Dictionary*. Accessed March 18, 2022 at https://www.oed.com/view/Entry/24523?rskey=2hNzIL&result=1&isAdvanced=false#eid.

7. Bert Kalmar, Harry Ruby, S. J. Perelman, and Will B. Johnstone, *Horse Feathers* [screenplay] (Hollywood, CA: Paramount Pictures, 1932). The song "Whatever It Is, I'm Against It" was written by Harry Ruby (music) and Bert Kalmar (lyrics). A few years later, the 1938 screwball comedy *Bringing Up Baby* was released; in it, Cary Grant played a spectacles-wearing paleontologist named David Huxley, an homage to the zoologist that, given Huxley's fame, would not be lost on the audience. Dudley Nichols and Hagar Wilde, *Bringing Up Baby* [screenplay] (Hollywood, CA: RKO Radio Pictures, 1938).
8. Andre Maurois, *Disraeli: A Picture of the Victorian Age*, trans. Hamish Miles (London: Penguin, 1940), 31.
9. L. Huxley, *Life and Letters of Thomas H. Huxley*, Vol. 1, 1–14.
10. Thomas H. Huxley, "Autobiography," in *Methods and Results: Collected Essays*, Vol. 1 (London: Macmillan, 1904), 1–17; L. Huxley, *Life and Letters of Thomas H. Huxley*, Vol. 1, 3– 4.
11. Cyril Bibby, *Scientist Extraordinary: The Life and Scientific Work of Thomas Henry Huxley, 1825–1895* (Oxford, UK: Pergamon Press, 1972), 1.
12. Hesketh Pearson, *Gilbert and Sullivan: A Biography* (London: Penguin Books, 1950), 10; Roderick Strange, ed., *John Henry Newman: A Portrait in Letters* (Oxford, UK: Oxford University Press, 2015), 25.
13. L. Huxley, *Life and Letters of Thomas H. Huxley*, Vol. 1, 6.
14. Huxley, "Autobiography," in *Methods and Results*, Vol. 1, 1–17; quotes are from pp. 5 and 6; Bibby, *Scientist Extraordinary*, 2; L. Huxley, *Life and Letters of Thomas H. Huxley*, Vol. 1 , 6.
15. *Annual Report for the Kent Lunatic Asylum, at Barming Heath Maidstone, for the Year 1854–1855, ending on July 5th* (Maidstone, UK: Walter Monckton, 1856), 30 ("May 17, G.H., male, an epileptic, found dead in bed. Inquest"); Bibby, *Scientist Extraordinary*, 1–2.
16. Huxley. "Autobiography," in *Methods and Results*, Vol, 1, 1–17; quotes are from pp. 5 and 6; L. Huxley, *Life and Letters of Thomas H. Huxley*, Vol. 1, 5–6.
17. Bibby, *Scientist Extraordinary*, 2; Bashford, *The Huxleys: An Intimate History of Evolution*.
18. Bibby. *Scientist* Extraordinary, 7.
19. Thomas H. Huxley, *Thoughts and Doings, 1840–1845*, Vol. 31 (174), Huxley Papers, Imperial College Archives, London; Huxley, "Autobiography," in *Methods and Results*, Vol. 1, 1–17. The underline is Huxley's.
20. Huxley, *Thoughts and Doings, 1840–1845*, [31. 174]; L. Huxley, *Life and Letters of Thomas H. Huxley*, Vol. 1, 11–12; Bashford, *The Huxleys: An Intimate History of Evolution*, 3–12.
21. Memoir written by Thomas Huxley, aboard the HMS *Rattlesnake*, Christmas, 1845. In L. Huxley, *Life and Letters of Thomas H. Huxley*, Vol. 1, 17.
22. Memoir written by Thomas Huxley, aboard the HMS *Rattlesnake*, Christmas, 1845.
23. Huxley, "Autobiography," in *Methods and Results*, Vol. 1, 6–7.

24. L. Huxley, *Life and Letters of Thomas H. Huxley*, Vol. 1, 18–19.
25. L. Huxley, *Life and Letters of Thomas H. Huxley*, Vol. 1, 19.
26. Thomas H. Huxley, "On a Hitherto Undescribed Structure in the Human Hair Sheath," *London Medical Gazette* 36 (1845): 1340–41.
27. Huxley, "Autobiography," in *Methods and Results*, Vol. 1, 6–7.
28. L. Huxley, *Life and Letters of Thomas H. Huxley*, Vol. 1, 25.
29. Thomas Huxley, *T. H. Huxley's Diary of the Voyage of H.M.S. Rattlesnake*, edited from the unpublished manuscript by Julian Huxley (London: Chatto and Windus, 1935).
30. T. H. Huxley, "On the Anatomy and the Affinities of the Family of the Medusae," *Philosophical Transactions of the Royal Society* 139 (1849): 413; T. H. Huxley, "Remarks upon the *Appendicularia* and *Doliolum*, Two Genera of the Tunicata," *Philosophical Transactions of the Royal Society* 2 (1851): 595–606; T. H. Huxley, "Observations upon the Anatomy and Physiology of *Salpa* and *Pyrosoma*," *Philosophical Transactions of the Royal Society* 2 (1851): 597–94; T. H. Huxley, "Notes on Medusae and Polyps," *Annals and Magazine of Natural History* 5 (1850): 66–67; T. H. Huxley, "Zoological Notes and Observations Made On Board the H.M.S. Rattlesnake," *Annals and Magazine of Natural History* 6 (1851): 304–306, 370–74; and 7 (1851): 433–442.
31. Thomas Huxley to Mrs. Elizabeth Scott, May 20, 1851; see: L. Huxley, *Life and Letters of Thomas H. Huxley*, Vol. 1, 80.
32. L. Huxley, *Life and Letters of Thomas H. Huxley*, Vol. 1, 80; Huxley, "Autobiography," in *Methods and Results*, Vol. 1, 1–17.
33. The term "mild hindoo" was an ironic but derogatory term from Charles Mackay's 1840 poem "The Hope of the World." In it, Mackay writes, "Taught by his creed, behold the mild Hindoo. Committing murders of the blackest hue." See Charles Darwin to Thomas Huxley (before November 12, 1857). (CUL. Classmark: Imperial College of Science, Technology and Medicine Archives, Huxley 5:58/ Letter # DCP-LETT-2166). In Burkhardt et al., *Correspondence*, Vol. 6, 484–85; see also Adrian Desmond and James Moore, *Darwin: The Life of a Tormented Evolutionist* (New York: W. W. Norton, 1991), 432–33; Adrian Desmond, *Archetypes and Ancestors in Victorian Paleontology, 1850–1875* (Chicago: University of Chicago Press, 1982), 20.
34. Thomas Huxley to Charles Darwin, April 14, 1874. (CUL. Classmark: DAR 166:333/ Letter #: DCP-LETT-9413). In Burkhardt et al., *Correspondence*, Vol. 22, 207.
35. L. Huxley, *Life and Letters of Thomas H. Huxley*, Vol. 1, 81–82; Huxley, "Autobiography," in *Methods and Results*, Vol. 1, 1–17.
36. L. Huxley, *Life and Letters of Thomas H. Huxley*, Vol. 1, 84.
37. L. Huxley, *Life and Letters of Thomas H. Huxley*, Vol. 1, 84.
38. L. Huxley, *Life and Letters of Thomas H. Huxley*, Vol. 1, 81; Huxley, "Autobiography," in *Methods and Results*, Vol. 1, 1–17.
39. Bibby, *Scientist Extraordinary*, 14.
40. L. Huxley, *Life and Letters of Thomas H. Huxley*, Vol. 1, 86.
41. T. H. Huxley to J. D. Hooker, July 6, 1855. In L. Huxley, *Life and Letters of Thomas H. Huxley*, Vol. 1, 128.
42. Charles Darwin to T. H. Huxley, September 29, 1855. (CUL. Classmark: Imperial College of Science, Technology and Medicine Archives, Huxley 5: 21; DAR 145:

222/ DCP-LETT-1757). In Burkhardt et al., *Correspondence*, Vol. 5, 441–44; Frederick Dyster to Thomas Huxley, December 3, 1860. [15.121–122]. Thomas H. Huxley Papers, Imperial College Archives, London.

43. L. Huxley, *Life and Letters of Thomas H. Huxley*, Vol. 1; quote is from p. 161 (letter from Thomas Huxley to his sister Lizzie, January 1, 1859); see also pp. 213, 220. Noel's scarlet fever progressed into an effusion of the brain (most likely streptococcal meningitis) and he died quickly after three days of illness; another child, their two-year-old daughter Jessie, was taken ill but only had a mild case and survived; their youngest child, Marian, who was still a baby, escaped illness. Between 1860 and 1866, they had five more children: Leonard, Rachel, Henrietta, Henry, and Ethel.
44. Charles Darwin to Thomas Huxley September 18, 1860. (CUL. Classmark: Imperial College of Science, Technology, and Medicine Archives, Huxley papers/Letter # DCP-LETT-2920B). In Burkhardt et al., *Correspondence*, Vol. 8, 365–66.
45. Henrietta's crisis of illness was from about October 20 to October 29, 1860. See Charles Darwin to Thomas Huxley, November 1 1860. (CUL. Classmark: Imperial College of Science, Technology, and Medicine Archives, Huxley Papers, 5:141/ Letter # DCP-LETT-2972). In Burkhardt et al., *Correspondence*, Vol. 8, 457–58; see also Charles Darwin to Asa Gray, October 31, 1860. (CUL. Classmark: Archives of the Gray Herbarium, Harvard University, [45 and 124a]/Letter #: DCP-LETT-2969). In Burkhardt et al., *Correspondence*, Vol. 8, 451.
46. A version of this phrase was originally used to describe Benjamin Disraeli ("he checkmated all the snobs of London") but it appears here, too, because it fits Huxley so well. Maurois, *Disraeli: A Picture of the Victorian Age*, 25.
47. L. Huxley, *Life and Letters of Thomas H. Huxley*, Vol. 1, 255–61.
48. Roy M. MacLeod, "The X-Club: A Social Network of Science in Late-Victorian England," *Notes and Records of the Royal Society of London* 24 (1970): 305–22.
49. L. Huxley, *Life and Letters of Thomas H. Huxley*, Vol. 1, 258.
50. E. F. Fiske, *The Life and Letters of John Fiske*, Vol. 1, ed. John Spencer Clark (Boston: Houghton Mifflin, 1917), 469.
51. J. Vernon Jensen, "Thomas Henry Huxley's Baptism into Oratory," *Notes and Records of the Royal Society of London* 30, no. 2 (1976): 181–207.
52. Leonard Huxley, *Life and Letters of Thomas H. Huxley*, Vol. 1, 297.
53. J.V. Jensen, "T. H. Huxley's Address at the Opening of the Johns Hopkins University in September 1876," *Notes and Records of the Royal Society of London* 47, no. 2 (1993): 257–69; "Prof. Huxley on University Education," *Nature* 14 (October 19, 1876): 546–50. Brunton was knighted by Queen Victoria in 1900 and is best known for prescribing amyl nitrate for those stricken with angina pectoris.
54. The quote in the text is from Gilman's journal, located at the Special Collections, Milton Eisenhower Library of the Johns Hopkins University Library, and cited in Fabian Franklin, *The Life of Daniel Coit Gilman* (New York: Dodd, Mead, and Co., 1910), 222–23. Upon his return to the United States, Gilman invited Huxley to Baltimore to deliver a major lectureship on modern education and science. Huxley did just that, on September 12, 1876, to an audience of more than two thousand people. The trustees decided to dispense with an opening benediction or prayer because it

was a scientific lecture and the university had already been blessed by clergymen when it was formally inaugurated the previous February 22. The scientists on the Johns Hopkins faculty praised Huxley's discourse. Offended by the absence of a prayer, many of the Catholic townspeople derided both the lecture and the lecturer. Huxley's many other gigs across "Yankee-land," as he called it, were far more successful. His tour lasted three months, and he gave full-house lectures everywhere he went, reaping thousands of pounds in lecture fees and book sales. See D. C. Gilman to T. H. Huxley, March 14, 1876. (HP. 17.51); D. C. Gilman to Mrs. Thomas Huxley, September 16, 1876. (H.P. 17.53), Thomas Huxley Papers, Archives, Imperial College, London; "Prof. Huxley on University Education," 546–50; William Peirce Randel, "Huxley in America," *Proceedings of the American Philosophical Society* 114, no. 2 (1970): 73–99.

55. Francis Darwin, *The Life and Letters of Charles Darwin*, Vol. 1 (London: John Murray, 1887), 222.
56. L. Huxley, *Life and Letters of Thomas H. Huxley*, Vol. 1, 176.
57. F. Darwin, *The Life and Letters of Charles Darwin*, Vol. 1, 222.
58. F. Darwin, *The Life and Letters of Charles Darwin*, Vol. 1, 222.
59. Thomas Huxley to Joseph D. Hooker, December 31, 1859. In L. Huxley, *Life and Letters of Thomas H. Huxley*, Vol. 1, 177.
60. The Francis Bacon reference is "For the inquiry of Final Causes is a barren thing, or as a virgin consecrated to God (produces nothing)." Anonymous [T. H. Huxley], "Darwin on the Origin of Species," *Times of London*, December 26, 1859, 8–9; see also T. H. Huxley, "The Darwinian Hypothesis," in *Collected Essays: Darwiniana*, Vol. 2 (London: Macmillan, 1907) 1–21; Francis Bacon, *The Advancement of Learning*, Chap. 5 (New York: The Colonial Press, 1900; originally published in 1605), 99.
61. Charles Darwin to Thomas Huxley, December 28, 1859. (CUL. Classmark: CUL. Classmark: Imperial College of Science, Technology, and Medicine Archives, Huxley Papers, 5:92/ Letter # DCP-LETT-2611). In Burkhardt et al., *Correspondence*, Vol. 7, 458–59.
62. Charles Darwin to Thomas Huxley, January 1, 1860. (CUL. Classmark: CUL. Classmark: Imperial College of Science, Technology, and Medicine Archives, Huxley Papers, 5:94/ Letter # DCP-LETT-2633). In Burkhardt et al., *Correspondence*, Vol. 8, 4–5.
63. Thomas H. Huxley, Preface, *Man's Place in Nature and Other Anthropological Essays*, authorized edition (New York: D. Appleton and Co., 1899; originally published in 1866), ix.
64. Charles Darwin, *On the Origin of Species by Means of Natural Selection, or the Preservation of Favoured Races in the Struggle for Life* (London: John Murray, 1859), 484.
65. Thomas H. Huxley, *Evidence as to Man's Place in Nature* (New York: D. Appleton and Co., 1863); see also Thomas H. Huxley, "On the Physical Basis of Life," *Fortnightly Review* 6 (new series) (1869):129–45.
66. Thomas H. Huxley, Preface. *Man's Place in Nature*, vi–vii.
67. Charles Darwin to T. H. Huxley, January 26, 1860. (CUL. Classmark: Imperial College of Science, Technology, and Medicine Archives, Huxley Papers 5:96/ Letter #

DCP-LETT-2646). In Burkhardt et al., *Correspondence*, Vol. 8, 52. The name of the poultry man was John Baily, Esquire Junior, of 113 Mount Street, Grosvenor Square, and the fee for the pigeons, including delivery by one man, was 2 shillings and 6 pence, and if two men were required, 5 shillings. Darwin instructed Huxley to be sure to write Baily "4 clear days" before the lecture to "tell him where & what hour the Pigeons must be sent" and to secure tickets for Baily to attend the lecture.

68. Weekly Evening Meeting, Friday, February 10, 1860 (Sir Henry Holland, MD, F.R.S., Vice-President in the Chair). Professor T. H. Huxley, F.R.S., "On Species and Races and Their Origin," *Notices of the Proceedings of the Meetings of the Members of the Royal Institution of Great Britain with Abstract of the Discourses Delivered at the Evening Meetings*, Vol. 3, 1858–1862 (London: William Clowes and Sons, 1862), 195–200; for discussion of the pigeons, see pp. 196–98; for a reported article on the lecture, see *Athenaeum*, March 3, 1860, 308.
69. Weekly Evening Meeting, Friday, February 10, 1860. Huxley, "On Species and Races and Their Origin," 195–200; quotes are from pp. 199–200.
70. Charles Darwin to Thomas H. Huxley, February 2, 1860. (CUL. Classmark: Imperial College of Science, Technology, and Medicine Archives, Huxley Papers, [5. 80]/ Letter # DCP-LETT-2679). In Burkhardt et al., *Correspondence*, Vol. 8, 64–65.
71. Charles Darwin to Charles Lyell, February 12, 1860. (CUL. Classmark: American Philosophical Society, Mss. B.D25.196/ Letter # DCP-LETT-2963). In Burkhardt et al., *Correspondence*, Vol. 8, 79–80.
72. Joseph D. Hooker to Asa Gray, March 16, 1860. Joseph Dalton Hooker Letters, 1844–1872. (HL Hook 1–3). Asa Gray Correspondence Files of the Gray Herbarium, Botany Libraries, Gray Herbarium Library, Harvard University Repository, Cambridge, MA.
73. Charles Darwin to Joseph D. Hooker, February 14, 1860. (CUL. Classmark: DAR 115:40/ Letter #: DCP-LETT-2696). In Burkhardt et al., *Correspondence*, Vol. 8, 84–85. Bold is Darwin's.
74. T. H. Huxley to Charles Darwin, November 23, 1859. (CUL. Classmark: DAR 98: B11–13/ DCP-LETT-2544). In Burkhardt et al., *Correspondence*, Vol. 7, 390–91. In his book review in the *Times* on December 26, 1859, Huxley wrote, "Nature does make jumps now and then and a recognition of that fact is of no small import in disposing of many minor objections to the doctrine of transmutation." Anonymous [T. H. Huxley], "Darwin on the Origin of Species," *Times of London*, December 26, 1859, 8–9; see also Sherrie L. Lyons, "The Origins of T. H. Huxley's Saltationism: History in Darwin's Shadow," *Journal of the History of Biology* 28, no. 3 (1995): 463–94; Stuart Mathieson. "Huxley: The Family that Championed Evolution," *Nature* 611 (2022): 228–30.
75. Evolutionary biologists today support both the slow continuous change Darwin described, or genetic drift, and so-called, discontinuous variation, or mutations. Because of the molecular structure of DNA, the jumps, or mutations, made at this level are on a discrete scale. That said, such mutations can have a big effect on the phenotypical expression of an organism. See Henry Fairfield Osborn, "Darwin's Theory of Evolution by the Selection of Minor Saltations," *American Naturalist* 46, no.

542 (1912): 76–82; R. G. Winther, "Darwin on Variation and Heredity," *Journal of the History of Biology* 33, no. 3 (2000): 425–55.

76. Thomas H. Huxley, "Time and Life: Mr. Darwin's 'Origin of Species,'" *Macmillan's Magazine*, 1, December 1859, 142–48.
77. T. H. Huxley, "The Origin of Species" [book review], *Westminster Review* 17 (n.s.) (April 1860): 541–70.
78. Huxley, "The Origin of Species" [book review], *Westminster Review*, April 1860; and Thomas H. Huxley, "The Origin of Species," in *Collected Essays: Darwiniana*, Vol. 2, 22–79. The Whitworth gun, a sniper rifle with a telescopic sight, was one of Britain's most advanced weapons of war at the time.
79. Huxley, "The Origin of Species" [book review], *Westminster Review*, April 1860, quotes are from pp. 567–69.
80. Charles Darwin to Charles Lyell, April 10, 1860. (CUL. Classmark: American Philosophical Society, Mss. B.D25.206/Letter #: DCP-LETT-2754). In Burkhardt et al., *Correspondence*, Vol. 8, 153–55.
81. Charles Darwin to Joseph D. Hooker, February 14, 1860. (CUL. Classmark: DAR 115:40/ Letter #: DCP-LETT-2696). In Burkhardt et al., *Correspondence*, Vol. 8, 84–85. Bold is Darwin's.
82. Sir Isaac Newton, *Principia Mathematica: The Mathematical Principles of Natural Philosophy*, trans. Andrew Motte (New York: Daniel Ader, 1846), 392.
83. Darwin, *On the Origin of Species*, 1st ed., Chap. 8, "Hybridism," 245–78; quote is on p. 245.
84. Charles Darwin to Thomas H. Huxley, January 14, 1862. (CUL Classmark: Imperial College of Science, Technology, and Medicine, Huxley Papers, 5:167/ Letter # DCP-LETT-3386). In Burkhardt et al., *Correspondence*, Vol. 10, 18–20. Darwin asked Huxley to read a paper he had just presented to the Linnean Society (November 21, 1861), which discussed this issue. Charles Darwin, "On the Two Forms, or Dimorphic Condition, in the Species of *Primula*, and on Their Remarkable Sexual Relations," *Journal of the Proceedings of the Linnean Society* (*Botany*) 6 (1862): 77–96; in Paul Barrett, ed., *Collected Papers of Charles Darwin*, Vol 2 (Chicago: University of Chicago Press, 1977), 45–63. In this paper, Darwin concludes, "Those who believe in the slow modification of specific forms will naturally ask themselves whether sterility may not have been slowly acquired for a distinct object, namely, to prevent two forms, whilst being fitted for distinct lines of life, becoming blended by marriage, and thus less well adapted for their new habits of life. But many great difficulties would remain, even if this view could be maintained." See also Charles Darwin to J. D. Hooker, September 28, 1861. In Burkhardt et al., *Correspondence*, Vol. 9, 283–86; T. H. Huxley to Charles Darwin January 20, 1862. (CUL. Classmark: DAR 166.2:291/ Letter # DCP-LETT-3396). In Burkhardt et al., *Correspondence*, Vol. 10, 33–35.
85. A. R. Wallace to Charles Darwin, March 1, 1868. (CUL. Classmark: DAR 106: B49–50, B53–5/ Letter #: DCP-LETT-5966). In Burkhardt, et al., *Correspondence*, Vol. 16, 219–23.
86. Charles Darwin to A. R. Wallace April 6, 1868. (CUL. Classmark: The British

Library, Add MS 46434: 125–9/ Letter #: DCP-LETT-6095). In Burkhardt et al., *Correspondence*, Vol. 16, 374–76.

## 7. The Dinosaur

1. Richard Owen, *On Parthenogenesis, or the Successive Production of Procreating Individuals from a Single Ovum. A Discourse Introductory to the Hunterian Lectures on Generation and Development for the Year 1849, Delivered at the Royal College of Surgeons* (London: John van Voorst, 1849), 3.
2. The best accounts of the critical reaction to Darwin's book can be found in Janet Browne, *Charles Darwin: The Power of Place*, Vol. 2 (London: Jonathan Cape; New York: Knopf, 2002), 83–125; and Adrian Desmond and James Moore, *Darwin: The Life of a Tormented Evolutionist* (New York: W. W. Norton, 1991), 485–518; see also Alvar Ellegard, "Public Opinion and the Press: Reactions to Darwinism," *Journal of the History of Ideas* 19, no. 3 (June 1958): 379–87; a list of the contemporary reviews of *On the Origin of Species*, circa 1859–1860, can be found in Frederick Burkhardt et al., eds., *The Correspondence of Charles Darwin*, Vol. 8 (Cambridge, UK: Cambridge University Press, in 30 volumes, 1985–2023), Appendix VII, 598–603, and Darwin Online, at www.darwin-online.org.uk.
3. Charles Darwin to T. H. Huxley, December 2, 1860. (CUL. Classmark: Imperial College of Science, Technology and Medicine Archives. Huxley 5:149/Letter DCP-LETT-3003.)
4. Huxley claimed that he coined the word at an 1869 meeting of the Metaphysical Club to describe his religious views. He later wrote, "[the word] simply means that a man shall not say he knows or believes that which he has no scientific grounds for professing to know or believe." He wrote three essays on the subject in 1889 for the February, March, and April issues of *The Nineteenth Century*. They were republished as "Agnosticism," "Agnosticism: A Rejoinder," and "Agnosticism and Christianity" in Thomas Huxley, *Collected Essays*, Vol. 5 (New York: D. Appleton and Co., 1902), 209–62, 263–308, 309–65, respectively.
5. For example, both Dickens and Darwin subscribed to and helped financially support the Jamaica Committee, protesting the suppression of Black Jamaicans revolting against white rule. Darwin did little more than send a check for £10. In 1866, Darwin warned Hooker that he would "shriek at me when you hear I have just subscribed to the Jamaica committee." By 1875, he told Herbert Spencer he could not even recall "whether [he] was nominally on the Committee of the Jamaica affair." See Charles Darwin to J. D. Hooker, November 20, 1866. (CUL. Classmark: DAR 115: 305; Letter #: DCP-LETT-5281). In Burkhardt et al., *Correspondence*, Vol. 14, 392–393; Charles Darwin to Herbert Spencer, November 13, 1875. (CUL. Classmark: University of London, Senate House Library/ MS. 791/111/ Letter #: DCP-LETT-10258). In Burkhardt et al., *Correspondence*, Vol. 23, 452–53; for a broader look at the affair, see Rande W. Kostal, *A Jurisprudence of Power: Victorian Empire and the Rule of Law* (Oxford, UK: Oxford University Press, 2009); Caroline Elkins, *Legacy of Violence: A History of the British Empire* (New York: Alfred A. Knopf, 2022); Darwin and Dick-

ens also knew and corresponded with several of the doctors working at the Great Ormond Street Children's Hospital, which was one of Dickens's favorite causes, and both were members of the Athenaeum Club.

6. Peter Ackroyd, *Dickens* (London: Sinclair-Stevenson, 1990), 663–65.
7. See, for example, Emma Wedgwood to Charles Darwin, December 26, 1838. (Cul. Classmark; DAR 204: 155/DCP-LETT-462). In Burkhardt et al., *Correspondence*, Vol. 2, 145–47.
8. Emma Darwin to William Darwin, May 6, 1859. (CUL. Classmark: DAR 210.6:44. Darwin Papers, Special Collections. Cambridge University Library.)
9. Emma Darwin to William Darwin, November 10, 1861. (CUL. Classmark: DAR 210.6:88. Darwin Papers, Special Collections. Cambridge University Library.)
10. Charles Dickens, "Our Nearest Relation," *All the Year Round* 1 (May 28, 1859), 112 (the weekly magazine conducted by Charles Dickens); Dickens and Owen were friends, and Owen was an occasional contributor to both of Dickens's magazines, *Household Words* and *All the Year Round*. See Adelene Buckland, "Charles Dickens, Man of Science," *Victorian Literature and Culture* 49, no. 3 (2021): 423–55; Howard Markel, "Charles Dickens' Work to Help Establish Great Ormond Street Hospital, London," *Lancet* 354 (1999): 673–75; Howard Markel, "Charles Dickens and the Art of Medicine," *Annals of Internal Medicine* 101 (1984): 408–11; Howard Markel, "The Childhood Suffering of Charles Dickens and His Literary Children," *Pharos* 48 (1985): 5–8.
11. Anonymous, "Species," *All the Year Round* 3 (1860): 174–77; Anonymous, "Natural Selection," *All the Year Round* 3 (1860): 293–99; Anonymous, "Transmutation of Species," *All the Year Round* 4 (1861): 519–21. In the first and second citation, Dickens was running Wilkie Collins's novel, *The Woman in White*, and in the last noted issue, Dickens was also running his latest (and for many, his best) novel, *Great Expectations*. Dickens scandalously left his wife Catherine in May 1858 and moved out of the family home.
12. Charles Darwin to Charles Lyell, June 14, 1860. (CUL. Classmark: American Philosophical Society, Mss. B. D25.216 / Letter #: DCP-LETT-2832). In Burkhardt et al., *Correspondence*, Vol. 8, 254–55.
13. Buckland, "Charles Dickens, Man of Science," 423–55.
14. Anonymous [Charles Dickens?], "Natural Selection," *All the Year Round* 3 (1860): 293–99; quote is from p. 299.
15. Robert FitzRoy (for the Great Britain Board of Trade), *Notes on Meteorology* (London: Printed by George E. Eyre and William Spottiswoode, for Her Majesty's Stationery Office, 1859); Robert FitzRoy, *Barometer Manual* (London: Printed by George E. Eyre and William Spottiswoode, for Her Majesty's Stationery Office, 1860); Robert FitzRoy, *The Weather Book: A Manual of Practical Meteorology* (London: Longman, Green, Longman, Roberts, and Green, 1863); Jim Burton, "Robert FitzRoy and the Early History of the Meteorological Office," *British Journal for the History of Science* 19, no. 2 (1986): 147–76.
16. Charles Darwin to William H. Dixon November 29, 1859. (CUL: Classmark: DAR 221, Darwin Papers, Special Collections, Cambridge University Library). In Bur-

khardt et al., *Correspondence*, Vol. 7, 414, n. 3; the original letter is missing, and this quote comes from another letter FitzRoy wrote to W. H. Dixon, editor of the *Athenaeum*, November 29, 1859.

17. [Robert FitzRoy], "Senex, Wise Old Man," *Times of London*, December 1, 1859, 8. It was Darwin's colleague, Leonard Horner, who made this estimate of human life in Egypt, in two papers that Darwin sponsored (rather than authored) for publication by the Royal Society. See Leonard Horner, "An Account of Some Recent Researches Near Cairo, Undertaken with the View of Throwing Light upon the Geological History of the Alluvial Land of Egypt" (Part I), *Philosophical Transactions of the Royal Society of London* 145 (1855): 105–38; and Leonard Horner, "An Account of Some Recent Researches Near Cairo, Undertaken with the View of Throwing Light upon the Geological History of the Alluvial Land of Egypt" (Part II), *Philosophical Transactions of the Royal Society of London* 148 (1858): 53–92.
18. Charles Darwin to Charles Lyell, December 3, 1859. (CUL. Classmark: American Philosophical Society, Mss. B.D25.182/ Letter #: DCP-LETT-2567). In Burkhardt et al., *Correspondence*, Vol. 7, 413–14.
19. Charles Darwin to Charles Lyell, December 3, 1859. (CUL. Classmark: American Philosophical Society, Mss. B.D25.182/ Letter #: DCP-LETT-2567). In Burkhardt et al., *Correspondence*, Vol. 7, 413–14; the italics are Darwin's; see also Richard Owen to Charles Darwin, November 12, 1859. (CUL. Classmark: Shrewsbury School, Taylor Library/ Letter #: DCP-LETT- 2526). In Burkhardt et al., *Correspondence*, Vol. 7, 373–74.
20. "Death of Sir Richard Owen," *Times of London*, December 19, 1892, 6. Owen published his first paper in 1826; his thesis was on urinary bladder stones, once the bane of men who dined on rich diets and when physicians still heeded the warnings of no less an authority than Hippocrates to "not use the knife, not even on sufferers from stone." They were instructed to leave that dicey procedure to only those who specialized in that craft. See Ludwig Edelstein, "The Hippocratic Oath: Text, Translation, and Interpretation," in *Ancient Medicine: Selected Papers of Ludwig Edelstein*, ed. O. Temkin and C. L. Temkin (Baltimore: Johns Hopkins University Press, 1967), 3–64; Howard Markel, "'I Swear by Apollo'—On Taking the Hippocratic Oath.," *New England Journal of Medicine* 350 (2004): 2026–29.
21. "Death of Sir Richard Owen," 6; Richard Owen, *The Life of Richard Owen* (London: John Murray, 1894).
22. "Death of Sir Richard Owen.," 6. Owen was initially assistant curator under William Clift, whose son was tapped to become the curator when the father retired. In 1834, after Clift Jr. fell out of a racing cab and died of head injuries, Owen promised the grieving father that "he would take the son's place and never desert him or the museum."
23. Owen. *The Life of Richard Owen*, Vol. 1, 26–58; J. H. Brooke, "Scientific Thought and Its Meaning for Religion: The Impact of French Science on British Natural Theology, 1827–1859," *Revue de Synthese* 110 (1989): 33–59; William Coleman. *Georges Cuvier: Zoologist* (Cambridge, MA: Harvard University Press, 1964).
24. "Death of Sir Richard Owen.," 6.

25. Scott F. Gilbert, "Owen's Vertebral Archetype and Evolutionary Genetics: A Platonic Appreciation," *Perspectives in Biology and Medicine* 23, no. 3 (1980): 475–88; R. Owen, "On the Anatomy of the Brachiopoda of Cuvier, and More Especially of the Genera *Terebratula* and *Orbicula*," *Transactions of the Zoological Society of London* 1 (1834): 145–64.
26. Richard Owen, "Report on British Fossil Reptiles, Part II," in *Annual Report of the Eleventh Meeting of the British Association for the Advancement of Science; Held at Plymouth in July 1841*, 60–204; Richard Owen, *A History of British Fossil Reptiles* (London: Cassell and Co., Ltd., 1849–1884).
27. Richard Owen, "Report on British Fossil Reptiles, Part II," 60–204; the naming of dinosaurs appears on p. 103.
28. Howard Markel, "The Origin of the Word 'Dinosaur'," *Science Friday*, NPR, July 6, 2015. Accessed 9 March, 2022, at https://www.sciencefriday.com/articles/the-origin-of-the-word-dinosaur/.
29. Charles Dickens, *Bleak House* (London: Bradbury and Evans, 1852), 1; see also Gowan Dawson, "Dickens, Dinosaurs and Design," *Victorian Literature and Culture* 44, no. 4 (2016): 761–78; Simon Guerrie, "Dickens and the Dinosaurs," *Lancet Psychiatry* 5, no. 8 (2018): e19; Henry Morley, "Our Phantom Ship on an Antediluvian Cruise," *Household Words* 3, no. 73 (August 16, 1851): 492–96. Conducted by Charles Dickens.
30. Steve McCarthy and Mick Gilbert, *The Crystal Palace Dinosaurs: The Story of the World's First Prehistoric Sculptures* (London: Crystal Palace Foundation, 1994). In 1842, Prime Minister Richard Peel placed Owen on the civil pension list (for an annual stipend of £200), and ten years later, 1853, Queen Victoria gave him a cottage in Richmond. Although Owen was offered knighthood in 1844, he did not accept the honor until 1884, which is why I do not refer to him as Sir Richard in this book.
31. G. H. Lewes, "Studies in Animal Life (Review of *On the Origin of Species*)," *Cornhill Magazine* 1 (January 1860): 438–47; quote is from p. 438.
32. Lewes, "Studies in Animal Life"; quote is from p. 447.
33. Charles Darwin to Richard Owen November 11, 1859. (CUL. Classmark: Shrewsbury School, Taylor Library/ Letter #: DCP-LETT-2515). In Burkhardt et al., *Correspondence*, Vol. 7, 365.
34. Charles Darwin to Charles Lyell, December 10, 1859. (CUL. Classmark: American Philosophical Society, Mss. B.D25. 184). In Burkhardt et al., *Correspondence*, Vol. 7, 421–23. The exclamation point and bold print are Darwin's.
35. Charles Darwin to Charles Lyell, December 10, 1859. (CUL. Classmark: American Philosophical Society, Mss. B.D25. 184). In Burkhardt et al., *Correspondence*, Vol. 7, 421–23. The exclamation point and bold print are Darwin's.
36. [Richard Owen], "Review of Darwin on the Origin of Species" [and other works: Alfred Wallace's 1859 Linnean Society paper, as well as papers by John Hooker, Flourens, Agassiz, Pouchet, the creationist Reverend Baden Powell, and some of Owen's recent work on paleontology and species], *Edinburgh Review* 111, no. 3, Article VIII (April 1860): 487–532; for an assessment of the *Review*'s influence, see John Clive, "*The Edinburgh Review*: 150 Years After," *History Today* 2, no. 12 (1952): 844–50.
37. Richard Owen, *On the Nature of Limbs: A Discourse* (London: John Van Voorst,

1849); facsimile version, edited by Ron Amundson (Chicago: University of Chicago Press, 2008), 3, 11, 59, 70, 112.

38. Richard Owen, *On Pathenogenesis, Or the Successive Production of Procreating Individuals from a Single Ovum* (London: John Van Voorst, 1849; Presidential Address, *Report of the Twenty-Eighth Meeting of the British Association for the Advancement of Science for 1858, Held at Leeds* (London: John Murray, 1859), xlix-cx; quote is from p. xc.

39. Nicholas Rupke, *Richard Owen: Victorian Naturalist* (Chicago: University of Chicago Press, 1994); Roy M. Macleod, "Evolutionism and Richard Owen, 1830–1868: An Episode in Darwin's Century," *ISIS* 56, no. 3 (1965): 259–280; Frank M. Turner, "The Victorian Conflict between Science and Religion: A Professional Dimension," *Isis* 69 (1978): 356–376; J. H. Brooke, "Why Did the English Mix Their Science and Their Religion?" in *Science and Imagination in Eighteenth-Century British Culture*, ed. Sergio Rossi (Milan: Edizioni Unicopli, 1987), 57–78.

40. Anonymous [Richard Owen], "Darwin on the Origin of Species," *Edinburgh Review* 111, no. 3 (1860):487–532; see also Richard Owen, "Report on the Archetype and Homologies of the Vertebrate Skeleton," *Report of the British Association for the Advancement of Science*, 1846, 169–340; Owen, *On Pathenogenesis, or the successive production of procreating individuals from a single Ovum*; Kevin Padian, "The Rehabilitation of Sir Richard Owen," *Bioscience* 47, no. 7 (1997: 446–453; Curtis N. Johnson, "Charles Darwin, Richard Owen, and Natural Selection: A Question of Priority," *Journal of the History of Biology* 52 (2019): 45–85; Evellene Richards, "A Question of Property Rights: Richard Owen's Evolutionism Reassessed," *British Journal for the History of Science* 20 (1987): 129–171; Richard Bellon, "Saints and Sinners: Sir Richard Owen," *Royal College of Surgeons of England Bulletin*. https://publishing.rcseng.ac.uk/doi/pdf/10.1308/003588413X13643054409180.

41. Thomas Huxley, *Man's Place in Nature and Other Anthropological Essays*, authorized edition (New York: D. Appleton and Co., 1899), 147; originally published in 1866.

42. Owen, Presidential Address, *Report of the Twenty-Eighth Meeting of the British Association for the Advancement of Science for 1858*, xlix-cx; quote is from p. lxxv.

43. [Owen], "Review of Darwin on the Origin of Species," 501–502, 522, 532.

44. [Owen], "Review of Darwin on the Origin of Species," 529–30.

45. Charles Darwin to Charles Lyell, December 10, 1859. (CUL. Classmark: American Philosophical Society, Mss. B.D25. 184). In Burkhardt et al., *Correspondence*, Vol. 7, 423.

46. M. Ruse, "Darwin's Debt to Philosophy: An Examination of the Influence of the Philosophical Ideas of John F. W. Herschel and William Whewell on the Development of Charles Darwin's Theory of Evolution," *Studies in History and Philosophy of Science* 6, no. 2 (1975): 159–81.

47. Michael Neve and Sharon Messenger, eds., *Charles Darwin: Autobiographies* (London: Penguin Books), 32, 36. The other influential book of his undergraduate days, he stated, was the German scientist and explorer Alexander von Humboldt's *Personal Narrative of Travels to the Equinoctial Regions of the New Continent during the Years 1799–1804*.

48. Charles Darwin to John Murray, April 9, 1860. (CUL. Classmark: National Library

of Scotland; John Murray Archive, Ms. 42152, ff. 90–91/DCP-LETT-2752). In Burkhardt et al., *Correspondence*, Vol. 8, 151–52.

49. Charles Darwin to Charles Lyell, April 10, 1860. (CUL. Classmark: American Philosophical Society, Mss. B. D25.206/ Letter #: DCP-LETT-2754). In Burkhardt et al., *Correspondence*, Vol. 8, 153–54. The italicized word "enjoyed" is Darwin's.
50. Charles Darwin to J. S. Henslow May 8, 1860. (CUL. Classmark: DAR 93: A70–71/ Letter #: DCP-LETT-2801). In Burkhardt et al., *Correspondence*, Vol. 8, 195–96.
51. Charles Darwin to Asa Gray, June 8, 1860. (CUL. Classmark: Gray Herbarium Archives, Harvard University, 50/ Letter #: DCP-LETT-2825). In Burkhardt et al., *Correspondence*, Vol. 8, 247–48.
52. Charles Darwin to John Hooker, March 4, 1874. (CUL. Classmark: DAR 95: 313–316/ Letter # DCP-LETT-9333). In Burkhardt et al., *Correspondence*, Vol. 22, 128–29.
53. Charles Darwin, *The Autobiography of Charles Darwin, 1809–1882*, ed. Nora Barlow (London: Collins, 1958), 104–105; see also Letter from Charles Darwin to Thomas Huxley, January 14, 1862. (CUL. Classmark: Imperial College of Science, Technology and Medicine Archives, Huxley 5: 167/ Letter #: 3386). In Burkhardt et al., *Correspondence*, Vol. 10, 18–20..
54. Owen, *The Life of Richard Owen*, Vol. 2, 269–73; quote is from p. 273.
55. "Death of Sir Richard Owen," *Times of London*, 6.

## 8. Soapy Sam

1. A. R. Ashwell, *Life of the Right Reverend Samuel Wilberforce, G.D., with Selections from His Diaries and Correspondence*, Vol. 1 (London: John Murray, 1880), xiii–xvi. The next two volumes of this biography were written by Wilberforce's son, Reginald, because Ashwell died after completing Volume 1.
2. Ambrose Bierce, *The Devil's Dictionary* (New York: Albert and Charles Boni, 1911), 235–36; originally published in 1906 as *The Cynic's Word Book*.
3. Reginald Wilberforce, *Life of the Right Reverend Samuel Wilberforce*, Vol. 2 (London: John Murray, 1881), 188.
4. Benjamin Disraeli, Speech on church policy and science delivered on Friday, November 25, 1864, at the third annual meeting of the Society for Augmenting Small Benefices in the Diocese of Oxford, held in the Sheldonian Theatre, at Oxford. Benjamin Disraeli, "On Church Policy, 1864," in *Church and Queen: Five Speeches Delivered by the Rt. Hon B. Disraeli, M.P., 1860–1864* (London: G. J. Palmer/Hamilton, Adams, & Co., 1865), 78.
5. Benjamin Disraeli, *Lothair* (London: Longmans, Green and Co., 1870); Andre Maurois, *Disraeli: A Picture of the Victorian Age*, trans. Hamish Miles (London: Penguin, 1940), 13–25, 124, 164, 210; Megan Dent, "There Must Be Design: The Threat of Unbelief in Disraeli's *Lothair*," *Victorian Literature and Culture* 44, no. 3 (2016): 671–86.
6. Letter from Samuel Wilberforce to Miss Marianne Thornton, May 28, 1870. In R. G. Wilberforce, *Life of Samuel Wilberforce*, Vol. 3, 344.
7. David Newsome, "How Soapy Was Sam? A Study of Samuel Wilberforce," *History Today* 13, no. 9 (September 1963): 624–32.

8. *Oxford English Dictionary*. Definition (No. 5a, slang) of "Soapy." Available online at https://www.oed.com/view/Entry/183682?redirectedFrom=soapy#eid. See also Diary entry for June 22, 1854. In Ellen Dwight Twistleton, *Letters of the Hon. Mrs. Edward Twistleton, Written to Her Family, 1852–1862* (London: John Murray, 1928), 202.
9. Wilberforce was hardly the only personage to be spoofed in this novel; for example, Trollope described Thomas Carlyle and Charles Dickens as "Dr. Pessimist Anticant" and "Mr. Popular Sentiment," respectively. Anthony Trollope, *The Warden*, new ed. (London: Longman, Green and Co., 1886), 78–79, 149–63.
10. R. G. Wilberforce, *Life of Samuel Wilberforce,* Vol. 3, 1–11.
11. Francis C. Legge, "Samuel Wilberforce." In *Dictionary of National Biography*, Vol. 61 (London: Smith, Elder and Co., 1885), 204–208.
12. The book, which was published four months after *Origin*, sold over twenty-two thousand copies in two years. The essays were written by churchmen Frederick Temple, Rowland Williams, Baden Power, Henry Bristow Wilson, Mark Pattison, and Benjamin Jowett and the layman Charles Wycliffe Goodwin. See John W. Parker, ed., *Essays and Reviews* (London: Longman, Green, Longman and Roberts, 1860); Anonymous [S. Wilberforce], Book Review of *Essay and Reviews*, in *Quarterly Review* 109 (1861): 248–301; Newsome, "How Soapy Was Sam?" 624–32; Adrian Desmond, *Huxley: From Devil's Disciple to Evolution's High Priest* (London; Penguin Books, 1997), 278.
13. *Pall Mall Gazette*, October 28, 1865, 5; Newsome, "How Soapy Was Sam?" 624–32. (Some have ascribed this witticism to Percy Smythe, the Lord Strangford, who may or may not have said it in 1845.)
14. Letter from William Wilberforce to Samuel Wilberforce, October 12, 1833. In Ashwell, *Life of the Right Reverend Samuel Wilberforce*, Vol 1, 21, 31. The italicized word is William Wilberforce's; see also Samuel and Robert Wilberforce, *Life of William Wilberforce* (London: John Murray, 1838).
15. Ashwell, *Life of the Right Reverend Samuel Wilberforce*, Vol. 1, 29.
16. Ashwell, *Life of the Right Reverend Samuel Wilberforce*, Vol. 1, 32–33.
17. Ashwell, *Life of the Right Reverend Samuel Wilberforce*, Vol. 1, 54.
18. Samuel's eldest brother, William Jr., remained an Anglican but proved an eternal embarrassment beginning with a raucous college career at Cambridge, engaging in a disastrous series of financial speculations, and for being unseated in Parliament over residency issues after serving only one year (1837–1838). In 1831, Junior had so many creditors, he fled to Naples. See Anne Stott, *Wilberforce: Family and Friends* (Oxford, UK: Oxford University Press, 2012), 260–61. Newman was canonized as a saint by the Catholic Church in 2009.
19. John Henry Newman to J. W. Bowden, November 6, 1838 (Vol. 6, 337); and John Henry Newman to H. A. Woodgate, September 22, 184. (Vol. 8, 27). In Stephen C. Dessain, ed., *Letters and Diaries of John Henry Newman* (Oxford, UK: Oxford University Press, 1978–2008).
20. Jerome Grosclaude, "'An Amazing Want of Christianity': Bishop Samuel Wilberforce and John Henry Newman's Tense Relationship," *Journal of Anglican Studies* 20, no. 1 (2022): 98–116.

21. Ashwell, *Life of the Right Reverend Samuel Wilberforce*, Vol. 1, 45.
22. Ashwell, *Life of the Right Reverend Samuel Wilberforce*, Vol 1, 53.
23. Ashwell, *Life of the Right Reverend Samuel Wilberforce*, Vol 1, 177–92.
24. Ashwell, *Life of the Right Reverend Samuel Wilberforce*, Vol 1, 176, 211, 241.
25. Arthur Burns, "Samuel Wilberforce." In *The Oxford Dictionary of National Biography* (Oxford, UK: Oxford University Press). This article was first published in print in 1900. Available online at https://doi.org/10.1093/odnb/9780192683120.013.29385.
26. Samuel (Wilberforce), Lord Bishop of Oxford, "Address III. The Sufficiency of the Holy Scriptures," in *Addresses to the Candidates for Ordination, on the Questions in the Ordination Service* (Oxford: J. H. and Jas. Parker; London: F. and J. Rivington, 1860), 37–55; quote is from p. 37.
27. Ashwell, *Life of the Right Reverend Samuel Wilberforce*, Vol. 1, 318–21; quote is from p. 319.
28. Ashwell, *Life of the Right Reverend Samuel Wilberforce*, Vol. 1, xvii.
29. John A. S. Abecasis-Phillips, "Prince Albert and the Church—Royal versus Papal Supremacy in the Hampden Controversy," in *Prinz Albert—Ein Wettiner in Großbritannien*, ed. John R, David, Vol. 22 in the Prince Albert Society Series (Munich: K. G. Saur, 2003), 95–110.
30. John William Burgon, "Samuel Wilberforce: The Remodeler of the Episcopate," in *The Lives of Twelve Good Men*, Vol. 2 (London: John Murray, 1888–1889); quotes are from pp. 46 and 48, respectively. The italics are Burgon's.
31. Ashwell, *Life of the Right Reverend Samuel Wilberforce*, Vol. 1, 419; Chapter 11 of this book is titled "The Hampden Controversy," 419–515.
32. Renn Dickson Hampden, *The Scholastic Philosophy Considered in Its Relation to Christian Theology in a Course of Lectures Delivered before the University of Oxford, in the year 1832 at the Lecture Founded by John Bampton* (Oxford, UK: S. Collingwood, Printer to the University for the Author and Sold by J. H. Parker, 1833), 39.
33. Ashwell, *Life of the Right Reverend Samuel Wilberforce*, Vol. 1, 421.
34. John Henry Newman, *History of My Religious Opinions* (London: Longman, Green, Longman, Roberts, and Green, 1865), 57.
35. Renn Dickson Hampden, *Observations on Religious Dissent* (Oxford, UK: S. Collingwood, Printer to the University for the Author and Sold by J. H. Parker, 1834).
36. Abecasis-Phillips, "Prince Albert and the Church," 95–110; see 104–105.
37. Letter from the Bishop of Oxford to Professor Renn Hampden, December 28, 1847. Reprinted in Ashwell, *Life of the Right Reverend Samuel Wilberforce*, Vol 1, 482–88, 494.
38. R. Wilberforce, *Life of the Right Reverend Samuel Wilberforce*, Vol. 2, 275; G. E. Buckle, ed., *The Letters of Queen Victoria*, Vol. 2, 2nd series (London: John Murray, 1926), 264.
39. Burgon, "Samuel Wilberforce: The Remodeler of the Episcopate," 17.
40. For a revisionist account of Wilberforce as "not being such a dinosaur after all," see R. Wrangham, "The Bishop of Oxford: Not so Soapy," *New Scientist* 83 (August 9, 1979): 450–51.

41. Anonymous [Samuel Wilberforce], "On the Origin of Species, by means of Natural Selection, or the Preservation of Favoured Races in the Struggle for Life. By Charles Darwin, M.A., F.R.S.," *Quarterly Review* 108 (1860): 226–64.
42. Anonymous [Wilberforce], "On the Origin of Species," 226.
43. Anonymous [Wilberforce], "On the Origin of Species," 231.
44. Anonymous [Wilberforce], "On the Origin of Species," 226–64; quote is from p. 258; see also R. Wrangham, "The Bishop of Oxford: Not so Soapy," 450–51.
45. Anonymous [Wilberforce], "On the Origin of Species," 234, 249.
46. Anonymous [Wilberforce], "On the Origin of Species," 253.
47. Anonymous [Wilberforce], "On the Origin of Species," 264. The italics are the author's (H.M.).
48. Charles Darwin Reprint Collection. (CUL. Classmark: DAR, Pamphlet R34), Darwin Papers, Cambridge University Library); Desmond and Moore, *Darwin: The Life of a Tormented Evolutionist*, 499.
49. Charles Darwin to J. D. Hooker, July 2, 1860. (CUL. Classmark: DAR 115: 64/ Letter #: DCP-LETT-2853). In Frederick Burkhardt et al., eds., *The Correspondence of Charles Darwin*, Vol. 8 (Cambridge, UK: Cambridge University Press, in 30 volumes, 1985–2023), 272–73.
50. Charles Darwin to Thomas Huxley, July 20, 1860. (CUL. Classmark: Huxley Papers, Archives, Imperial College, London/DCP-LETT- 2873). In Burkhardt et al., *Correspondence*, Vol. 8, 294–96.
51. Charles Darwin to Asa Gray, July 22, 1860. (CUL. Classmark: Gray Herbarium Archives, Harvard University, 30/ Letter #: DCP-LETT-2876) In Burkhardt et al., *Correspondence*, Vol. 8, 298–99.
52. Charles Darwin to Charles Lyell, July 30, 1860. (CUL. Classmark: American Philosophical Society, Mss. B.D25.222/ Letter #: DCP-LETT-2881). In Burkhardt et al., *Correspondence*, Vol. 8, 306–307.
53. Charles Darwin to John Murray, August 3, 1860. (CUL. Classmark: National Library of Scotland, John Murray Archive, Ms. 42152, ff. 74–75/ Letter #: DCP-LETT-2895). In Burkhardt et al., *Correspondence*, Vol. 8, 309.

## 9. A Mysterious Malady

1. Charles Darwin to Dr. Henry Bence Jones, January 3, 1866. (CUL. Classmark DAR 249:86/Letter # DCP-LETT-4968A). In Frederick Burkhardt et al., eds., *The Correspondence of Charles Darwin*, Vol. 14 (Cambridge, UK: Cambridge University Press, in 30 volumes, 1985–2023), 3–4.
2. See, for example, Janet Browne, *Charles Darwin: Voyaging. Volume I of a Biography* (London: Jonathan Cape; New York: Knopf, 1995); Janet Browne, *Charles Darwin: The Power of Place. Volume II of a Biography* (London: Jonathan Cape; New York: Knopf, 2002).
3. Charles Darwin, Notes for His Physician, John Chapman. Albert and Shirley Small Special Collections Library, Darwin Evolution Collection, University of Virginia, Charlottesville, VA (UVa-Darwin-Evolution-3314–1.43).

4. Charles Darwin to J. D. Hooker, June 17, 1847. (CUL. Classmark: DAR 114: 96/Letter # DCP-LETT-1098). In Burkhardt et al., *Correspondence*, Vol. 4, 51.
5. Charles Darwin, "Diary of Health, 1849–1855." Archives of Down House/Darwin Museum, Down, Kent, UK. This diary has been reprinted in the most extensive account of Darwin's many ills: Ralph Colp Jr., *Darwin's Illness* (Gainesville, FL: University Press of Florida, 2008), 187–257, which is the revised edition of Dr. Colp's earlier volume, *To Be an Invalid: The Illness of Charles Darwin* (Chicago: University of Chicago Press, 1977); see also Mike Dixon and Gregory Radick, *Darwin in Ilkley* (Stroud, UK: The History Press, 2009).
6. Francis Darwin, ed., *Life and Letters of Charles Darwin*, Vol. 1 (London: John Murray, 1887–1888), 135; Henrietta Litchfield, ed., *Emma Darwin: A Century of Family Letters, 1792–1896* (London: John Murray, 1915). Henrietta, or Etty, was a most sickly child and adult who long modeled her beloved father's symptoms.
7. Gwen Raverat, *Period Piece: A Cambridge Childhood* (London: Faber and Faber, 1952), 121–22.
8. James Johnson, *An Essay on the Morbid Sensibility of the Stomach and Bowels, as the Proximate Cause or Characteristic Condition of Dyspesy, Nervous Irritability, Mental Despondency, Hypochondriasis, and Many other Ailments of Body and mind to which are added, Condition on the Diseases and Regimen of Invalids on Their Return from Hot and Unhealthy Climates* (London: S. Highley, 1826), 1. This book went through several editions during the nineteenth century.
9. John Henry Clarke, *Indigestion: Its Causes and Cure*, 3rd ed. (London: James Epps and Co., 1886), 8; Clarke was also well known as a virulent anti-Semite who headed a nasty group of florid Jew-haters known as the Britons.
10. Andre Maurois, *Disraeli: A Picture of the Victorian Age*, trans. Hamish Miles (London: Penguin, 1940), 64.
11. Colp, *To Be an Invalid*, 12, 22, 37, 45–46, 65, 76, 78–80.
12. Charles Darwin to Joseph D. Hooker, March 28, 1849. (CUL. Classmark DAR 114:113/Letter # DCP-1236). In Burkhardt et al., *Correspondence*, Vol. 4, 227–29.
13. James M. Gully, *The Water Cure in Chronic Disease: An Exposition of the Causes, Progress and Terminations of Various Chronic Diseases of the Digestive Organs, Lungs, Nerves, Limbs, and Skin; and of Their Treatment by Water and Other Hygienic Means* (London: J. Churchill, 1846). This book went through many editions over the last half of the nineteenth century.
14. Dixon and Radick, *Darwin in Ilkley*, 36–43, 47, 49, 94, 122–23; Janet Browne, "Spas and Sensibilities: Darwin at Malvern," *Medical History*, Supplement 10 (1990): 102–13.
15. James Johnson, *Pilgrimages to the Spas in Pursuit of Health and Recreation; with an Inquiry into the Comparative Merits of Different Mineral Waters: The Maladies to Which They Are Applicable and Those in Which They Are Injurious* (London: S. Highley, 1841). This was the same Dr. Johnson who wrote a treatise on the stomach and digestion (see footnote 19); see also Richard Metcalfe, *The Rise and Progress of Hydropathy in England and Scotland* (London: Simpkin, Marshall, Hamilton, Kent & Co., 1906); Howard Markel, *The Kelloggs: The Battling Brothers of Battle Creek* (New York: Pantheon Books, 2017).

16. G. H. Lewes, *The Physiology of Common Life*, Vol. 1 (Edinburgh and London: William Blackwood and Sons, 1859), 190; see also, R. E. Smith, "George Henry Lewes and His *Physiology of Common Life*, 1859," *Proceedings of the Royal Society of Medicine* 53, no. 7 (1960): 569–74; A. Barrat, "*Physiology of Common Life* by G. H. Lewes: A Contribution to the Scientific Edification of the Victorians," *George Eliot, George Henry Lewes Newsletter*, no. 18/19 (1991): 10–21.
17. Charles Darwin to Joseph Hooker, March 28, 1849. (CUL. Classmark DAR 114:113/ Letter # DCP-1236). In Burkhardt et al., *Correspondence*, Vol. 4, 227–29.
18. Charles Darwin to W. D. Fox, July 7, 1849. (CUL. Classmark: Christ's College Library, Cambridge/MS 53, Fox 74/Letter # DCP-LETT-1249). In Burkhardt et al., *Correspondence*, Vol. 4, 246–47.
19. Charles Darwin to J. D. Hooker, October 12, 1849. (CUL. Classmark: DAR 114:116/ Letter # DCP-LETT-1260). In Burkhardt et al., *Correspondence*, Vol. 4, 268–71; see also F. Darwin, ed., *The Life and Letters of Charles Darwin*, Vol. 1, 346–47.
20. Dixon and Radick, *Darwin in Ilkley.*
21. Charles Darwin to J. D. Hooker, April 29, 1857. (CUL. Classmark: DAR 114:194/Letter # DCP-LETT-2084). In Burkhardt et al., *Correspondence*, Vol. 6, 384–85.
22. Charles Darwin to Emma Darwin, April 28, 1858. (CUL. Classmark: DAR 210.8:34/ Letter # DCP-LETT-2261). In Burkhardt et al., *Correspondence*, Vol. 7, 84.
23. Dixon and Radick, *Darwin in Ilkley*, 116–18.
24. John Chapman, *Functional Diseases of the Stomach. Part I: Seasickness: It's Nature and Treatment* (London: Trübner and Co., 1864), 5–6.
25. Darwin paid £10 and 10 shillings for the consultation. Charles Darwin, Notes for His Physician, John Chapman, May 21, 1865 (Emma Darwin's hand). (UVa-Darwin-Evolution-3314–1.43). Albert and Shirley Small Special Collections Library, Darwin Evolution Collection, University of Virginia, Charlottesville, VA; Charles Darwin to John Chapman, May 16, 1865. University of Virginia Library, Special Collections (3314, 1; 42); see also Charles and Emma Darwin to J. D. Hooker, July 10, 1865. (CUL. Classmark: DAR 115 272/ Letter # DCP-LETT-4868). In Burkhardt et al., *Correspondence*, Vol. 13, Supplement 1822–1864, 194–95.
26. The most complete analysis of Darwin's health can be found in Ralph Colp Jr., *To be an Invalid. The Illness of Charles Darwin* (Chicago: University of Chicago Press, 1977), first edition; which was updated and republished as *Darwin's Illness* (Gainesville, FL: University of Florida Press, 2008).
27. Colp, *To Be an Invalid*, updated as *Darwin's Illness*; Janet Browne, "I Could Have Retched All Night. Charles Darwin and His Body," in C. Lawrence, S. Shapin, eds., *Science Incarnate: Historical Embodiments of Natural Knowledge*, ed. C. Lawrence and S. Shapin (Chicago: University of Chicago Press, 1998), 240–87; Browne, *Charles Darwin: The Power of Place*; Browne, *Charles Darwin: Voyaging*; Desmond and Moore, *Darwin: The Life of a Tormented Evolutionist*; David Quammen, *The Reluctant Mr. Darwin: An Intimate Portrait of Charles Darwin and the Making of His Theory of Evolution* (New York: W. W. Norton/Atlas Books, 2006); see also Anonymous, "Charles Robert Darwin FRS, Obituary," *Lancet* i (1882): 714; W. W. Johnston, "The Ill Health of Charles Darwin: Its Nature and Its Relation to His Work," *American Anthropolo-*

*gist* 3 (1901):157; G. M. Gould, "Charles Darwin," *Biographic Clinics* 1 (1903): 103; G. G. Simpson, "Charles Darwin in Search of Himself," *Scientific American* 199 (1958): 117–22; Saul Adler, "Darwin's Illness," *Nature* 124 (1959): 1102–3; D. Stetten, "Gout," *Perspectives in Biology and Medicine* 2, no. 2 (winter 1959): 194–95; L. A. Kohn, "Charles Darwin's Chronic Ill Health," *Bulletin of the History of Medicine* 37 (1963): 239–56; H. J. Roberts, "Reflections on Darwin's illness," *Journal of Chronic Disease* 19 (1966): 723–25; J. H. Winslow, *Darwin's Victorian Malady: Evidence for Its Medically Induced Origin* (Philadelphia: American Philosophical Society, 1971); F. Smith, "Charles Darwin's Ill Health," *Journal of the History of Biology* 23 (1990): 443–59; F. Smith, "Charles Darwin's Health Problems: The Allergy Hypothesis," *Journal of the History of Biology* 25 (1992): 285–306; S. B. Matthews, J. P. Waud, A. G. Roberts, et al., "Systemic Lactose Intolerance: A New Perspective on an Old Problem," *Postgraduate Medical Journal* 81 (2005): 167–73; E. Kempf, "Charles Darwin—The Affective Sources of His Inspiration and Anxiety," *Neurosis Psychoanalytical Review* 5 (1918): 167; D. Hubble, "Charles Darwin and Psychotherapy," *Lancet* 241, no. 6231 (1943): 129–33; W. C. Alvarez, "The Nature of Charles Darwin's Lifelong Ill-Health," *New England Journal of Medicine* 261 (1959): 1109–12; A. K. Campbell, "What Darwin Missed in Fred Hoyle's Universe: A Memorial Symposium to Sir Fred Hoyle," *Astrophysics Space Science* 285 (2004): 1–5; A. W. Woodruff, "Darwin's Health in Relation to His Voyage to South America," *BMJ* 1 (1965): 745–50; A. W. Woodruff, "Darwin's Illness," *Israel Journal of Medical Science* 26, no. 3 (1990): 163–64; A.W. Woodruff, "The Impact of Darwin's Voyage to South America on His Work and Health," *Bulletin of the New York Academy of Medicine* 44, no. 6 (1968): 661–672; J. Bolby, *Charles Darwin* (London: Hutchinson, 1990); J. Bowlby, "Darwin's Health," *British Medical Journal* 1 (1956): 999; R. E. Bernstein, "What Was Charles Darwin's Illness?" *South African Medical Journal* 61 (1982): 939; T. J. Barloon and R. Noyes, Charles Darwin and Panic Disorder," *Journal of the American Medical Association* 277 (1997): 138–41; F. Darwin, ed., *Life and Letters of Charles Darwin* (London: John Murray, 1887; New York: D. Appleton, 1889); G, Raverat. *Period Piece* (London: Faber and Faber, 1952), 19; R. Keynes, *Annie's Box: Charles Darwin, His Daughter and Human Evolution* (London: Fourth Estate, 2001), 110; George Pickering, *Creative Malady: Illness in the Lives and Minds of Charles Darwin, Mary Baker Eddy, Sigmund Freud, Florence Nightingale, Marcel Proust, and Elizabeth Barret Browning* (Oxford: Oxford University Press, 1974); Buckston Browne, "Darwin's Health," *Nature* 151 (January 2, 1943): 14–15; Ralph Colp, "More on Darwin's Illness," *History of Science* 38 (2000): 219–36; F. Orrego and C. Quintana, "Darwin's Illness: A Final Diagnosis," *Notes and Records of the Royal Society of London.* 61 (2007): 23–29; Louis Heyse-Moore, "Charles Darwin's (1809–1882) Illness—The Role of Post-Traumatic Stress Disorder," *Journal of Medical Biography* 27, no. 1 (2019): 13–25; D. A. Young, "Darwin's Illness and Systemic Lupus Erythematosus," *Notes and Records of the Royal Society of London* 51: (1997): 77–86; A. G. Gordon, "The Dueling Diagnoses of Darwin," *Journal of the American Medical Association* 16 (1997): 1276–77; Claudia Kalb, "Evolution and Angst: Charles Darwin Was a Worrier," excerpt in *ScientificAmerican.com* (February 11, 2016), from C. Kalb, *Andy Warhol Was a Hoarder: Inside the Minds of History's Great Personalities* (Washington, DC: National Geographic Books, 2016).

28. Charles Darwin, *Journal of Researches into the Natural History and Geology of the Countries Visited during the Voyage of H.M.S. Beagle Round the World, under the Command of Captain R.N. Fitz Roy*, 2nd ed. (London: John Murray, 1845), 330.
29. Adler, "Darwin's Illness," 1102–3; see also, Woodruff, "Darwin's Health in Relation to His Voyage to South America," 745–50; J. H. Goldstein, "Darwin, Chagas' Mind and Body," *Perspectives in Biology and Medicine* 32, no. 4 (1989): 586–601; Peter Salwen, "Chagas Disease Claimed an Eminent Victim," Letter to the Editor, *New York Times*, June 15, 1989, A30; C. Botto-Mahan and R. Medel, "Was Chagas Disease Responsible for Darwin's Illness? The Overlooked Eco-Epidemiological Context in Chile," *Revista Chilena de Historia Natural* 94 (2021): 7.
30. John Hayman, "Charles Darwin's Mitochondria," *Genetics* 194 (2013): 21–25; Winslow, *Darwin's Victorian Malady: Evidence for Its Medically Induced Origin* (Philadelphia: American Philosophical Society, 1971); Smith, "Charles Darwin's Ill Health," 443–59; Smith, "Charles Darwin's Health Problems: The Allergy Hypothesis," 285–306; Matthews, Waud, Roberts, et al, "Systemic Lactose Intolerance: A New Perspective on an Old Problem," 167–73; Kemf, "Charles Darwin—The Affective Sources of His Inspiration and Anxiety," 167.
31. Hubble, "Charles Darwin and Psychotherapy," 129–33; Alvarez, "The Nature of Charles Darwin's Lifelong Ill-Health," 1109–12.
32. Charles Darwin to Robert FitzRoy, October 1, 1846. (CUL. Classmark: DAR 144: 119/ DCP-LETT-1002). In Burkhardt et al., *Correspondence*, Vol. 3, 344–45; in 1863, he described his life as that of a "hermit." See Charles Darwin to C. T. Whitley, June 20, 1863. (CUL. Classmark: Shrewsbury School, Taylor Library/ Letter # DCP-LETT-4217A). In Burkhardt et al., *Correspondence*, Vol. 11, 500–501.
33. Francis Darwin, "Reminiscences of My Father's Everyday Life" (Chapter 3), in *The Life and Letters of Charles Darwin*, Vol. 1, 85–136; R. B. Freeman, *Charles Darwin. A Companion* (Hamden, CT: Archon Books, 1978), 88–89.
34. Browne, *Charles Darwin: The Power of Place*, 10, 165, 209.
35. Charles Darwin to Henry Fawcett, September 18, 1861. (CUL. Classmark: Karpeles Library Museum/DCP-LETT-3257). In Burkhardt et al., *Correspondence*, Vol. 9, 269.
36. Leonard Darwin, "Memories of Down House," *The Nineteenth Century* 106 (1929): 118–23; quote is from p. 120.
37. Charles Darwin to J. D. Hooker, December 29, 1860. (CUL. Classmark: DAR 115:83/ Letter # DCP-LETT-3034). In Burkhardt et al., *Correspondence*, Vol. 8, 541.
38. Charles Darwin, "Autobiography, 1876," in *The Life and Letters of Charles Darwin*, ed. Francis Darwin, Vol. 1, 85.
39. William Allingham, "Recollections of Darwin," in *A Diary*, ed. Helen Paterson Allingham and Dollie Radford (London: Macmillan, 1907), 184–85.
40. Letter to Joseph D. Hooker, February 20–22, 1864. (CUL. Classmark: DAR 115: 221a-c/Letter # CP-LETT-4412). In Burkhardt et al., *Correspondence*, Vol. 12, 56–59.
41. Emma Darwin, Mrs. Charles Darwin's Recipe Book at Down. (CUL. Classmark: DAR 214–0–157. Special Collections, Cambridge University Library. This manuscript has been republished as Dusha Bateson and Weslie Janeway, eds, *Mrs. Charles Darwin's Recipe Book, Revived and Illustrated* (New York: Glitterati, 2008); the burnt

cream recipe can be found on p. 120 and in a facsimile of her handwriting in Appendix I, p. xi.

42. Charles Darwin to Susan Elizabeth Darwin, November 27, 1844 [?]. (CUL. Classmark: DAR 92: A9–10/ Letter # DCP-LETT-833). In Burkhardt et al., *Correspondence*, Vol. 3, 86–87.
43. Charles Darwin to Susan Elizabeth Darwin, March 19, 1849. (CUL. Classmark: DAR 92: A7–A8. Letter # DCP-LETT-1234). In Burkhardt et al., *Correspondence*, Vol. 4, 224–26.
44. Charles Darwin to Henry Bence Jones, January 3, 1866. (CUL. Classmark: DAR 249: 86/DCP-LETT-4968A). In Burkhardt et al., *Correspondence*, Vol. 14, 3–4.
45. A. K. Campbell and S. B. Matthews, "Darwin's Illness Revealed," *Postgraduate Medical Journal* 81 (2005): 248–51; Matthews, Waud, Roberts, and Campbell, "Systemic Lactose Intolerance: A New Perspective on an Old Problem," 167–73; Dixon and Radick, *Darwin in Ilkley*.
46. A. Dahlqvist, J. B. Hammond, R. K. Crane, J. V. Dunphy, and A. Littman, "Intestinal Lactase Deficiency and Lactose Intolerance in Adults: Preliminary Report," *Gastroenterology* 45 (1963): 488–91; P. Cuatrecasas, D. H. Lockwood, and J. Caldwell, "Lactase Deficiency in the Adult," *Lancet* 285, no. 7375 (1965): 14–18; H. B. McMichael, J. Webb, and A. M. Dawson, "Lactase Deficiency in Adults: Cause of "Functional" Diarrhoea," *Lancet* 1 (1967): 717–20.
47. Anne Ferguson and J. D. Maxwell, "Genetic Aetiology of Lactose Intolerance," *Lancet* 290, no. 7508 (1967): 188–91.
48. Yuval Itan, Adam Powell, Mark A. Beaumont, Joachim Burger, and Mark G. Thomas, "The Origins of Lactase Persistence in Europe," *PLOS Computational Biology* 5, no. 8 (2009): e1000491.
49. Email correspondence to author (H.M.) from Katrina Dean, Keeper of Manuscripts and Curator of Scientific Collections, Cambridge University Library, April 19, 2022. The lock of Annie's hair is filed under the CUL. Classmark: MS DAR 210.13:27; similar requests have been made over the years to examine the remains of Charles Darwin, buried at Westminster Abbey, which were all appropriately refused by the Abbey's curator. For the ethical implications of such grave robbing, see Howard Markel, "King Tutankhamun, Modern Medical Science, and the Expanding Boundaries of Historical Inquiry," *Journal of the American Medical Association* 303, no. 7 (2010): 667–68.
50. Charles Dickens, *A Christmas Carol: Being a Ghost Story of Christmas* (London: Chapman and Hall, 1845).
51. Francis Darwin, ed., *The Life and Letters of Charles Darwin*, Vol. 1, 135–36.

## Part IV: Oxford

1. Benjamin Disraeli, Speech on church policy and science delivered on Friday, November 25, 1864, at the third annual meeting of the Society for Augmenting Small Benefices in the Diocese of Oxford, held in the Sheldonian Theatre, at Oxford. Benjamin Disraeli, "On Church Policy, 1864," in *Church and Queen: Five Speeches Delivered by*

*the Rt. Hon B. Disraeli, M.P., 1860–1864* (London: G. J. Palmer/Hamilton, Adams, & Co., 1865), 78.

## 10. The Association

1. Charles Darwin to George Rolleston (physician, former attending pediatrician at the Great Ormond Street Children's Hospital, and Linacre Professor of Physiology at the University of Oxford from 1860 to 1881; he was also president of the subsection on physiology for the upcoming meeting of the British Association for the Advancement of Science at Oxford that year), June 6, 1860. (CUL. Classmark: Wellcome Collection, MS. 6119/Letter #: DCP-LETT-2822A). In Frederick Burkhardt et al., eds., *The Correspondence of Charles Darwin*, Vol. 8 (Cambridge, UK: Cambridge University Press, in 30 volumes, 1985–2023), 244–45.
2. Charles Darwin to Charles Lyell, June 25, 1860. (CUL. Classmark: American Philosophical Society, Mss. B.D25.220/Letter #: DCP-LETT-2843). In Burkhardt et al., *Correspondence*, Vol. 8, 265–56.
3. Evelyn Waugh, *Brideshead Revisited: The Sacred and Profane Memoirs of Captain Charles Ryder* (London: Chapman and Hall, 1945), 20.
4. Waugh, *Brideshead Revisited*, 20–21.
5. A. D. Orange, "The Origins of the British Association for the Advancement of Science," *British Journal for the History of Science* 6, no. 2 (1972): 152–76.
6. Ruth Barton, "'Men of Science': Language, Identity and Professionalization in the Mid-Victorian Scientific Community," *History of Science* 41, no. 1 (2003): 73–119.
7. *Report of the Thirtieth Meeting of the British Association for the Advancement of Science, Held at Oxford in June and July, 1860* (London: John Murray, 1861), xvii–xxx. (Hereafter referred to as *BAAS Report.*) The Archives of the BAAS are held in the Bodleian Library Archives and Manuscript Collection, Oxford University. Papers related to the 1860 meeting can be found under: Dep. B.A.A.S. Printed Materials, 147, Press-cuttings, 405. They contain a rich trove of details on the order of the talks and the organization of the meeting as well as recreational lectures or tours and housing options for the attendees.
8. Janet Browne, *Charles Darwin: The Power of Place. Volume II of a Biography* (London: Jonathan Cape; New York: Knopf, 2002), 119.
9. Press-cuttings of the 1860 BAAS Meeting. Dep. B.A.A.S. 405, BAAS Papers, Bodleian Library Archives and Manuscript Collection, Oxford University.
10. Quote is from: Press-cuttings of the 1860 BAAS Meeting, p. 3. Dep. B.A.A.S. 405, BAAS Papers. Bodleian Library Archives and Manuscript Collection, Oxford University; see also "Literature and Art," *The Press* (London), July 7, 1860, 656.
11. David Layton, "Lord Wrottesley, F.R.S., Pioneer Statesman of Science," *Notes and Records of the Royal Society of London* 23, no. 2 (1968): 230–46; quote is from p. 230.
12. "Presidential Address by the Right Honorable Lord Wrottesley," *BAAS Report*, lv–lxxv; quote is from pp. lxxiv–lxxv. The italics are his.
13. "Literature and Art," *The Press* (London), July 7, 1860, 656; Howard Markel, "Doctors Still Argue about This Prince's Early Death," *PBS NewsHour*, December 15, 2017.

14. *Oxford University Herald*, July 7, 1860, 4.
15. *BAAS Report*, i–xxx.
16. George Rolleston to T. H. Huxley, [?] December 1860. [25.150]. Thomas H. Huxley Papers, Imperial College Archives, London.
17. Francis Darwin, ed., *The Life and Letters of Charles Darwin*, Vol. 2 (New York: D. Appleton, and Co., 1889), 113.
18. Charles Daubeny, *A Description of Active and Extinct Volcanos, of Earthquakes, and of Thermal Springs* (London: W. Phillips; Oxford, UK: Joseph Parker, 1826, revised and augmented in 1848).
19. Robert T. Gunther, *Oxford Gardens Based upon Daubeny's Popular Guide to the Physick Garden of Oxford* (Oxford, UK: Parker and Son, 1912), 76–77, 80–81.
20. Charles Daubeny, "Remarks on the Final Causes of Sexuality of Plants with Particular Reference to Mr. Darwin's Work 'On the Origin of Species by Natural Selection,'" *BAAS Report* 1861, 109–10.
21. Daubeny, "On the Final Causes of Sexuality of Plants," 109–10.
22. Thomas H. Huxley to Henrietta Huxley, June 27, 1860. Second series, 475. Thomas Huxley Papers, Archives, Imperial College, London.
23. Henrietta Huxley to Thomas Huxley, June 28, 1860. Second series, 477. Thomas Huxley Papers, Archives, Imperial College, London.
24. "Thursday Session of Section D—Zoology and Botany, including Physiology," *Athenaeum*, July 7, 1860, 25–26.
25. *BAAS Report*, 110.
26. F. Darwin, *The Life and Letters of Charles Darwin*, Vol. 2, 113–14; *Athenaeum*, July 7, 1860, 25–26.
27. *Athenaeum*, July 7, 1860, 25–26.
28. Charles Darwin to Joseph D. Hooker, July 5, 1857. (CUL. Classmark: DAR 114: 203/Letter #: DCP-LETT-2117). In Burkhardt et al., *Correspondence*, Vol. 6, 420–21. The paper Darwin is referring to was published as: Richard Owen, "On the Characters, Principles of Division and Primary Groups of the Class Mammalia," *Proceedings of the Linnean Society (Zoology)* 2 (June 1857): 1–37. The paper was read at the February and April 1857 meetings. Years later, anatomists did, in fact, locate a hippocampal structure in the brains of gorillas. See, for example: Charles C. Gross, "Hippocampus Minor and Man's Place in Nature; A Case Study in the Social Construction of Neuroanatomy," *Hippocampus* 3 (1993): 403–15; Charles C. Gross, "Huxley versus Owen; The Hippocampus Minor and Evolution," *Trends in Neuroscience* 16 (1993): 493–98; as well as S. K. Barks, M. E. Calhoun, W. D. Hopkins, et al., "Brain Organization of Gorillas Reflects Species Differences in Ecology, *American Journal of Physical Anthropology* 156, no. 2 (February 2015): 252–62; R .E. Clark and L. R. Squire, "Similarity in Form and Function of the Hippocampus in Rodents, Monkeys, And Humans," *Proceedings of the National Academy of Sciences of the USA* 110, Suppl. 2 (2013): 10365–70.
29. "Thursday Session of Section D—Zoology and Botany, including Physiology," *Athenaeum*, July 7, 1860, 25–26.

30. "Thursday Session of Section D—Zoology and Botany, including Physiology," *Athenaeum*, July 7, 1860, 26.
31. Here, Huxley was citing the work of Friedrich Tiedemann, Heidelberg neuroanatomist and acolyte of Owen's mentor, Georges Cuvier. Tiedemann argued against racist tropes that the brain "of the Negro" was smaller or less capable than the brains of white European men and that "negroes were not," as many once claimed, "a subspecies of (white) human beings or apes in "On the Brain of the Negro, Compared with That of the European and Orang-Outang," *Philosophical Transactions of the Royal Society of London* 126 (1836): 497–527. Huxley was also referring to Owen's 1835 *Descriptive and Illustrated Catalogue on the Physiological Series of the Hunterian Collection*, where he found similarities between baboon and chimpanzee brains.
32. "Thursday Session of Section D—Zoology and Botany, including Physiology," *Athenaeum*, July 7, 1860, 26. The neuroanatomical structures determining expression and understanding of speech would not be identified until 1861 and 1874, respectively. In 1861, Pierre Broca identified "Broca's area" in the posterior inferior frontal gyrus—which is linked to the production of speech; and in 1874, Carl Wernicke found an area in the posterior section of the superior temporal gyrus that allows for the comprehension of speech and language. A destructive lesion to the former part of the brain causes expressive aphasia and to the latter, receptive aphasia.
33. F. Darwin, *The Life and Letters of Charles Darwin*, Vol. 2, 114.
34. *Athenaeum*, July 7, 1860, 19. The battle at Farnborough was an illegal prize fight on April 17, 1860. It became so violent that it had to be broken up by the local police. The Volunteer Review refers to the immense gathering of soldiers in Hyde Park, London, and elsewhere.
35. Thomas Huxley to Henrietta Huxley, June 28, 1860. Series 2, 476. Thomas Huxley Papers. Archives, Imperial College, London.
36. Henrietta Huxley to Thomas Huxley, June 28, 1860. Series 2 477. Thomas Huxley Papers, Archives, Imperial College, London.
37. Thomas Huxley to Henrietta Huxley, June 28, 1860. Series 2, 476. Thomas Huxley Papers. Archives, Imperial College, London.
38. Thomas Huxley to Henrietta Huxley, June 28, 1860. Series 2, 476. Thomas Huxley Papers. Archives, Imperial College, London; Adrian Desmond, *Huxley: From Devil's Disciple to Evolution's High Priest* (London; Penguin Books, 1997), 277.
39. Henrietta Huxley to Thomas Huxley, June 29, 1860. T. H. Huxley-Henrietta Huxley Correspondence, Huxley Papers, Archives, Imperial College, London; Desmond, *Huxley: From Devil's Disciple to Evolution's High Priest*, 277.
40. L. Huxley, *Life and Letters of Thomas H. Huxley*, Vol. 1, 187–88; Notes and Abstracts of the Miscellaneous Communications to the Sections, *BAAS Report*, 136–40.
41. Thomas Huxley to Henrietta Huxley June 29, 1860. Series 2, 479. Thomas Huxley Papers. Archives, Imperial College, London.
42. Thomas Huxley's Pocket Diary for 1860. June 30, 1860. Thomas Huxley Papers. Archives, Imperial College, London. Recorded on this page, he also considered two later trains at 4:43 and 5:15 p.m.
43. Desmond, *Huxley: From Devil's Disciple to Evolution's High Priest*, 276–77; Anony-

mous [Robert Chambers], *Vestiges of the Natural History of Creation* (London: John Churchill, 1844).

44. Draper noted that his talk was scheduled for Monday, July 2. James C. Ungureanu, "A Yankee at Oxford: John William Draper at the British Association for the Advancement of Science at Oxford, 30 June 1860," *Notes and Records of the Royal Society of London* 70 (2015): 135–50; F. Darwin, *The Life and Letters of Charles Darwin*, Vol. 2, 114; see also *BAAS Report*, 1860, Notices and Abstracts, 115–16; *Athenaeum*, July 7, 1860, 19; July 14, 1860, 64.
45. L. Huxley, *Life and Letters of Thomas H. Huxley*, Vol. 1, 187–88; see also T. H. Huxley to Frederick D. Dyster, September 9, 1860. T. H. Huxley Papers, Volume XV, Folios 115–118, Archives, Imperial College, London.
46. Desmond, *Huxley: From Devil's Disciple to Evolution's High Priest*, p. 277.
47. Thomas Huxley to Henrietta Huxley, June 29, 1860. Series 2, 479. Thomas Huxley Papers. Archives, Imperial College, London.
48. Joseph D. Hooker to Charles Darwin, July 2, 1860. (CUL. Classmark: DAR 115: 64/ Letter #: DCP-LETT-2853). In Burkhardt et al., *Correspondence*, Vol. 8, 270–71.
49. F. Darwin, *The Life and Letters of Charles Darwin*, Vol. 2, 114; *Athenaeum*, July 7, 1860, 19.

## 11. Pax Interruptus

1. *The Press* (London), July 7, 1860, 656; see also Ian Hesketh, *Of Apes and Ancestors: Evolution, Christianity, and the Oxford Debate* (Toronto: University of Toronto Press, 2009), 76; J. Vernon Jensen, "Return to the Wilberforce-Huxley Debate," *British Journal for the History of Science* 21, no. 2 (1988): 161–79.
2. Frank A. J. L. James, "An 'Open Clash between Science and the Church'? Wilberforce, Huxley and Hooker on Darwin at the British Association, Oxford, 1860," in *Science and Beliefs: From Natural Philosophy to Natural Science, 1700–1900*, ed. David M. Knight and Matthew D. Eddy (Aldershot, UK: Ashgate, 2005), 171–93.
3. Josef L. Altholz, "The Huxley-Wilberforce Debate Revisited," *Journal of the History of Medicine and Allied Sciences* 35 (1980): 313–16.
4. There is a superb secondary literature on the "Oxford debate," including James, "An 'Open Clash Between Science and the Church'? 171–93; Jensen, "Return to the Huxley-Wilberforce Debate," 161–79; Hesketh, *Of Apes and Ancestors*; William Irvine, *Apes, Angels, and Victorians: The Story of Darwin, Huxley, and Evolution* (New York: McGraw-Hill, 1955); P. J. Bowler, *Monkey Trials and Gorilla Sermons: Evolution and Christianity from Darwin to Intelligent Design* (Cambridge, MA: Harvard University Press, 2007); see also, Altholz, "The Huxley-Wilberforce Debate Revisited," 313–16; Cyril Bibby, "The Huxley-Wilberforce Debate: A Postscript" [Letter to the Editor], *Nature* 176, no. 4477 (August 20, 1955): 363; John Hedley Brooke. "The Wilberforce-Huxley Debate: Why Did it Happen?" *Science and Christian Belief* 13 (2001): 127–41; John Hedley Brooke, "Samuel Wilberforce, Thomas Huxley, and Genesis," in *The Oxford Handbook of the Reception History of the Bible*, ed. M. Leib, E. Mason, and J. Roberts, C. Rowland (Oxford, UK: Oxford University Press, 2011); Edward Cau-

dill, *Darwinian Myths: The Legends and Misuses of a Theory* (Knoxville: University of Tennessee Press, 1997); F. Darwin, ed., *The Life and Letters of Charles Darwin*, (London: John Murray, 1887; New York: D. Appleton, 1889); Gowan Dawson, *Darwin, Literature, and Victorian Respectability* (Cambridge, UK: Cambridge University Press, 2007); Desmond, Adrian, *Archetypes and Ancestors: Palaeontology in Victorian London, 1850–1875* (Chicago: University of Chicago Press, 1982); Adrian Desmond, *Huxley: The Devil's Disciple* (London: Michael Joseph, 1994); Adrian Desmond, *The Politics of Evolution: Morphology, Medicine, and Reform in Radical London* (Chicago: University of Chicago Press, 1989); John W. Draper, *History of the Conflict of Science and Religion* (New York: Appleton, 1874); John W. Draper, *A History of the Intellectual Development of Europe* (New York: Harper, 1863); John W. Draper, *Human Physiology, Statical and Dynamical* (New York: Harper, 1856); James C. Ungureanu, "A Yankee at Oxford: John William Draper at the British Association for the Advancement of Science at Oxford, 30 June 1860," *Notes and Records of the Royal Society of London* 70 (2015): 135–50; Sheridan Gilley, "The Huxley-Wilberforce Debate: A Reconsideration," in *Religion and Humanism*, ed. Keith Robbins (Oxford, UK: Blackwell, for the Ecclesiastical History Society, 1981); Studies in Church History, No. 17, 325–40; Stephen Jay Gould, "Knight Takes Bishop?" in *Bully for Brontosaurus: Reflections in Natural History* (New York: W. W. Norton, 1991), 385–401; Larry A. Witham, "The Great Debate" (Chapter 12), in *Where Darwin Meets the Bible: Creationists and Evolutionists in America* (New York: Oxford University Press, 2002), 212–26; Stephen Greenblatt et al., eds., *The Norton Anthology of English Literature*, 9th ed., Vol. 2 (New York: W. W. Norton, 2012); Leonard Huxley, ed., *Life and Letters of Sir Joseph Dalton Hooker* (London: John Murray, 1918); Leonard Huxley, ed., *Life and Letters of Thomas Henry Huxley* (London: Macmillan, 1900); Bernard Lightman, *Victorian Popularizers of Science: Designing Nature for New Audiences* (Chicago: University of Chicago Press, 2007); "Literature and Art," *The Press* (London), July 7, 1860, 656; David N. Livingstone, "Myth 17. That Huxley Defeated Wilberforce in Their Debate over Evolution and Religion," in *Galileo Goes to Jail and Other Myths about Science and Religion*, ed. Ronald L. Numbers (Cambridge, MA: Harvard University Press, 2009), 152–60; J. R. Lucas, "Wilberforce and Huxley: A Legendary Encounter," *Historical Journal* 22 (1979): 313–30; Nanna K. L Kaalund, "Oxford Serialized: Revisiting the Huxley-Wilberforce Debate through the Periodical Press," *History of Science* 52, no. 4 (2014): 429–53; Lynn A. Phelps and Edwin Cohen, "The Wilberforce-Huxley Debate," *Western Speech* 37 (Winter 1973): 56–64; James R. Moore, *The Post-Darwinian Controversies* (Cambridge, UK: Cambridge University Press, 1979); Jack Morrell and Arnold Thackray, *Gentlemen of Science: Early Years of the British Association for the Advancement of Science* (Oxford, UK: Clarendon Press of Oxford University Press, 1981); Anonymous [Owen, Richard], "Darwin on the Origin of Species," *Edinburgh Review* 111, no. 3, (1860): 487–532; *Report of the Thirtieth Meeting of the British Association for the Advancement of Science* (London: John Murray, 1861); Nicolaas Rupke, *Richard Owen: Biology without Darwin* (Chicago: University of Chicago Press, 2009); "Science—British Association," *Athenaeum*, July 7 and 14, 1860, 18–32, 59–69; James A. Secord, *Victorian Sensation: The Extraordinary Publication, Reception, and Secret*

*Authorship of Vestiges of the Natural History of Creation* (Chicago: University of Chicago Press, 2000); Steven Shapin, *The Scientific Revolution* (Chicago: University of Chicago Press, 1996); Frank M. Turner, *Contesting Cultural Authority: Essays in Victorian Intellectual Life* (Cambridge, UK: Cambridge University Press, 1993); Andrew Dickson White, *A History of the Warfare of Science with Theology in Christendom* (New York: D. Appleton, 1896); A. R. Ashwell, *Life of the Right Reverend Samuel Wilberforce, D.D.*, Vol. 1 (London: John Murray, 1880); Reginald G. Wilberforce, *Life of the Right Reverend Samuel Wilberforce, D.D.*, Vol. 2 (London: John Murray, 1881); Reginald G. Wilberforce, *Life of the Right Reverend Samuel Wilberforce, D.D.*, Vol. 3 (London: John Murray 1882); [Samuel Wilberforce], "Darwin's *On the Origin of Species*," *Quarterly Review* 108 (1860): 225–64; Carla Yanni. *Nature's Museums: Victorian Science and the Architecture of Display* (Baltimore: Johns Hopkins University Press, 1999); Jonathan Smith, "The Huxley-Wilberforce 'Debate' on Evolution, 30 June 1860," 2013, at *BRANCH: Britain, Representation and Nineteenth-Century History*, ed. D. F. Felluga.

5. L. Huxley, *Life and Letters of Thomas H. Huxley*, Vol. 1, 181.
6. Eartha Kitt, with Hugo Winterhalter's Orchestra. *Proceed with Caution*. Music and lyrics by Wilson Stone. RCA Records, January 1958.
7. Charles Darwin to Asa Gray, May 22, 1860. (CUL. Classmark: Gray Herbarium of Harvard University, 26 and 37a/ DCP-LETT-2814). In Frederick Burkhardt et al., eds., *The Correspondence of Charles Darwin*, Vol. 8 (Cambridge, UK: Cambridge University Press, in 30 volumes, 1985–2023), 223–26.
8. Asa Gray, "Review of On the Origin of Species," *Atlantic Monthly* 6 (July, August, and October 1860): 109–16, 229–39, 406–25. These articles were published as a pamphlet in 1861 under the title *Natural Selection Not Inconsistent with Natural Theology. A Free Examination of Darwin's Treatise 'On the Origin of Species' and of Its American Reviewers* (Boston: Ticknor and Fields, 1861); they are most easily found in Asa Gray, *Darwiniana: Essays and Reviews Pertaining to Darwinism* (New York: D. Appleton and Co., 1876, reprint in 1889); quotes are from this version, on pp. 152–53. Twelve years later, Gray remained just as cautious when describing the coexistence of religious faith and science. During his 1872 presidential address to the American Association for the Advancement of Science meeting at Dubuque, Iowa, he opined, "Faith in an order, which is the basis of science, will not—as it cannot reasonably—be dissevered from faith in an Ordainer, which is the basis of religion." See Asa Gray, "Sequoia and Its History; The Relations of North American to Northeast Asian and to Tertiary Vegetation. Presidential Address to the American Association for the Advancement of Science, at Dubuque, August 1872," in *Darwiniana*, 235.
9. George Rolleston to T. H. Huxley April 13, 1860. [25.142–143]. Thomas H. Huxley Papers, Imperial College Archives, London. Soon after, Huxley supported Rolleston's appointment to become the Linacre Professor of Physiology at Oxford. He was Huxley's host during the 1860 BAAS meeting and was supposed to arrange rooms for Darwin until Darwin canceled to go to Ilkley. The First and Second articles of the Apostles' Creed are: "I believe in God, the Father almighty, maker of heaven and earth; And in Jesus Christ, his only Son, our Lord. . . ."

10. Stephen Burt and Tim Burt, *Oxford Weather and Climate Since 1767* (Oxford, UK: Oxford University Press, 2019), 230.
11. F. Darwin, *The Life and Letters of Charles Darwin*, Vol. 2, 114. Unfortunately, Francis Darwin does not identify the eyewitness here—but this description is quite similar to a letter Charles Lyell wrote to Charles Bunbury after the event. It should be noted that in the letter, Lyell wrote, "I was not able to attend the section of Zoology and Botany (Henslow in the chair), when first Owen and Huxley, and on a later day the Bishop of Oxford and Huxley. . . ." He then went on to give a full account of the proceedings. See Charles Lyell (edited by his sister-in-law, Mrs. Katherine M. Lyell), *Life, Letters and Journals of Sir Charles Lyell, Bart*, Vol. 2 (London: John Murray, 1881), 335.
12. L. Huxley, *Life and Letters of Thomas H. Huxley*, Vol. 1, 182; see also F. Darwin, *The Life and Letters of Charles Darwin*, Vol. 2, 113–16.
13. L. Huxley, *Life and Letters of Thomas H. Huxley*, Vol. 1, 182. Among the undergraduates was the future philosopher, reformer, and then fellow of Balliol College, Thomas Hill Green, who "listened but took no part in the cheering."
14. William Tuckwell, *Reminiscences of Oxford*, 2nd ed. (New York: E. P. Dutton, and Co., 1908), 56; see also William Tuckwell, *Reminiscences of a Radical Parson* (London: Cassell and Co., Ltd., 1905).
15. Thomas Huxley to Henrietta Huxley, June 29, 1860. Series 2, 479. Thomas Huxley Papers. Archives, Imperial College, London.
16. Specifically, he called it "the B. of O." Charles Darwin to T. H. Huxley, July 3, 1860. (CUL. Classmark: Imperial College of Science, Technology, and Medicine Archives, London. Huxley 5.121/Letter: DCP-LETT-2854.). In Burkhardt et al., *Correspondence*, Vol. 8, 277–78.
17. Alison Bashford, *The Huxleys: An Intimate History of Evolution* (Chicago: University of Chicago Press, 2021), 378.
18. Apparently, Daubeny exposed vegetable matter to temperatures exceeding 300°F. The boiled vegetables were "subsequently brought in contact with nothing" but distilled water and air. Hours later, upon inspecting the flasks containing the vegetable matter, Daubeny claimed to have found evidence of new, spontaneous, organic life. This turned out to be a wrongheaded conclusion to the famous debates then going on at the French Academy of Sciences, in which Louis Pasteur demonstrated the fallacy of spontaneous generation and the existence of bacteria, much to the disapproval of Félix Pouchet. See *BAAS Report*, 1860, 115; and John Farleyand Gerald Geison, "Science, Politics and Spontaneous Generation in Nineteenth-Century France: The Pasteur-Pouchet Debate," *Bulletin of the History of Medicine* 48, no. 2 (1974): 161–98; Nils Roll-Hansen, "Experimental Method and Spontaneous Generation: The Controversy between Pasteur and Pouchet, 1859–1864," *Journal of the History of Medicine and Allied Sciences* 34, no. 3 (1979): 273–92.
19. Quote is in L. Huxley, *Life and Letters of Thomas Huxley*, Vol. 1, 182.
20. Donald Fleming, *John William Draper and the Religion of Science* (Philadelphia: University of Pennsylvania Press, 1950), 66–73; see also "Dr. Draper's Lecture on Evolution: Its Origin, Progress, and Consequences," *Popular Science Monthly* 12 (December 1877): 175–92; George F. Barker, "Memoir of John William Draper, 1811–

1882: Read Before the National Academy, April 21, 1886," *Memoirs of the National Academy* (Washington DC: National Academy of Sciences, 1886); Sketch of Dr. J. W. Draper. *Popular Science Monthly* 4 (January 1874): 361–67; "John William Draper," *Daily Graphic*, December 13, 1877, 285; "John William Draper," *Scientific American* 37 (December 15, 1877): 378; "The Late Professor Draper," *Manchester Guardian*, January 6, 1882, 5; "John William Draper," *Nature* 25 (1882): 274–75; "Dr. John William Draper," *Athenaeum*, January 14, 1882, 60–61.

21. Henry J. S. Smith to John William Draper, March 21, 1860. Container 7, John William Draper Family Papers, Library of Congress, Washington DC.
22. John Draper, *History of the Intellectual Development of Europe* (New York: Harper and Brothers, 1863); and John Draper, *The History of the Conflict between Religion and Science* (New York: D. Appleton and Co., 1874); Ungureanu, "A Yankee at Oxford," 135–50.
23. *BAAS Report*, Notes and Reports, 116.
24. John William Draper, *Human Physiology: Statical and Dynamical, or The Conditions and Course of the Life of Man* (New York: Harper and Brothers, 1856), 457–58.
25. Hesketh, *Of Apes and Ancestors*, 79; L. Huxley, *Life and Letters of Thomas H. Huxley*, Vol. 1, 182; Leslie Stephen, ed., *The Letters of John Green* (New York: Macmillan, 1901), 44; Isabel Sidgwick, "A Grandmother's Tales," *Macmillan's Magazine* 78 (1898): 425–35.
26. Sidgwick, "A Grandmother's Tales," 425–35; quote is from p. 433. The italics are the author's (H.M.).
27. John Green to W. Boyd Dawkins, July 3, 1860. In Leslie Stephen, ed., *Letters of John Green* (New York and London: Macmillan, 1901), 45.
28. Alfred Newton, "Early Days of Darwinism," *Macmillan's Magazine* 57 (February 1888): 242–49; quote is on p. 248.
29. J. D. Hooker to Charles Darwin, July 2, 1860. (CUL. Classmark: DAR 100: 141–142/ Letter #: DCP-LETT-2852). In Burkhardt et al., *Correspondence*, Vol. 8, 270–71. Herbert Spencer was the founding father of social Darwinism. Thomas H. Buckle was the author of *History of Civilization in England*, in 3 volumes (London: Parker, Son, and Bourne, 1857, 1861, and 1868, respectively).
30. J. W. Draper to his family, July 6, 1860. Container 43. J.W. Draper Family Letters, Library of Congress; also see Ungureanu, "A Yankee at Oxford," 135–50.
31. J. W. Draper to his family, July 6, 1860. Container 43. J.W. Draper Family Letters, Library of Congress; also see Ungureanu, "A Yankee at Oxford," 135–50.
32. John W Draper to his family, July 6, 1860, J.W. Draper Family Letters, Library of Congress; also see Ungureanu, "A Yankee at Oxford," 135–50; *Oxford Chronicle and Berks and Bucks Gazette*, July 21, 1860, 3.
33. Ungureanu, "A Yankee at Oxford," 135–50.
34. Ungureanu, "A Yankee at Oxford," 135–50.

## 12. Mawnkey! Mawnkey!

1. Edward Clodd, *Thomas Henry Huxley* (New York: Dodd, Mead, and Co., 1902), 21–22. Clodd was a banker, writer, and amateur anthropologist who knew Huxley,

Spencer, and many other eminent Victorians. Here he is using the "gold pencil-case" analogy recorded by W. H. Freemantle in 1892. See the account of the Oxford meeting by the Reverend W. H. Freemantle in Francis Darwin, ed., *Charles Darwin, His Life Told in an Autobiographical Chapter, and in a Selected Series of his Published Letters* (London: John Murray, 1908), 238. Reprinted in L. Huxley, *Life and Letters of Thomas H. Huxley*, Vol. 1 (London: Macmillan, 1900), 186–87.

2. Letter from John S. Henslow to John Hooker May 10, 1860. (CUL. Classmark: R.A. Hooker/ Letter #: DCP-LETT-2794). In Frederick Burkhardt et al., eds., *The Correspondence of Charles Darwin*, Vol. 8 (Cambridge, UK: Cambridge University Press, in 30 volumes, 1985–2023), 200–201; Letter to the Editor from Professor Henslow, in *Macmillan's Magazine* 3, (1861): 336; Janet Browne, *Charles Darwin: The Power of Place. Volume II of a Biography* (London: Jonathan Cape; New York: Knopf, 2002), 117–25.
3. United States Congress. "Act to Establish the 'Smithsonian Institution' for the Increase and Diffusion of Knowledge among Men," Public Law 76, 29th Congress, 1st Session. Signed into Law by President James K. Polk on August 10, 1846.
4. Russel L. Carpenter, ed., *Memoirs of the Life and Work of Philip Pearsall Carpenter, Chiefly Derived from His Letters* (London: Kegan Paul, 1880), 244.
5. *Athenaeum*, Section D, July 14, 1860, 64–65; Richard England, "Censoring Huxley and Wilberforce: A New Source for the Meeting that the *Athenaeum* 'Wisely Softened Down.'" *Notes and Records of the Royal Society of London* 71 (2017): 371–84. For more than a century and a half, the chief contemporaneous sources for describing what happened next was the official report in the July 14, 1860, issue of the *Athenaeum* and a shorter account in *Jackson's Oxford Journal*, July 7, 1860, 2; in 2017, Richard England found a far longer report in the liberal, anti-Tractarian, penny weekly *Oxford Chronicle and Berks and Bucks Gazette*, July 21, 1860, 3, and July 30, 1860, 3. These three sources are the main contemporary versions, albeit not precise transcriptions, of the discourse that followed Draper's lecture on June 30, 1860; and they have been supplemented with the biographies that both Huxley's and Darwin's sons wrote and several other of the after-the-fact reminiscences as cited below. See also the press accounts in *John Bull*, July 6, 1860, 422; "Literature and Art, *The Press*, July 7, 1860, 656; *Morning Chronicle*, July 9, 1860, 7; *Glasgow Herald*, July 4, 1860; *Evening Star*, July 2, 1869, 3; *Oxford University Herald*, July 7, 1860, 4–10; *Guardian*, July 4, 1960, 593.
6. *Oxford Chronicle and Berks and Bucks Gazette*, July 21, 1860, 3; *Athenaeum*, Section D, July 14, 1860, 64–65; *Jackson's Oxford Journal*, July 7, 1860, 4, and July 14, 1860, 2; *John Bull*, July 6, 1860, 422; "Literature and Art," *The Press*, July 7, 1860, 656; *London Morning Chronicle*, July 9, 1860, 7; *Glasgow Herald*, July 4, 1860, 4; *London Morning Star*, July 2, 1869, 3; *Reading Mercury*, July 7, 1860, 10.
7. F. Darwin, *The Life and Letters of Charles Darwin*, Vol. 2, 113–15; quote is on p. 114.
8. Joseph Hooker to Charles Darwin, July 1860. In Burkhardt et al., *Correspondence*, Vol. 8, 270–71.
9. A. S. Farrar to Leonard Huxley, July 12, 1899 (Vol. 16:13–19), and A. S. Farrar to T. H. Huxley, September 2, 1899 (Vol. 16:20). Thomas Huxley Papers, Archives, Imperial College, London; A. S. Farrar, *Science in Theology: Sermons Preached in St. Mary's,*

*Oxford, before the University* (London: John Murray, 1859); see also L. Huxley, *Life and Letters of Thomas H. Huxley*, 182, which includes this description as well. In 1864, Farrar was appointed Professor of Divinity at the University of Durham. He also delivered the Bampton Lectures at Oxford in 1862; in 1899, Farrar wrote Leonard Huxley a point-by-point letter of his perspective on what happened that day, and most of his account corresponds to many other people's versions. No other account, however, includes the parson from Brompton. Although Francis Darwin included the comments in his biography of his father, he was skeptical of them. In 1898, he wrote Leonard Huxley, "I have a perfectly unfounded prejudice against Canon Farrar which would extend itself to his notes." Francis Darwin to Leonard Huxley, December 3, 1898. [13.80]. Thomas Huxley Papers, Archives, Imperial College, London. See also Farrar, *Science in Theology*.

10. *BAAS Report*, 1860, 4, 60, 184–85, 202, 222; "Journal of Sectional Proceedings, No. 3, Issued Friday Morning, June 29, at 8 am," Dep. B.A.A.S. 147, British Association for the Advancement of Science Archives, Bodleian Library Archives and Manuscript Collection, Oxford University; *Journal of the Society of Arts*, July 13, 1860, 648–63; Reverend Dr. James Booth, "On the Principles of an Income Tax," *Journal of the Statistical Society of London* 23, no. 4 (1860): 455–564; J. W. L. Glaisher, "James Booth," *Monthly Notices of the Royal Astronomical Society* 39 (1879): 219–25; C. Sutton and F. Foden, "Booth, James (1806–1878), Mathematician and Educationist," *Oxford Dictionary of National Biography* (Oxford, UK: Oxford University Press, 2004).
11. Sutton and Foden, "Booth, James (1806–1878), Mathematician and Educationist."
12. A. S. Farrar to Leonard Huxley, July 12, 1899 (Volume 16:13–19), and A. S. Farrar to T. H. Huxley, September 2, 1899 (Volume 16:20). Thomas Huxley Papers, Archives, Imperial College, London; List of Resident and Non-resident Members, 30th Meeting of the BAAS, Printed Materials for the 1860 BAAS meeting. Dep. B.A.A.S. 147, British Association for the Advancement of Science Archives, Bodleian Library Archives and Manuscript Collection, Oxford University. In most of the press accounts, Greswell is described as the first to speak. In several secondary accounts, he is referred to as Richard Cresswell, but nowhere in the BAAS report is there a reference to this phantom individual. In the *Athenaeum* (Section D, July 14, 1860, 64–65), for example, his name is misspelled as Cresswell, and this spelling is used by many secondary sources; but after reviewing the *Oxford Chronicle* and various letters and memoirs, it appears that the man in question is Richard Greswell, of Oxford University. The matter becomes more confusing in that there was a Dr. Richard Cresswell with a degree from Oxford around this time, but there is no record of him attending the BAAS meeting in 1860.
13. *BAAS Report*, 1860, lviii, 249; L. Huxley, *Life and Letters of Thomas H. Huxley*, Vol. 1, 182.
14. John William Burgon, "Samuel Wilberforce: The Remodeler of the Episcopate" and "Richard Greswell: The Faithful Steward," in *The Lives of Twelve Good Men*, Vol. 2 (London: John Murray, 1888–1889) 1–70, 94–121; quote is from p. 117.
15. Burgon, "Samuel Wilberforce: The Remodeler of the Episcopate" and "Richard Greswell: The Faithful Steward," 1–70, 94–121.

16. A Derby dog is a "masterless dog" named for the canines who run out onto a racecourse before an important horse race. William Tuckwell, *Reminiscences of Oxford*, 2nd ed. (New York: E. P. Dutton and Co., 1908), 54.
17. Quote is from *Oxford Chronicle and Berks and Bucks Gazette*, July 21, 1860, 3; see also *Athenaeum*, Section D, July 14, 1860, 64–65; *Jackson's Oxford Journal*, July 7, 1860, 4, and July 14, 1860, 2.
18. *Oxford Chronicle and Berks and Bucks Gazette*, July 21, 1860, 3.
19. *BAAS Report*, 1860, Geology Section, p. 77; the year before, Dingle presented a jumbled assessment of the constitution, or components, of the earth below the crust, in which he "regarded mathematical reasoning as inadequate to the solution of the question." *Report of the 29th Meeting of the BAAS, 1859* (London: John Murray, 1860), 102–103.
20. John Dingle, *The Harmony of Revelation and Science: A Series of Essays on Theological Questions of the Day* (London: Bell and Daldy; Cambridge: Deighton, Bell and Co., 1863); John Dingle, *The Universal Obligation of the Fourth Commandment Demonstrated, from a Consideration of the Nature of the Jewish Dispensation* (London: Hamilton, Adams, and Co., 1850); see also John Dingle, *The War in Connection with the Prophesies: A Sermon Preached in Lanchester Church, on the Day of Thanksgiving* (London: Wertheim and Macintosh, 1856); John Dingle, *Memoir of the Rev. John M. Dingle* (South Shields, UK: Northern Press, 1925); Dingle was also an early advocate of the scientific applications of the stereoscope. John Dingle, Letter to the Editor, *The Reader* 7, no. 176 (May 12, 1866): 473.
21. Carpenter, *Memoirs of the Life and Work of Philip Pearsall Carpenter*, Vol. 1, 244.
22. L. Huxley, *Life and Letters of Thomas H. Huxley*, 182. The italics are the author's (H.M.).
23. Carpenter, *Memoirs of the Life and Work of Philip Pearsall Carpenter*, 244–45.
24. L. Huxley, *Life and Letters of Thomas H. Huxley*, 182–83. The italics are the author's (H.M.).
25. Carpenter, *Memoirs of the Life and Work of Philip Pearsall Carpenter*, 244–45; Hesketh, *Of Apes and Ancestors*, 80.
26. *Oxford Chronicle and Berks and Bucks Gazette*, July 21, 1860, 3.
27. Carpenter, *Memoirs of the Life and Work of Philip Pearsall Carpenter*, 245.
28. R. E. Graves, "Pellegrini, Carlo [Ape] (1839–1889)." In *Dictionary of National Biography, 1885–1900*, Vol. 44 (London: Smith, Elder and Co., 1900), 265; Leonard Naylor, *The Irrepressible Victorian: The Story of Thomas Gibson Bowles: Journalist, Parliamentarian, and Founder Editor of the Original Vanity Fair* (London: Macdonald and Co., 1965).
29. Carlo "Ape" Pelligrini, "Statesman No. 25. Samuel Wilberforce," *Vanity Fair*, July 24, 1869, 50. Equally brilliant was Pelligrini's caricature of Thomas Huxley, *Vanity Fair*, January 28, 1871 ("Men of the Day No. 19," with the caption "A great Med'cine-Man among the inquiring Redskins"), 306.
30. Tuckwell. *Reminiscences of Oxford*, 54.
31. [Samuel Wilberforce], "On the Origin of Species, by Means of Natural Selection;

or the Preservation of Favoured Races in the Struggle for Life. By Charles Darwin, M.A., F.R.S," *Quarterly Review* 108 (1860): 226–64.

32. *Oxford Chronicle and Berks and Bucks Gazette*, July 21, 1860, 3; see also *Athenaeum*, Section D, July 14, 1860, 64–65; *Jackson's Oxford Journal*, July 7, 1860, 4, and July 14, 1860, 2.
33. *Oxford Chronicle and Berks and Bucks Gazette*, July 21, 1860, 3; see also *Athenaeum*, Section D, July 14, 1860, 64–65; *Jackson's Oxford Journal*, July 7, 1860, 4, and July 14, 1860, 2.
34. *Oxford Chronicle and Berks and Bucks Gazette*, July 21, 1860, 3.
35. L. Huxley, *Life and Letters of Thomas H. Huxley*, Vol. 1, 183; F. Darwin, *The Life and Letters of Charles Darwin*, Vol. 2, 114–15; quote is from p. 114.
36. *Oxford Chronicle and Berks and Bucks Gazette*, July 21, 1860, 3.
37. *Oxford Chronicle and Berks and Bucks Gazette*, July 21, 1860, 3; see also L. Huxley, *Life and Letters of Thomas H. Huxley*, Vol. 1, 183; F. Darwin, *The Life and Letters of Charles Darwin*, Vol. 2, 114–15; quote is from p. 114.
38. [Samuel Wilberforce], "On the Origin of Species," 226–64; quote is from p. 258.
39. J. D. Hooker to Charles Darwin, July 2, 1860. (CUL. Classmark: DAR 115: 64/ Letter #: DCP-LETT-2853). In Burkhardt et al., *Correspondence*, Vol. 8, 270–71; a revised version of these comments appears in F. Darwin, *The Life and Letters of Charles Darwin*, Vol. 2, 114–15. See also Janet Browne, "The Charles Darwin-Joseph Hooker Correspondence: An Analysis of Manuscript Resources and Their Use in Biography," *Journal of the Society for the Bibliography of Natural History* 8 (1978): 351–66; Sheridan Gilley, "The Huxley-Wilberforce Debate: A Reconsideration," in *Religion and Humanism*, ed. Keith Robbins (Oxford, UK: Blackwell, 1981), 25–40; J. R. Lucas, "Wilberforce and Huxley: A Legendary Encounter," *Historical Journal* 22 (1979): 313–30.
40. F. Darwin, *The Life and Letters of Charles Darwin*, Vol. 2, 114–15; see also J. D. Hooker to Charles Darwin, July 2, 1860. (CUL. Classmark: DAR 115: 64/ Letter #: DCP-LETT-2853). In Burkhardt et al., *Correspondence*, Vol. 8, 270–71.
41. Tuckwell, *Reminiscences of Oxford*, 54–55.
42. L. Huxley, *Life and Letters of Thomas H. Huxley*, Vol. 1, 181. This last statement was challenged by Adam Storey Farrar, who wrote Leonard to tell him: "The ladies there were few. The tale of them waving their handkerchiefs is a legend." See A. S. Farrar to Leonard Huxley, July 18, 1899. (Volume 16:13–19), and A. S. Farrar to T. H. Huxley, September 2 1899 (Volume 16:20). Thomas Huxley Papers, Archives, Imperial College, London.
43. *Oxford Chronicle and Berks and Bucks Gazette*, July 21 1860, 3; see also *Athenaeum*, Section D, July 14, 1860, 64–65; *Jackson's Oxford Journal*, July 7, 1860, 4, and July 14, 1860, 2.
44. *Athenaeum*, Section D, July 14, 1860, 64–65.
45. Jensen, "Return to the Huxley-Wilberforce Debate," 161–79; Hesketh, *Of Apes and Ancestors*; James, "An Open Clash between Science and the Church?," 171–93.
46. F. Darwin, *The Life and Letters of Charles Darwin*, 114–15.

47. The quote is found in the *Oxford Chronicle and Berks and Bucks Gazette*, July 21, 1860, 3; the shock in the room is described in several articles, including Henry Fawcett, "A Popular Exposition of Mr. Darwin on the Origin of Species," *Macmillan's Magazine* 3, no. 14 (December 1860): 81–92.
48. Stephen, *The Letters of John Green*, 45–46. For example, the bootblack Sam Weller is referred to as Samivel in Dickens's *The Pickwick Papers* (London: Penguin Books, 2000), 308, 358–59, 433, 463, 575–77, 596–97; the London accent used "a 'v' for 'w' and 'w' for 'v' and so forth." See George Orwell, *Down and Out in Paris and London* (London: Penguin Books, 1950), 177.
49. Stephen, *The Letters of John Green*, 45–46.
50. Tuckwell, *Reminiscences of Oxford*, 55.
51. T. H. Huxley to Francis Darwin, June 27, 1891. In Leonard Huxley, *Life and Letters of Thomas H. Huxley*, Vol. 1, 187–89; quote is on p. 188; Book of Samuel I, 23:7.
52. Stephen, *The Letters of John Green*, 45. Adam S. Farrar, on the other hand, heatedly disputed Green's version (especially the use of the word "equivocal"), calling it "spiteful" and "hardly worthy of him." See A. S. Farrar to Leonard Huxley, July 12, 1899 (Volume 16:13–19), and A. S. Farrar to T. H. Huxley, September 2, 1899 (Volume 16:20). Thomas Huxley Papers, Imperial College, London. The underline is Green's and the italicized "equivocal" was inserted into the text by the author (H.M.) for clarity. Years later, Julius Victor Carus, the German zoologist, corroborated to Leonard Huxley that Thomas Huxley did not use the word "equivocal" to describe Wilberforce's accomplishments. Leonard Huxley, *Life and Letters of Thomas H. Huxley*, Vol. 1 (London: Macmillan, 1900), 187–89.
53. Letter from A. G. Vernon-Harcourt to Leonard Huxley, undated. In L. Huxley, *Life and Letters of Thomas H. Huxley*, Vol. 1, 185.
54. George Johnstone Stoney to Francis Darwin, May 17, 1895. (CUL. Classmark: MS DAR 107 A36–A39), Charles Darwin Papers, Special Collections and Manuscripts, Cambridge University Library, Cambridge, UK. The underlining is Stoney's; I am indebted to Frank Bowles, archivist at the Cambridge University Library, for helping me to find this source. At the time of the debate, Stoney was Secretary of Queen's University of Ireland in Dublin. See J. G. O'Hara, "George Johnstone Stoney, F.R.S. and the Concept of the Electron," *Notes and Records of the Royal Society of London* 29 (1975): 265–76.
55. Isabel Sidgwick, "A Grandmother's Tales," *Macmillan's Magazine* 78 (1898): 425–35; quote is from p. 433; see also F. Darwin, *The Life and Letters of Charles Darwin*, Vol. 2, 115.
56. Thomas H. Huxley to Frederick Dyster, September 9, 1860. [HP 15.117] Thomas H. Huxley Papers, Archives, Imperial College, London; see also Jensen, "Return to the Huxley-Wilberforce Debate," 161–79; Appendix VI: Report of the British Association meeting in Oxford, June 26–July 3, 1860. In Burkhardt et al., *Correspondence*, Vol. 8, 590–97; D. J. Foskett, "Wilberforce and Huxley on Evolution," *Nature* 172 (November 14, 1954): 920.
57. Thomas H. Huxley to Frederick Dyster, September 9, 1860. [HP 15.117] Thomas H. Huxley Papers, Archives, Imperial College, London; see also Jensen, "Return to the Huxley-Wilberforce Debate," 161–79; Appendix VI: Report of the British Association

meeting in Oxford, June 26–July 3, 1860. In Burkhardt et al., *Correspondence*, Vol. 8, 590–97; Foskett, "Wilberforce and Huxley on Evolution," 920.

58. Tuckwell, *Reminiscences of Oxford*, 55.
59. William Shakespeare, *Hamlet*. Act I, scene v, lines 105–109, spoken by Hamlet after his first encounter with the ghost of his father.
60. Carpenter, *Memoirs of the Life and Work of Philip Pearsall Carpenter*, 245. Carpenter was, undoubtedly, referring to the lynching of African Americans by white supremacists in the United States.
61. Sidgwick, "A Grandmother's Tales," 425–35, quote is from p. 433; Hooker identifies the "lady" as Lady Brewster; see letter from J. D. Hooker to Charles Darwin, July 2, 1860. In Burkhardt et al., *Correspondence*, Vol. 8, 270.
62. Fawcett, "A Popular Exposition of Mr. Darwin on the Origin of Species," 81–92; quote is from p. 88.
63. Charles Lyell to Charles Bunbury, July 4, 1860. In K. M. Lyell, ed., *Life, Letters, and Journals of Sir Charles Lyell, Bart*, Vol. 2 (London: John Murray, 1881), 335. The parentheses are Lyell's; the brackets are the author's (H.M.).
64. George Rolleston to T. H. Huxley, [?]December 1860. [25.150]. Thomas H. Huxley Papers, Imperial College Archives, London.
65. J. D. Hooker to Charles Darwin, July 2, 1860. (CUL. Classmark: DAR 115: 64/ Letter #: DCP-LETT-2853). In Burkhardt et al., *Correspondence*, Vol. 8, 270–71. The italics are Hooker's.
66. George Johnstone Stoney to Francis Darwin, May 17, 1895. (CUL. Classmark: MS DAR 107 A36–A39), Charles Darwin Papers, Special Collections and Manuscripts, Cambridge University Library, Cambridge, UK.
67. *Oxford Chronicle and Berks and Bucks Gazette*, July 21, 1860, 3.
68. *Oxford Chronicle and Berks and Bucks Gazette*, July 21, 1860, 3.
69. *Oxford Chronicle and Berks and Bucks Gazette*, July 21, 1860, 3.
70. L. Huxley, *Life and Letters of Thomas H. Huxley*, Vol. 1, 189; see also Gerald Geison, *Michael Foster and the Cambridge School of Physiology: The Scientific Enterprise in Late Victorian Society* (Princeton, NJ: Princeton University Press, 1978).
71. See, for example, Charles Darwin to Charles Lyell, December 10, 1859. (CUL. Classmark: American Philosophical Society, Mss. B.D25. 184). In Burkhardt et al., *Correspondence*, Vol. 7, 423; William Whewell to Charles Darwin, January 2, 1860. (CUL. Classmark: DAR 98 (series 2): 19/Letters #: DCP-LETT-2634). In Burkhardt et al., *Correspondence*, Vol. 8, 6.
72. These quotes are from *Oxford Chronicle and Berks and Bucks Gazette*, July 21, 1860, 3.
73. *Oxford Chronicle and Berks and Bucks Gazette*, July 21, 1860, 3.

## 13. The Rebuttals

1. Charles Darwin to Charles Lyell, November 23, 1859. (CUL. Classmark: American Philosophical Society, Mss. B. D25.176/ Letter #: DCP-LETT-2543). In Frederick Burkhardt et al., eds., *The Correspondence of Charles Darwin*, Vol. 7 (Cambridge, UK: Cambridge University Press, in 30 volumes, 1985–2023), 391–92.

2. *BAAS Report*, 1860, Chemistry Section, 66.
3. G. H. Brown, *Lives of the Fellows of the Royal College of Physicians of London, 1826–1925* (London: Royal College of Physicians, 1955), 101–102.
4. James Bird, *Historical Researches On the Origin and Principles of the Bauddha and Jaina Religions: Embracing the Leading Tenets of Their System, As Found Prevailing in Various Countries; Illustrated by Descriptive Accounts of the Sculptures in the Caves of Western India, with Translations of the Inscriptions of Kanari, Karli, Ajanta, Ellora, Nasik, &c. Which Indicate Their Connection with the Coins and Topes of the Panjab and Afghanistan* (Bombay: American Mission Press, 1847); Ali Muhammad Khan, *The Political and Statistical History of Gujarat: To Which Are Added, Copious Annotations, And a Historical Introduction*, trans. James Bird (London: Richard Bentley, 1835).
5. *The British Museum Catalogue*. Accessed November 11, 2022, at https://www.britishmuseum.org/collection/term/BIOG209138.
6. "Obituary. James Bird," *Lancet* 2 (July 23, 1864): 109. Bird was an active member of the Statistical Society of London, which, like the BAAS, enjoyed the patronage of Prince Albert. See Masthead of Officers, *Journal of the Statistical Society of London* 23, no. 1 (1860): iv; for many years, Bird was also foreign secretary for India of the Epidemiological Society of London, founded in 1850 and also known as the Royal Society of Medicine's Epidemiological Society; see also David E. Lilienfeld, "The Greening of Epidemiology: Sanitary Physicians and the London Epidemiological Society (1930–1870)," *Bulletin of the History of Medicine* 52, no. 4 (1978): 503–28.
7. Reports of Societies and Academies. Westminster Medical Society. John Snow, "Pathology and Mode of Communication of Cholera," *London Journal of Medicine* 1 (October 13, 1849): 1077–1084; quote by Bird is on page 1082. See also James Bird, Contributions to the Pathology and Treatment of Cholera," *London Journal of Medicine* 1 (1849): 313–21, 829–44; James Bird, *The Laws of Epidemic and Contagious Diseases and the Importance of Preventive Medicine* (London: John Churchill, 1854), 5–7; James Bird, *Contributions to the Pathology of Cholera, Embracing Its History, Modifications, Stages, and Treatment, as the Disease Appeared in the Bombay Presidency, from 1818 to 1842* (London: John Churchill, 1849).
8. John Snow, *On the Mode of Communication of Cholera* (London: John Churchill, 1849; 2nd ed, 1854); Howard Markel, "On John Snow," in *Literatim: Essays at the Intersection of Medicine and Culture* (New York: Oxford University Press, 2019), 75–78.
9. *Oxford Chronicle and Berks and Bucks Gazette*, July 21, 1860, 3.
10. William Tuckwell. *Reminiscences of Oxford*, 2nd ed. (New York: E. P. Dutton, and Co., 1908), 55.
11. *BAAS Report*, 1860, Meteorology Section, p. 39; July 7, 1860, p. 4. The Met Office was founded in 1854. The now world famous "Daily Weather Report" was inaugurated on September 3, 1860. *Daily Weather Report, 1860–1870*. (MET/2/4/2/1/1)/. Meteorological Office. Board of Trade. Accessed August 3, 2022 at https://library.metoffice.gov.uk/Portal/Default/en-GB/recordview/index/626295.
12. Leonard Huxley, *Life and Letters of Thomas H. Huxley*, Vol. 1 (London: Macmillan,

1900), 187; although this account appears in the 1887 British version of Francis Darwin's *Life and Letters of Charles Darwin*, on p. 238, it was not included in the 1889 American version.

13. Stoney had the venue wrong—it was held at the museum, not at the Sheldonian Theatre, and admitted he was paraphrasing FitzRoy, to some extent. George Johnstone Stoney to Francis Darwin, May 17, 1895. (CUL. Classmark: MS DAR 107 A36–A39), Charles Darwin Papers, Special Collections and Manuscripts, Cambridge University Library, Cambridge, UK.
14. George Johnstone Stoney to Francis Darwin, May 17, 1895. MS DAR 107 A36–A39, Charles Darwin Papers, Special Collections and Manuscripts, Cambridge University Library, Cambridge, UK; see also, J. V. Carus to Charles Darwin, November 15, 1866. (CUL. Classmark: DAR 161: 54/ Letter #: DCP-LETT-5279). In Burkhardt et al., *Correspondence*, Vol. 14, 382–84; the German biologist wrote Darwin after FitzRoy's death, "I shall never forget that meeting of the combined sections of the British Association when at Oxford 1860, where Admiral FitzRoy expressed his sorrows for having given you the opportunities of collecting facts for such a shocking theory as yours. The poor man is gone, and yet we thank him."
15. Tuckwell, *Reminiscences of Oxford*, 55.
16. *Oxford Chronicle and Berks and Bucks Gazette*, July 21, 1860, 3.
17. Charles Darwin to John S. Henslow, July 16, 1860. (CUL. Classmark: DAR 93: A74–75/ Letter #: DCP-LETT-2869). In Burkhardt et al., *Correspondence*, Vol. 8, 289–90.
18. Charles Shaw to Charles Darwin, October 5, 1865. (CUL. DAR 177 : 147/ Letter # : DCP-LETT-4908). In Burkhardt et al., *Correspondence*, Vol. 13, 258–59; Darwin recorded this payment to the Admiral FitzRoy Testimonial Fund on October 4, 1865, under the heading "Charities" in his Classed Account Book. Archives of Down House, Down, Kent, UK.
19. Brown, *Lives of the Fellows of the Royal College*, 100.
20. *Oxford Chronicle and Berks and Bucks Gazette*, July 21, 1860, 3.
21. Brown, *Lives of the Fellows of the Royal College*, 100.
22. *Oxford Chronicle and Berks and Bucks Gazette*, July 21, 1860, 3.
23. E. Benton, "Vitalism in 19th Century Scientific Thought: A Typology and Reassessment," *Studies in History and Philosophy of Science, Part A* 5, no. 1 (1974): 17–48; quote is on p. 18.
24. *Oxford Chronicle and Berks and Bucks Gazette*, July 21, 1860, 3.
25. *Oxford Chronicle and Berks and Bucks Gazette*, July 21, 1860, 3; Fred Somkin, "The Contributions of Sir John Lubbock, F.R.S. to the 'Origin of Species': Some Annotations to Darwin," *Notes and Records of the Royal Society of London* 17, no. 2 (1962): 183–91.
26. In 1871, as a member of the House of Commons, Lubbock introduced the Bank Holidays Act, which set four public holidays for England, Wales and Ireland, and Scotland. They unofficially became known as "St Lubbock's Days" to honor the man who created these statutory days of rest. J. F. M. Clark, "John Lubbock, Science, and the Liberal Intellectual," *Notes and Records of the Royal Society of London* 68, no.

1 (2014): 65–87; J. F. M. Clark, "The Science of John Lubbock," *Notes and Records of the Royal Society of London* 68, no. 1 (2014): 3–6; Alison Pearn, "The Teacher Taught? What Charles Darwin Owed to John Lubbock," *Notes and Records of the Royal Society of London* 68, no. 1 (2014): 7–19.

27. Leslie Matthew "Spy" Ward, "Bankers and Financiers. The Bank Holiday. Sir John Lubbock," *Vanity Fair* (UK) February 23, 1878. Edison's light bulb was "invented" at Menlo Park in January 1879.
28. *Oxford Chronicle and Berks and Bucks Gazette*, July 21, 1860, 3; see also *Athenaeum*, Saturday session, Section D, Zoology, July 14, 1860, 64–65.
29. Hooker's comments are reported in the *Athenaeum*, Saturday session, Section D, Zoology, July 14, 1860, 64–65, and are reproduced in Burkhardt et al., *Correspondence*, Vol. 8, 590–97; Hooker's comments appear on pp. 596–97.
30. Hooker was referring to the distinction between a species—the basic unit of classification, whose members share similar traits and offspring—and a variety, or subspecies, whose members share traits because of geographical or environmental conditions but remain true to their species. *Athenaeum*, Saturday session, Section D, Zoology, July 14, 1860, 64–65; for a study of Hooker's science, see, for example, Jim Endersby, "Lumpers and Splitters: Darwin, Hooker, and the Search for Order," *Science* 326, no. 5959, (2009): 1496–99.
31. *Athenaeum*, Saturday session, Section D, Zoology, July 14, 1860, 64–65.
32. *Athenaeum*, Saturday session, Section D, Zoology, July 14, 1860, 64–65.
33. Charles Lyell to Charles Bunbury, July 4, 1860. In K. M. Lyell, ed., *Life, Letters, and Journals of Sir Charles Lyell, Bart*, Vol. 2 (London: John Murray, 1881), 335.
34. The Amalekites were enemies of the people of Israel. (Genesis 24:12–16; Samuel 1:15); Joseph D. Hooker to Charles Darwin, July 2, 1860. (CUL. Classmark: DAR 100: 141–2/DCP-LETT-2852). In Burkhardt et al., *Correspondence*, Vol. 8, 270–71; see also Leonard Huxley, *The Life and Letters of Joseph Dalton Hooker*, Vol. 1 (London: John Murray, 1918), 525–27.
35. Tuckwell, *Reminiscences of Oxford*, 56; L. Huxley, *The Life and Letters of Joseph Dalton Hooker*, Vol. 1, 527.
36. *Oxford University Herald*, July 7, 1860, 8.
37. Undated letter from Isabel Sidgwick to Leonard Huxley. In L. Huxley, *Life and Letters of Thomas Huxley*, Vol. 1, 189.
38. Isabel Sidgwick, "A Grandmother's Tales," *Macmillan's Magazine* 78 (1898): 425–35; quote is from pp. 433–34.
39. Thomas H. Huxley to Frederick Dyster, September 9, 1860. [15: 115–118]. Thomas H. Huxley Papers, Archives, Imperial College, London; J. Vernon Jensen, "Return to the Huxley-Wilberforce Debate," *British Journal for the History of Science* 21 (1988): 161–79.
40. F. Darwin, *The Life and Letters of Charles Darwin*, Vol. 2, 116. In a letter from Francis Darwin to Leonard Huxley, on December 3, 1898, Francis lets him know that Hooker was "the eyewitness" who asked him "not to publish his name tho' I can't imagine why. . . . It is most curious how different accounts vary" [13.80–81]. Thomas H. Huxley Papers, Archives, Imperial College, London.

41. Joseph Hooker to Francis Darwin, Royal Botanic Gardens, Kew, Hooker Outgoing Letters 3, 153, and 6, 151; as cited in Janet Browne, "The Charles-Darwin-Joseph Hooker Correspondence: An Analysis of Manuscript Resources and Their Use in Biography," *Journal of the Society for the Bibliography of Natural History* 8, no. 4 (1978): 351–66 (see p. 362).
42. L. Huxley, *Life and Letters of Thomas Huxley*, Vol. 1, 179.

## 14. The Dogs Bark but the Caravan Moves On

1. W. S. Gilbert, *Princess Ida, or Castle Adamant* (London: G. Bell and Sons, Ltd., 1912), 27. This was sung by a character named "Lady Psyche, Professor of Humanities."
2. Henry Fawcett, "A Popular Exposition of Mr. Darwin on the Origin of Species," *Macmillan's Magazine* 3, no. 14 (December 1860): 81–92; quote is from p. 88.
3. Alfred Newton, "Early Days of Darwinism," *Macmillan's Magazine* 3, no. 340 (February 1888): 242–49; quote is on p. 249.
4. Account of the Oxford meeting by the Reverend W. H. Freemantle, in F. Darwin, ed., *Charles Darwin, His Life Told in an Autobiographical Chapter, and in a Selected Series of His Published Letters* (London: John Murray, 1908), 238. Reprinted in Leonard Huxley, *Life and Letters of Thomas H. Huxley*, Vol. 1 (London: Macmillan, 1900), 186–87.
5. L. Huxley, *Life and Letters of Thomas H. Huxley*, Vol. 1, 189.
6. C. J. Cornish, *William Henry Flower: A Personal Memoir* (London: Macmillan, 1904), 66.
7. On October 3, 1862, at the BAAS meeting, Section D, Zoology and Botany, including Physiology, Richard Owen presented a paper titled "On the Zoological Significance of the Cerebral and Pedal Characters of Man." The paper and resulting discussion were reported in *The Times of London*, October 4, 1862, 7. At this meeting, Owen again defended his work that man was a separate subspecies from apes because of the hippocampal and bipedal issues he discussed in 1860. He further agreed with Tiedemann that the "Negro's brain . . . was as large as the average one of the Caucasian . . . [and] the weight of a full-grown male Gorilla being one-third more than that of an average-sized Negro" (p. 117). At the same meeting, Owen also read a paper, "On the Characters of the 'Aye-Aye' as a Test of the Lamarckian and Darwinian Hypotheses of the Transmutation and Origin of Species," in which he concluded that "Darwin seems to be as far from giving a satisfactory explanation . . . as Lamarck." (See *BAAS Report*, 1862, 32, Part II, pp. 114–18.) Huxley (who was president of Section D for 1862), George Rolleston, W. H. Flower, and Alfred Newton argued against Owen's concepts at the session. See also: T. H. Huxley to Charles Darwin, October 9, 1862. (CUL. Classmark: DAR 162.2: 294/Letter #: DCP-LETT-3755). In Frederick Burkhardt et al., eds., *The Correspondence of Charles Darwin*, Vol. 10 (Cambridge, UK: Cambridge University Press, in 30 volumes, 1985–2023), 55–56; J. E. Gray to Charles Darwin, January 29, 1862. In Burkhardt et al., *Correspondence*, Vol. 10, 450–51.
8. "Weasand" is an archaic term for "throat." T. H. Huxley to Charles Darwin, Octo-

ber 9, 1862. (CUL. Classmark: DAR 162.2: 294/Letter #: DCP-LETT-3755). In Burkhardt et al., *Correspondence*, Vol. 10, 55–56; see also *The Times of London*, October 3, 1862, 5, and October 4, 1862, 7; similarly, in a letter to Charles Lyell on August 17, 1862, Huxley referred to a poem by the paleontologist Philip Egerton: "I do not think you will find room to complain of any want of distinctness in my definition of Owen's position touching the Hippocampus question. I mean to give the whole history of the business in a note, so that the paraphrase of Sir Ph. Egerton's line 'To which Huxley replies that Owen he lies,' shall be unmistakable." See L. Huxley, *Life and Letters of Thomas Henry Huxley*, Vol. 1, 200.

9. Newton, "Early Days of Darwinism," 242–49; quote is on p. 249.
10. Frederick Temple, *The Present Relations of Science to Religion: A Sermon Preached on Act Sunday, July 1, 1860, Before the University of Oxford during the Meeting of the British Association* (Oxford and London; J. H. and Jas. Parker, 1860), 15.
11. Arthur J. Munby, *Munby, Man of Two Worlds: The Life and Diaries of Arthur J. Munby, 1828–1910*, ed. Derek Hudson (London: John Murray, 1972), 64.
12. Munby, *Munby, Man of Two Worlds*, 64.
13. Alfred Newton, *A Dictionary of Birds* (London: A & C Black, 1896).
14. T. R. Birkhead and P. T. Gallivan, "Alfred Newton's Contribution to Ornithology: A Conservative Quest for Facts Rather Than Grand Theories," *IBIS: International Journal of Avian Science* 154, no. 94 (2012): 887–905.
15. J. P. Hume, A. S. Cheke, and A. McOran-Campbell, "How Owen 'Stole' the Dodo: Academic Rivalry and Disputed Rights to a Newly Discovered Subfossil Deposit in Nineteenth Century Mauritius," *Historical Biology: An International Journal of Paleobiology* 21, nos. 1/2 (2009): 33–49.
16. Newton, "Early Days of Darwinism," 242–49; quotes are on pp. 248 and 249; for a more contemporaneous version of Newton's experiences at the 1860 meeting, see A. F. R. Wollaston, *Life of Alfred Newton* (New York: E. F. Dutton and Co., 1921), 118–20.
17. Newton, "Early Days of Darwinism," 242–49; quotes are on pp. 248 and 249.
18. F. W. Farrar to Leonard Huxley, July 12, 1899. Thomas H. Huxley Papers, Vol. 16, folios 13–19, Special Collections, Archives, Imperial College, London; see also L. Huxley, *Life and Letters of Thomas H. Huxley*, Vol. 1, 183–84, footnote 2. The underlining is Farrar's, and the italics are the author's (H.M.).
19. Frank A. J. L. James, "An Open Clash between Science and the Church? Wilberforce, Huxley and Hooker on Darwin at the British Association, Oxford, 1860," in *Science and Beliefs: From Natural Philosophy to Natural Science, 1700–1900*, ed. David M. Knight and Matthew D. Eddy (Aldershot, UK: Ashgate, 2005); quote is from p. 382; the most complete analysis of the press reports on the meeting is Nanna K. L Kaalund, "Oxford Serialized: Revisiting the Huxley-Wilberforce Debate through the Periodical Press," *History of Science* 52, no. 4 (2014): 429–53.
20. *Athenaeum*, July 7, 1860, 19. The political views of the newspapers cited in this section can be found in *The Newspaper Press Directory and Advertisers' Guide* (London: C. Mitchell and Company, 1860) and Alvar Ellegård, "The Readership of the Periodical

Press in Mid-Victorian Britain: II: Directory," *Victorian Periodicals Newsletter* 4, no. 3 (1971): 3–22.

21. *London Morning Star*, July 2, 1860, 3.
22. "Literature and Art," *The Press* (London), July 7, 1860, 656.
23. *John Bull*, July 6, 1860, 422.
24. *Christian Remembrancer* 40 (October 1860): 237–61; quotes are from pp. 238–40, 243, and 261.
25. *Guardian*, July 4, 1860, 593.
26. Samuel Wilberforce Diary, 29 June 1860, p. 117, Dep. +e + 327, Samuel Wilberforce Papers, Bodleian Library, Oxford University.
27. Samuel Wilberforce Diary, 30 June 1860, p. 117, Dep. +e + 327, Samuel Wilberforce Papers, Bodleian Library, Oxford University; Adrian Desmond, *Huxley: From Devil's Disciple to Evolution's High Priest* (London: Penguin Books, 1997), 281.
28. Samuel Wilberforce to Sir Charles Anderson, July 3, 1860. MS. Wilberforce d.29; folios 30–32; Samuel Wilberforce Papers, Bodleian Library, Oxford University.
29. Balfour Stewart to J. David Forbes, July 4, 1860. J.D. Forbes Papers, (GB 227 msdep7). Special Collections, University of St Andrews; quoted in Stephen Jay Gould, "Knight Takes Bishop," in *Bully for Brontosaurus: Reflections in Natural History* (New York: W. W. Norton, 1992), 385–401; quote is on p. 389.
30. Charles Darwin to Joseph D. Hooker, July 2, 1860. (CUL. Classmark: DAR 115: 64/ Letter #: DCP-LETT-2853). In Burkhardt et al., *Correspondence*, Vol. 8, 272–73.
31. Charles Darwin to Thomas H. Huxley, July 3, 1860. (CUL. Classmark: Imperial College of Science, Technology and Medicine, Archives (Huxley 5: 121)/ Letter #: DCP-LETT-2854). In Burkhardt et al., *Correspondence*, Vol. 8, 277. Here Darwin is referring to Huxley, Hooker, Lyell, and the naturalist and physician William Benjamin Carpenter, whose work on comparative physiology was of use to Darwin when writing his book. He states these credits in a letter to Asa Gray on July 22, 1860. (CUL. Classmark: Archives, Gray Herbarium, Harvard University, [30]/ Letter #: DCP-LETT-2876.) In Burkhardt et al., *Correspondence*, Vol. 8, 299. Other letters declaring victory, to Charles Lyell, John S. Henslow, and others, appear in the same volume, pp. 270–300. The bold is Darwin's.
32. Charles Darwin to Thomas H. Huxley, July 5, 1860. (CUL. Classmark: Imperial College of Science, Technology and Medicine, Huxley Papers, 5: 123/ Letter #: DCP-LETT-2861). In Burkhardt et al., *Correspondence*, Vol. 8, 280. Unfortunately, the long letter Huxley wrote Darwin, prompting this letter, had been lost.
33. "Elèves" is from the French for student. Hugh Falconer to Charles Darwin, July 9, 1869. (CUL. Classmark; DAR 164.1:5; DC-LETT-2863). In Burkhardt et al., *Correspondence*, Vol. 8, 281–82.
34. Charles Darwin to T. H. Huxley, July 20, 1860. (CUL. Classmark: Imperial College of Science, Technology and Medicine, Huxley Papers, 5: 125/ Letter #: DCP-LETT-2873). In Burkhardt et al., *Correspondence*, Vol. 8, 294–95; see also Darwin to J. D. Hooker, July 20 [?]1860. (CUL. Classmark: DAR 115: 33a/ Letter #: DCP-LETT-2875). In Burkhardt et al., *Correspondence*, Vol. 8, 293; Charles Darwin to

John Lubbock, July 20, 1860. (CUL. Classmark: DAR 263: 40a[EH 88206447]/ DCP-LETT-2874). In Burkhardt et al., *Correspondence*, Vol. 8, 296–97.

35. Charles Darwin to T. H. Huxley, July 30, [?]1860. (CUL. Classmark: DAR 145/Letter #: DCP-LETT-2887). In Burkhardt et al., *Correspondence*, Vol. 8, 302.
36. Mountstuart E. Grant Duff, *Notes from a Diary, 1851–1872*, Vol. 1 (London: John Murray, 1897), 139. Milnes was a notable champion of women's rights, a poet, and a jilted suitor of Florence Nightingale.
37. Thomas H. Huxley to Charles Kingsley. In L. Huxley, *Life and Letters of Thomas H. Huxley*, Vol. 1, 221; see Charles Kingsley, *Water-Babies: A Fairy Tale for a Land Baby* (London: Macmillan, 1863). Within the text of this novel is a spoof on the "great hippocampus question," addressing the long-running debate between Huxley and Owen.
38. Thomas H. Huxley to Samuel Wilberforce, January 3, 1861. (c.13, folios 1–2). Samuel Wilberforce Papers, Bodleian Library, Oxford University; Cyril Bibby, "The Huxley-Wilberforce Debate: A Postscript," *Nature* 176 (1955): 363; the paper Huxley sent was Thomas H. Huxley, "On the Zoological Relations of Man with the Lower Animals," *Natural History Review*, new series 1 (1861): 67–84. Huxley co-owned this journal with Rolleston, John Lubbock. George Busk, and William Carpenter—all Darwinists. See also Charles C. Gross, "Hippocampus Minor and Man's Place in Nature; A Case Study in the Social Construction of Neuroanatomy," *Hippocampus* 3 (1993): 403–15; it is hard to believe that Huxley and Wilberforce did not see each other, at least occasionally, in the dining room of the Athenaeum. For a history of the Athenaeum Club, see Michael Wheeler, *The Athenaeum: More Than Just Another London Club* (New Haven, CT: Yale University Press, 2020).
39. Samuel Wilberforce to Thomas H. Huxley, January 30, 1861. Vol. 29, Folio 25. Archives, Imperial College, London.
40. L. Huxley, *Life and Letters of Thomas H. Huxley*, Vol. 2, 190.
41. Thomas H. Huxley to Charles Darwin, January 20, 1862. (CUL. Classmark: DAR 166.2: 291/ Letter #: DCP-LETT-3396). In L. Huxley, *Life and Letters of Thomas H. Huxley*, Vol. 1, 196; also in Burkhardt et al., *Correspondence*, Vol. 10, 33–35.
42. For information on Bishop John Colenso, who supported Darwinism and wrote of errors he found in the Old Testament, see John W. Colenso, *First Lessons in Science. Designed for the Use of Children* (London: Ridgeway, 1861); John W. Colenso, *The Pentateuch and Book of Joshua Critically Examined* (London: Longman, Green, Longman, Roberts and Green, 1862); Timothy Larsen, "Bishop Colenso and His Critics: The Strange Emergence of Biblical Criticism in Victorian Britain," *Scottish Journal of Theology* 50, no. 4 (1997): 433–58.
43. *London Daily Telegraph*, April 10, 1863, 4; L. Huxley, *Life and Letters of Thomas H. Huxley*, Vol. 1, 201–202; see also Colenso, *First Lessons in Science*.
44. L. Huxley, *Life and Letters of Thomas H. Huxley*, Vol. 1, 202.
45. R. W. Church, "Death of Bishop Wilberforce," *Manchester Guardian*, July 23, 1873; "Fatal Accident to the Bishop of Winchester," *Sunday Times* (London), July 27, 1873, 8.
46. Thomas H. Huxley to John Tyndall, July 30, 1873. Volumes 8.151 (typed letter) and 9.72 (handwritten). The italics are the author's (H.M.); Thomas H. Huxley Papers,

Archives, Imperial College, London. A version of this letter appears in L. Huxley, *Life and Letters of Thomas Huxley*, Vol. 1, 394–95, but the sentences so harshly describing Wilberforce's death are struck. Only one year later, 1874, Tyndall would deliver his famous BAAS presidential address at Belfast on the superiority of scientific authority over religious thought. See John Tyndall, *Presidential Address Delivered before the British Association of the Advancement of Science Assembled at Belfast, 1874* (London: Longmans, Green, 1874).

47. Reginald Wilberforce, *Life of Samuel Wilberforce*, Vol. 2 (London: John Murray, 1881), 450–51. One year later, the errata for Vol. 2 appear, in tiny agate font, on p. xiv of Vol. 3 (London: John Murray, 1882). The error concerning Huxley still does not mention him by name but reads, "Page 451, line 7, 'If I had to choose between being descended from an ape or a man who would use his great powers of rhetoric to crush an argument, I should prefer the former.'"
48. Thomas Huxley, "On the Reception of the *Origin of Species*" (Chapter 14), in *The Life and Letters of Charles Darwin*, ed. Francis Darwin, Vol. 1 (New York: D. Appleton and Co., 1889), 533–58.
49. Francis Darwin, *Life and Letters of Charles Darwin*, Vol. 1 (New York: D. Appleton and Co., 1889), 536–39.
50. R. G. Wilberforce, "Professor Huxley and the *Life and Letters of Charles Darwin*" (Letter to the Editor), *Times of London*, November 29, 1887, 10. Wilberforce was mistaken about the location of the debate; it was, as noted several times on these pages, at the Oxford Museum of Natural History. Wilberforce also quoted Darwin in his letter as stating, "[Wilberforce's review was] uncommonly clever, it picks out with skill all the most conjectural parts and brings forward well all the difficulties." This snippet, however, is selective and misleading. The full quote from that letter read: "I have just read Quarterly R. It is uncommonly clever; picks out with skill all the most conjectural parts, & brings forwards well all difficulties.— It quizzes me quite splendidly by quoting the Anti-Jacobin versus my grandfather.— You are not alluded to; nor, strange to say, Huxley, & I can plainly see here & there Owen's hand.— The concluding pages will make Lyell shake in his shoes. By Jove if he sticks to us, he will be a real Hero." Charles Darwin to J. D. Hooker, July 20, 1860. (CUL. Classmark: DAR 115: 33a/ Letter #: DCP-LETT-2875). In Burkhardt, et al., *Correspondence*, Vol. 8, 294.
51. T. H. Huxley, "Bishop Wilberforce and Professor Huxley," *Times of London*, December 1, 1887, 8; see also J. Vernon Jensen, "Return to the Huxley-Wilberforce Debate," *British Journal for the History of Science* 21 (1988): 161–79.

## Epilogue: After Myth

1. Charles Darwin, *On the Origin of Species by Means of Natural Selection, or the Preservation of Favoured Races in the Struggle for Life*, 6th ed. (London: John Murray, 1872), 421.
2. *BAAS Report*, 1862; 32, Part II, pp. 114–118; see also T. H. Huxley to Charles Darwin, October 9, 1862. (CUL. Classmark: DAR 162.2: 294/Letter #: DCP-LETT-3755). In Frederick Burkhardt et al., eds., *The Correspondence of Charles Darwin*, Vol. 10 (Cam-

bridge, UK: Cambridge University Press, in 30 volumes, 1985–2023), 55–56; J. E. Gray to Charles Darwin, January 29, 1862. In Burkhardt et al., *Correspondence*, Vol. 10, 450–51.

3. *Christian Remembrancer*, 40 (1860): 237–61; quote is from p. 239.
4. Francis Darwin, ed., *The Life and Letters of Charles Darwin* (London: John Murray, 1887, in 3 volumes; New York: D. Appleton and Co., 1889, in 2 volumes); Leonard Huxley, *Life and Letters of Thomas H. Huxley* (London: Macmillan, 1900).
5. James A. Secord, *Victorian Sensation: The Extraordinary Publication, Reception, and Secret Authorship of Vestiges of the Natural History of Creation* (Chicago: University of Chicago Press, 2000), 526; Morse Peckham, ed., *The Origin of Species by Charles Darwin: A Variorum Text* (Philadelphia: University of Pennsylvania Press, 1959), Appendix II, 775–87. Another literary milestone occurred in 1901, forty-two years after the original *Origin* first appeared and the British copyright ran out. At this point, John Murray—and several other publishers—were free to produce cheap copies of the book, which sold in the many tens of thousands, or more, every year thereafter. British copyrights in Darwin's time were guided by the Copyright Act of 1842 and lasted for the lifetime of the author plus seven years or for forty-two years after its first appearance, whichever was longer. In Darwin's case, the forty-two-year rule, from 1859, was best suited to his advantage.
6. Thomas Huxley, "Owen's Position in the History of Anatomical Science," in Richard Owen, *The Life of Richard Owen*, Vol. 2 (London: John Murray, 1894), 273–332.
7. Henrietta Litchfield, ed., *Emma Darwin: A Century of Family Letters, 1792–1896* (London: John Murray, 1915); Reginald Wilberforce, *Life of Samuel Wilberforce*, Vol. 2 (London: John Murray, 1881), 450–51; Leonard Huxley, *Life and Letters of Sir Joseph Dalton Hooker, O.M., G.C.S.I., Based on Materials Collected and Arranged by Lady Hooker* (London: John Murray, 1918); Rev. Richard Owen, *The Life of Richard Owen* (London: John Murray, 1894).
8. William Carew Hazlitt, *English Proverbs and Proverbial Phrases Collected from the Most Authentic Sources: Alphabetically Arranged and Annotated, with Much Matter Not Previously Published* (London: Reeves and Turner, 1907), 173.
9. Howard Markel, "*The Principles and Practice of Medicine*: How a Textbook, a Former Baptist Minister, and an Oil Tycoon Shaped the Modern American Medical and Public Health Industrial-Research Complex," *Journal of the American Medical Association* 299, no. 10 (2008): 1199–1201; Howard Markel, "Abraham Flexner and His Remarkable Report on Medical Education: A Century Later," *Journal of the American Medical Association* 303, no. 9 (2010): 888–90.
10. Jerome Lawrence and Robert E. Lee, *Inherit the Wind: The Powerful Drama of the Greatest Courtroom Clash of the Century* (New York: Ballantine books, reprint edition, 2003).
11. Edward J. Larson, *Summer of the Gods. The Scopes Trial and America's Continuing Debate over Science and Religion* (Cambridge, MA: Harvard University Press, 1998), 17, 22.
12. Colin Gauld, "Wilberforce, Huxley, and the Use of History in Teaching about Evolution," *American Biology Teacher* 54, no. 7 (1992): 406–10; Larry A. Witham, "The

Great Debate" (Chapter 12), in *Where Darwin Meets the Bible: Creationists and Evolutionists in America* (New York: Oxford University Press, 2002), 212–26.

13. See, for example, Georgina Ferry, *A Wonderland of Natural History: A Souvenir Guide* (Oxford, UK: Oxford Museum of Natural History, 2011), 9; "The Great Debate," Oxford University Museum of Natural History, accessed October 8, 2022, at https://oumnh.ox.ac.uk/great-debate. See also Darwin Exhibition, 2006, at the American Museum of Natural History, accessed October 8, 2022, at https://www.amnh.org/exhibitions/darwin; "The World Reacts," https://www.amnh.org/exhibitions/darwin/a-life-s-work/the-world-reacts; *Darwin Educator's Guide*, www.amnh.org/education/resources/exhibitions/darwin.
14. Garrison Keillor, *The Writer's Almanac* for Thursday, June 30, 2022. Originally aired on National Public Radio June 30, 2011. https://www.garrisonkeillor.com/radio/twa-the-writers-almanac-for-june-30–2022/.
15. *The Man Who Shot Liberty Valance*, directed by John Ford, screenplay by James Warner Bellah and Willis Goldbeck, based on a short story by Dorothy M. Johnson. Paramount Pictures, 1962.
16. Whig historians are often charged with using the present to interpret the past and presenting triumphalist accounts of progress over time at the risk of oversimplification of the past. For the classic text explaining this line of inquiry, see Herbert Butterfield, *The Whig Interpretation of History* (New York: W. W. Norton, 1965).
17. The Nobel Prize in Physiology or Medicine 2022. NobelPrize.org. Nobel Prize Outreach AB 2022. Friday, October 7, 2022. https://www.nobelprize.org/prizes/medicine/2022/press-release/. See also Carl Zimmer, "Oldest Known DNA Offers Glimpse of a Once-Lush Arctic," *New York Times*, December 8, 2022, Section A, 1; Elie Dolgen, "Ashkenazi Jews Have Become More Genetically Similar over Time," *New York Times*, December 6, 2022, Section D, 4.
18. Emma Darwin, "Reminiscences of Charles Darwin's Last Years," 1882. Charles Darwin Papers. (Classmark: MS No. CUL-DAR210.9); Special Collections and Archives, Cambridge University Library, Cambridge, UK.
19. J. Charles Deduchson to T. H. Huxley, January 24, 1887. [13.137]. Huxley responded, on February 12, 1887, with Francis Darwin's permission, "I have the best authority in informing you that the statement . . . is totally false and without foundation." [13.138] Thomas Huxley Papers, Archives, Imperial College, London.
20. Alison Pearn, "The Teacher Taught? What Charles Darwin Owed to John Lubbock," *Notes and Records of the Royal Society of London* 68, no. 1 (2014): 7–19.
21. Charles Darwin to Charles Lyell, December 10, 1859. (CUL. Classmark: American Philosophical Society, Mss. B.D25. 184). In Burkhardt et al., *Correspondence*, Vol. 7, 423.
22. Herbert Spencer to Charles Darwin, February 22, 1860. (Classmark: University of London, Senate House Library, MS. 791/51; Letter #: DCP-LETT-3126). In Burkhardt et al., *Correspondence*, Vol. 8, 98–99. Spencer had been articulating this notion for some time even before *On the Origin of Species* was published; see Herbert Spencer, "The Development Hypothesis," in *Essays: Scientific, Political, and Specu-*

*lative* (London: Longman, Brown, Green, Longmans, and Roberts, 1858), 389–94. (Originally published March 20, 1852, in *The Leader*.)

23. Charles Darwin to Alfred R. Wallace, July 5, 1866. (CUL. Classmark; British Library, Add 46434, f. 70/ DCP-LETT-5145). In Burkhardt et al., *Correspondence*, Vol. 14, 235–37; see also Herbert Spencer, *The Principles of Biology*, Vol. 1, 2nd ed. (London: Williams and Norgate, 1864–1867), 444; Herbert Spencer, *First Principles*, 2nd ed. (London: Williams and Norgate, 1867).
24. The quote can be found in T. H. Huxley, *The Romanes Lecture of 1893: Evolution and Ethics* (London: Macmillan, 1893), 32; see also Alison Bashford, *The Huxleys: An Intimate History of Evolution* (Chicago: University of Chicago Press, 2022), 300–306; see also, for conflicting Huxley views, an earlier essay written at the end of the American Civil War. Arguing for the abolition of one of humankind's most inhumane acts, he nonetheless echoed Darwin's racist opinions on African Americans: "The highest places in the hierarchy of civilization will assuredly not be within the reach of our dusky cousins, though it is by no means necessary that they should be restricted to the lowest." See Thomas H. Huxley, "Emancipation—Black and White (1865)," in *Collected Essays: Science and Education*, Vol. 3 (New York: D. Appleton and Co., 1895), 66–75; quote is on pp. 66–67.
25. Francis Galton, *Hereditary Genius: An Inquiry into Its Laws and Consequences* (London: Macmillan, 1869), 202–203.
26. Charles Darwin, *The Descent of Man, and Selection in Relation to Sex*, Vol 1, 1st ed. (London: John Murray, 1871), 110–11.
27. Darwin, *The Descent of Man, and Selection in Relation to Sex*, Vol. 2, 1st ed. (London: John Murray, 1871), 403. It is interesting to note that Darwin worried about passing on his ill health by inheritance to his children, especially Henrietta and Leonard. To his cousin William Fox, he wrote in September 1859, "I very much fear that my eldest girl Etty will never be strong; & Lenny [his fourth son and eighth child] has been often ailing for last 2 years with intermittent (but only symptomatically so) pulse.—My poor constitution like everything else is transmitted by inheritance." See Charles Darwin to William D. Fox, September 23, 1859. (CUL. Classmark: Library, Cambridge/MS 53 Fox 122/Letter # DCP-LETT-2493). In Burkhardt et al., *Correspondence*, Vol. 7, 335–36.
28. Galton also coined the phrase "nurture vs. nature." He and Charles Darwin were both grandsons of the same Birmingham physician, Erasmus Darwin, but by different wives (grandmothers). See Frances Galton, *Inquiries into Human Faculty and Its Development* (London: Macmillan, 1883), 17, 24–25, 44; Frances Galton, *Hereditary Genius: An Inquiry into its Laws and Consequences* (London: Macmillan, 1869); Frances Galton, "On Men of Science: Their Nature and Their Nurture," *Proceedings of the Royal Institution of Great Britain* 7 (1874): 227–36; Frances Galton, *Natural Inheritance* (London: Macmillan, 1889); Frances Galton, *Essays in Eugenics* (London: The Eugenics Education Society, 1909).
29. The author (H.M.) discusses eugenics at length in Howard Markel, *Quarantine! East European Jewish Immigrants and the New York City Epidemics of 1892* (Baltimore: Johns Hopkins University Press, 1997; updated edition, 2021), 179–82; Howard

Markel, *When Germs Travel: Six Major Epidemics That Have Invaded America and the Fears They Unleashed* (New York: Vintage, updated edition, 2020), 81–110; Howard Markel, *The Kelloggs: The Battling Brothers of Battle Creek* (New York: Pantheon Books, 2017), 298–321; Howard Markel, *The Secret of Life; Rosalind Franklin, James Watson, Francis Crick, and the Discovery of DNA's Double Helix* (New York: W. W. Norton, 2021), 22–34.

30. Eventually, eugenics impregnated Adolf Hitler's "Final Solution," although had he not had Galton's theories, *Der Führer* would have surely found some other rationale to justify his extermination of millions of Jews, Hungarian Gypsies, homosexuals, and disabled people. See Richard J. Evans, *The Third Reich Trilogy*, Vol. I: *The Coming of the Third Reich*; Vol. II: *The Third Reich in Power*; Vol. III: *The Third Reich at War* (London: Penguin Books, 2003, 2005, 2008); Richard Weikart, *From Darwin to Hitler: Evolutionary Ethics, Eugenics and Racism in Germany* (London: Palgrave Macmillan, 2004); Paul Crook, *Darwin's Coattails: Essays on Social Darwinism* (New York: Peter Lange, 2007); Richard Hofstadter, *Social Darwinism in American Thought, 1860–1915* (Philadelphia: University of Pennsylvania Press, 1944).
31. Nancy Stepan, *The Idea of Race in Science: Great Britain, 1800–1960* (London: Palgrave Macmillan, 1982), 47–139; Markel, *The Kelloggs: The Battling Brothers of Battle Creek*, 298–321.
32. Darwin, *The Descent of Man*, Vol. 2, 385.
33. The spokesman was Rev. Malcolm Brown, and he published his "unofficial" statement on the church's website. Jordan Lite, "Better Late Than Never? Clergyman Says Church Owes Darwin an Apology," *Scientific American* 16 (September 2008). The same week, the Vatican noted that while it never banned *Origin* and would not formally apologize to a dead man, "Darwin's theories are compatible with the Bible." Phillip Pullella, "Evolution Fine But No Apology to Darwin: Vatican," *Reuters News*, September 16, 2008.
34. Howard Markel, Science Diction: The Origin of the Word "Evolution." National Public Radio's Science Friday/Talk of the Nation, August 13, 2010.
35. Darwin, *On the Origin of Species* (1st. ed.), 488–89.
36. Darwin, *On the Origin of Species* (1st. ed.), 490. Italics are the author's (H.M.).
37. Emily Anthes, " 'Anthropause' during Pandemic Healed Nature, but Hurt It, Too," *New York Times*, July 17, 2022, A1.

# ILLUSTRATION CREDITS



### Introduction: A Temple of Science



## Part I: Down



### 1. The Letter

### 2. First



### 3. Survival of the Fittest



## Part II: The Book



### 4. The Devil's Chaplain

### 5. Best Seller



## Part III: Friends and Foes



### 6. Darwin's Bulldog



### 7. The Dinosaur



### 8: Soapy Sam

### 9: A Mysterious Malady



## Part IV: Oxford



### 10. The Association



### 11. Pax Interruptus



### 12. Mawnkey! Mawnkey!



### 13: Rebuttals

### 14. The Dogs Bark but the Caravan Moves On



### Epilogue: After Myth

# INDEX

Page numbers in *italics* indicate illustrations or tables.
Endnotes are indicated by *n* after the page number.